Spatio-Temporal Narratives

Spatio-Temporal Narratives: Historical GIS and the Study of Global Trading Networks (1500-1800)

Edited by

Ana Crespo Solana

CAMBRIDGE SCHOLARS

PUBLISHING

Spatio-Temporal Narratives: Historical GIS and the Study of Global Trading Networks (1500-1800)
Edited by Ana Crespo Solana

This book first published 2014

Cambridge Scholars Publishing

12 Back Chapman Street, Newcastle upon Tyne, NE6 2XX, UK

British Library Cataloguing in Publication Data
A catalogue record for this book is available from the British Library

ISBN (10): 1-4438-5542-1, ISBN (13): 978-1-4438-5542-6

This investigation has been sponsored by the project:
"GlobalNet", Spanish Ministry of Science and Innovation, Reference: HAR2011-27694

Contents

PREFACE

A Flemish scholar from Haarlem, Michael Van der Veen, disciple of Johan Huizinga, gave a precise account of the punctual appearances on several places on the planet since 1614, of "the Flying Dutchman" (Cunqueiro, 1998: 121–123). In 1731 the Dutchman enters the port of Genoa on his ship, and an old sailor from Liguria recognizes him because they had been drinking together in Lisbon in 1689. Forty-two years later, the Dutchman is still a young-looking, pitch-black haired fellow of an intriguingly melancholic sort. Rumours of witchcraft reach the Genoese authorities, and at dawn on a stormy morning the Dutchman flees. Then in 1718 he sets foot on Saint-Maló and wins the love of the young daughter of a member of the Chamber of Commerce. He runs off with her only to abandon her some months later on a beach near Boulogne. The girl will remain insane during her short life, ceaselessly shouting "the burning man is coming, the burning man!" In 1736 he is again to be found in Lisbon just arrived from New Spain, where he visits a woman and gives her news of her husband who is an innkeeper in Veracruz. But then, once again, the Dutchman is gone with the outward wind. Then in 1751 he is seen in Naples where an aristocratic lady dates him and then tells the story of a Dutch captain that cannot live on land for longer than nine days, and those nine days come at a price of nine months of ceaseless navigation; and he can be killed only by fire; and if you ask him for his name he will reply, "Call me *Foreigner!*"

The very last documented account of the stay of the Dutchman on land has him stay in London, then Marseille, in 1819; in the latter he met an envoy from Napoleon. Before that he had stayed in Lübeck some time during the blockade years in Europe; he had been seen as well in the Bermudas, Amsterdam, Java, and sailing around the Cape of Good Hope. This piece of sailors' imaginary describes the ceaseless wandering of Willem van der Decken from Holland, a Jewish captain, as he was sentenced to an eternity of aimless sailing that would see him join political or military causes, frequent taverns and make women fall for him, or engage in business of an uncertain kind. Almost every man who met him turned mad, and no-one ever knew how to put an end to such long and miserable peregrination. This story has been repeatedly depicted in art, music and literature, and its protagonist has been identified with a real man by the name of Bernard Fokke (Falkenburg in Dutch literature). He was the captain of a vessel notorious for its cruise speed when travelling between Holland and Java; he was thought to have struck a deal with the Devil some time between 1641 and 1680. There are differences between the various versions of this famous example of the sailors' imaginary in which the sea plays the part of a frontier world across which seamen lead a life of constant wandering between port cities (Fitzball, 1826; Marryat, 1837; Irving, 1855; Neale, 1840).

This is a reference to changes in time although strongly linked to space since space is neither static nor immutable; what is more, this is a clear reference to travelling narratives. The Flying Dutchman seemed to have the gift of ubiquity, he was everywhere and nowhere. Time and movement are dialectic concepts in space and they are related to the dynamic systems and their evolution. Just as in sailors' tales, the spatial frontier where historical progression takes place is a dynamic, non-linear reality where consequences may not be proportional to their causes (Puu, 2003). Besides, it is deterministic and described in terms of three distinct elements: the configuration space, the agent making the change, and the whole of final states. Just as in any good narrative, the space where events take place acts as a category needed for the description of events. Space makes history immutable and ever-changing at the same time as can be seen in the materialization of the various forms of existence, in policy making, in socio-historic organizations and in theoretic paradigms. Every time we construct a narrative it manifests itself in space as space is essential to human action, especially when it comes to thinking historically as advised by Pierre Vilar, because every historical moment has its own way of thinking, and the evil of anachronism must be avoided as Lucien Febvre warned us (quoted in Vilar, 2004).

This book is a collective dream that has come true. It is the result of years of research conducted by the authors of this volume as part of a project that has meant a new turn of historical studies as a scientific discipline that is vital so we can understand the evolution of human behaviour, the interactions among individuals and groups, and the resulting societies shaped by this behaviour. That project was named "Dynamic Complexity of Cooperation-Based Self-Organizing Commercial Networks in the First Global Age" and it was funded by the European Science Foundation. This project ran between 2007 and 2011 as part of the EUROCORES "The Evolution of Cooperation and Trading" and under the DynCoopNet acronym (Owens, 2008: 23–35). Its object was the expansion of the available data on the evolution of cooperative relationships responsible for the self-organizing commercial networks that created and sustained the first global economy (1400–1800) by integrating disciplinary methods such as Geographic Information Systems (GIS) and other information technologies. The perspective was multinational and multidisciplinary and it was innovative in its conceptual focus and the unusual disciplinary combination of history and other social science methods. As far as historical sciences are concerned, this project intended to refute the classic theory that greater human cooperation in trading became possible with the increased effectiveness of the state, which was accompanied by the rise, primarily in Europe after 1500, of capitalism and the individualism characteristic of modernity. This theory will not stand up to the scrutiny of the analyses of the processes of spatial economic integration, by using actual historical data, on agents, forms of network behaviour, and the mechanisms these agents and networks developed in order to communicate through trade the areas of production and markets along the intercontinental sea routes. These international routes greatly expanded from the 15th century onwards. And this era of so-called merchant capitalism was central to the understanding of the emergence of new economic/financial and social behaviours as determined by how the world was integrated and spatially linked. This dynamic complexity, as defined by Andre Gunder Frank, requires a wealth of historical data as well as powerful tools that are capable of handling massive amounts of information (Frank, 1998). The DynCoopNet project has gone a great length in revealing the mechanisms of cooperation that permitted merchants and others to establish the often long-distance trading networks of the first global age. This is particularly relevant when studying the expansion of the Iberian empires of Portugal and Spain; these were much more interconnected than would appear on first sight even when the dynastic union between the Kingdom of Portugal and the Spanish Monarchy between 1580 and 1640 was no longer in force. This interconnectedness has been proven in other works where the special features of these corporate and composite monarchies were discussed, and the impact of the activities conducted by social agents was heavier than that from the political and even the military powers. Between the 16th and the 18th centuries, the specificities of certain shadow empires (Winius, 1983: 83–101) made up of a multitude of different agents conditioned Spain and Portugal to became the first imperial powers on a worldwide scale, and although they lost their political primacy in the 17th century, both monarchies survived and were able to enjoy a relative success until the early 19th century. The networks in the empire did definitely condition this expansion as well as other "expansions" in England, France or the Low Countries since all these expansions had to adapt to how other territories in Europe, Asia and America would socially and economically fit in these Iberian empires. Besides, the analysis of the social networks and their spatial links shows how important these networks and their links were on the circulation of ideas, cultures, religious confessions, identities and other forms of institutional organisation. These studies are being continued in various projects run by former DynCoopNet researchers[1]. Also, further works along similar lines point out those relations between networks and institutions or between networks and ways of commercial cooperation. Having said this, what is really characteristic of the post-DynCoopNet works conducted by these researchers is their increasing interest in spatial history (Fujita, Krugman & Venables, 1999; Hewings, Sonis & Boyce, 2001). This spatial history claims the need to tell narratives not only with text and language but also with images so history can be visualized (Shnayder, 2010), as well as analysed from the fashionable digital humanities perspective. Beyond technical meanings, the specialists in spatial history focus on the space – rather than on the time – aspect as a coordinate that is central to approaching how heavily spatial changes affected behaviours throughout history. It is worth mentioning that this approach is a long-standing aspiration for historians, as it was already present in Fernand Braudel's definition of geohistory as "the history that the environment imposes on men through its constants, what happens most often, or through slight variations in those constants, when they have consequences at human level" (Braudel, 2002: 78), and in the Annales school when the road to a new scientific quality in history was open through empirical-quantitative analyses and the intensive use of historical databases.

Consequently, a cartographical perception of landscape and the geographic framework where historical events take place are central to the spatial turn analysis. This is a new approach to the need to create "mental maps" by historians, but also a return to the *ever-so-important-in regional-*

---

[1] The projects are: "Understanding social networks within complex, nonlinear systems: geographically-integrated history and dynamics GIS" (SOCNET), which is administered by the NSF Office of Cyberinfrastructure (J. B. Owens and May Yuan); "Una ruta Global: análisis histórico con bases de datos y geovisualización espacio-temporal del comercio marítimo internacional (1717-1850)" (GlobalNet) Reference number: HAR2011-27694 (Ana Crespo Solana) and *Geografía Fiscal y Poder financiero en Castilla en un siglo de transición. Análisis mediante Sistemas de Información Geográfica* Ref. HAR2010-15168 (David Alonso García), both funded by the Spanish Ministry of Science and Innovation (MICINN); and Dynamic Complexity of Cooperation-Based Self-Organizing Networks in the First Global Age, FCT - TECH/0002/2007 (http://www.dyncoopnet-pt.org/) funded by the Portuguese government (Amélia Polonia Da Silva).

studies spatio-temporal analysis, now framed in ever wider contexts, whether national, transnational or, above all, global. This review of the spatial paradigm is very important when it comes to delimiting the social and economic spaces analysed in terms of their relationships, such as in Atlantic history (Pieper & Schmidt, 2005: 17). Although this review is in fact a revival of an approach previously seen in certain classic studies by thinkers and scientists such as Jules Michelet or Fernand Braudel, the notion of "space" allows us to pinpoint geographic areas that condition historic evolution. Certain specialists in Atlantic history take into consideration this relationship with "space" as a fundamental aspect in order to understand the dynamics between empires, or the formation of hierarchical structures around monopolies, or either the "peripheral" or the "leading" nature shown by certain regions. This has been described in studies of the various "expansions" observed from a large-scale treatments perspective that claims that the Atlantic must be regarded as a historical subject, and that analysing how far the influences of the spatial interchanges go is of great importance (Greene & Morgan, 2009). Space as a historic category allows us to highlight a specific event and attach to it further data that greatly add to this event's knowledge; hence the use of historic GISs means an IT-based added value as this technology is placed at the very frontier of knowledge (Middell, 2010: 149–170).

In fact, one of the main basic assumptions of the DynCoopNet project was that the history of any place, no matter how large (the Atlantic), is shaped by the way it is connected to other places and all places are connected to an ultimately global system that shapes the histories of all places (Owens, 2007: 2014–2040). This is why this book insists on a new turn for the studies on expansion and integration in the Modern era; these should be conducted with an interdisciplinary methodology in mind and the use of data-processing technologies that combine historical databases, tools for spatial visualisation and network analysis. The commercial networks to be studied involved a high level of cooperation among people of various ethnic, linguistic, and religious backgrounds. They served to move goods and people within a highly open system over an expanding geographic space. These self-organizing networks, and the strategies employed to achieve cooperation among a large number of widely-dispersed people which made them possible, served as the source of the creativity and innovation necessary to respond in a flexible manner to the endemic disruptions to transportation and capital flows occasioned by war, disease epidemics, arbitrary government actions, environmental changes, and the transportation problems associated with weather, distance, and the available technologies. Without the diffusion of authority and the widespread cooperation among merchants and others, often overcoming apparently serious cultural and political barriers, the emergence and spatial intensification of the first global economy would have been impossible, but existing research on such trading activity has been too limited and fragmentary to grasp how such networks of cooperating individuals emerged and how they were sustained and evolved for hundreds of years. Therefore studies involve the gathering, organizing, and sharing of data on commercial relationships. In part, this will involve micro-studies identifying these networks and the people involved in them. The historians will find out how, within their social and cultural environments, individual merchants maintained the "creditworthiness" necessary for such a remarkable degree of cooperation over often great distances and with people they sometimes did not know. They will focus on the cultural environment of ways of understanding the world and on the types of cooperation at various scales: family, small firm, rural community, political and economic urban centre, and geographically extensive trade routes (Crespo Solana & Alonso García, 2012).

This book attempts to explore new methods and techniques for research about merchant networks and maritime routes of trade during the First Global Age by using Geographic Information Systems as a tool for visualizing the formation of trading systems, for database management, and for cartographic and spatio-temporal analysis in historical GIS. In order to achieve this goal we focus on key issues for understanding the birth of the so-called First Global Age (16th to 18th centuries): integration of spatial economies, regionalization of markets, the organization of maritime trade routes and the evolution of self-organizing networks of merchants, producers, communities, and other social agents during the age of expansion. The collections of essays will deal with relevant information about spatial and social network analysis, the use of digital cartography and the possibilities of digital humanities: sharing databases via the Internet, metadata and possibilities, problems and new design by using geographic information systems (GIS) for the organization and visualization of data and spatial analysis. This volume consists of twelve chapters dealing with various subjects devoted to the study of the mechanisms by which the socio-economic spaces integrated the Iberian monarchies; specific examples from the Spanish Atlantic colonial trade routes are studied (Crespo Solana); how transnational commercial networks operated, from the specific case of Castilian merchant Simón Ruiz and Portuguese networks (Polonia Da Silva, Pinto, Ribeiro, Morais Barros). These works highlight the process of integration between network and space. The authors intend to argue, with empirical evidence, how this still innovative methodological approach to the field of history, combined with a renewed theoretical framework, enabled us to generate new historiographical perspectives with significant impact upon the historiography of the First Global Age.

Other papers are devoted to the detailed description, at scientific and technical levels, of the methodology used in a GIS; also the social network analysis and the use of cartography as a representation and visualization method are clearly detailed (Yuan). Three papers explain how spatio-temporal GISs can greatly add to historical studies. J. B. (Jack) Owens is the intellectual father of this research team; he has made a truly outstanding contribution to the theoretic configuration of historic studies of spatial integration, and has had a truly remarkable impact on the development of new methodological devices for spatio-temporal GISs (Owens and Wachowikz). Together with Kantabutra he has now produced an article in which they explain what an ILE (intentionally-linked entity) is: a database system for representing dynamic social networks, narrative geographic information, and general abstractions of reality. In their work, Guerrero Nieto and Saurí describe the development of tools for data transfer from a semantic web perspective. This is also a major innovation that will greatly contribute in the future to the processing of spatio-temporal information contained in historical sources. The paper written by Pérez Asensio, Salas Tovas, Del Bosque, Maestre and Crespo analyzes the application of Social Network Analysis software to Spatial Analysis. The authors study what these technologies offer in terms of understanding and visualizing the relationships between merchants and ships. Analysis is based on the data collected by the DynCoopNet project and made publicly available from the Digital Repository of the Spanish National Research Council (CSIC).[2]

Cartography has contributed from a twofold approach: cartography and networks (Picazo Muntaner), and cartography and GIS (Crespo Sanz). The elaboration of network cartography and the subsequent visualization greatly add to our ability to conduct in-depth network analyses and perceive the main systemic nodes, the network's geographic expanse and configuration as a "small world", and assess its fragility. Two characteristic examples of the global network can be visualized with this tool showing trans-Pacific commerce and its impact on continental America, the Caribbean area, and even the various metropolises. This type of study leads to a cartographic visualization of the port networks including the main connecting ports, and allows for a better observation of

[2] Available at http://digital.csic.es/handle/10261/28394 and http://hdl.handle.net/10261/23414.

the movements of the merchant networks at local level as well as in their connections with other geographic areas. In addition to this, the combined use of GIS and cartographic data offers cartographers new tools to examine the characteristics of historical maps, their distortions, and their functions, as well as making it possible to compare different historical maps through new overlapping technologies. Crespo Sanz uses GIS technologies to study the *Atlas de El Escorial* (c. 1538), a manuscript of the Iberian Peninsula consisting of an index map (1:2,600,000) and twenty detailed maps (1:400,000). The geographical and spatial details of this Renaissance manuscript allow scholars to use three distinct, yet complementary, methods of metric analysis. First, traditional mathematical and statistical analysis will be used to obtain the necessary data to appraise the map – determine its scale, accuracy, the degree length, geographical and hydrographical details, territorial limits, and so forth. Such analysis, however useful, generates many cumbersome tables and datasets, making it difficult for scholars to use. GIS with its analytical capabilities will allow scholars to manipulate such large datasets more easily, and thereby gain greater insights into early modern maps and their making. Second, GIS makes cartographical georeferencing possible. This allows us to compare geographical coordinates between various old maps as well as between old and modern maps. That is, we can overlay maps or points on maps to determine more accurately the distortions and other differences between maps. Although there are many georeferencing programs, this study uses Google Earth, which is free and readily available to scholars. Third, MapAnalyst is another tool for cartographical analysis that combines mathematical and visual analysis. This program is user-friendly for those with limited mathematical and computer programming skills; you can compare all kind of maps estimating the distortion between them, watching the values as vectors that express the direction and length of the deformation. By applying these three methods of analysis to the Atlas of the Escorial, this study will show how these approaches complement each other and define a general methodology for using GIS and metric analysis on any historical maps. Along this line Mukherjee discusses the construction of spaces in her work on Srihatta/Sylhet in the medieval Bay of Bengal world. Finally, as added assets to this book I wish to mention the works focusing on the analysis of the links between social networks and the spaces of interaction in which cooperation amongst agents takes place; how the latter take advantage of the strengths and weaknesses present in the systems (Alonso García), and the impact of these agents on the economic structure of ports and maritime routes (Scheltjens).

The authors of these papers have travelled to distant ports in their search for the spatio-temporal relations conducted by the protagonists in history. Technology has assisted us in trying to understand a teleological narrative which, in turn, we can attempt to make available to our readers.

> I pass, like night, from land to land;
> I have strange power of speech;
> that moment that his face I see,
> I know the man that must hear me:
> to him my tale I teach.
>
> (Samuel Taylor Coleridge, *The Rime of the Ancient Mariner*, 1799).

Ana Crespo Solana

CHAPTER ONE

THE WIDER WORLD: SPATIAL EXPANSION AND INTEGRATION IN THE HISPANIC ATLANTIC, 16TH TO 18TH CENTURIES[1]

ANA CRESPO SOLANA[2]

1. The Spatial History of the Hispanic Atlantic World

In the 16th century the Atlantic was not perceived as a unit of historical analysis, and there was no Mediterranean Sea to be found in the Americas (Games, 2006: 741–757; Greene & Morgan, 2009). However, it was at that time when the Atlantic World began to materialize as a "social environment" in which new forms of human behaviour emerged as a consequence of the challenge faced by those historical actors in that new social, economic, and political environment. The Spanish colonization of the Atlantic world was instituted in La Hispaniola in 1493 by Christopher Columbus on his second trip; this colonization led to the "Hispanization" of vast American territories that were linked through the maritime routes of the Atlantic Ocean. This maritime region was meteorologically ruled by the trade winds that linked the area comprising the Canaries and the Azores with the Lesser Antilles first, and later with the whole of the Caribbean space, including the Gulf of Mexico. That Atlantic space was defined by Pierre Chaunu as a parallelogram drafted around these routes: "lying between the ebb of the trade wind and the flow from the mid-way latitude, this was an immense front that was well marked out between the oceanic coast of Andalusia – a highly privileged point on earth from which the westward, trade winds are born in summer and in which the eastward,

[1] This investigation has been sponsored by the GlobalNet project of the Spanish Ministery of Science and Innovation, Reference: HAR2011-27694.

[2] Instituto de Historia, CSIC.

return winds culminate in winter – and the immense arch of the 'Mediterranean Caribbean' spreading from Florida to Trinidad" (Chaunu, 1973). This vast area affected by the trade winds soon became the main way for integrating the Atlantic and the Pacific oceans since that meteorological phenomenon was discovered by the Portuguese sailors in relation to the Atlantic in the 15th century, and later by Andrés de Urdaneta in 1565 in relation to the Pacific (Hayes, 2001).

Between the "First America" (as David Brading calls the Spanish America in which the first Atlantic world emerged as a result of the conquest and first colonization by the Iberians) (Bernard, 1994), and the port cities of Seville, Huelva and Cadiz, new channels of exchange were opened that was ever-changing in terms of time and space. And in those early days of the hispanization process, in the American Atlantic three issues merged together: race, religion, and language; and these are still widespread in large parts of the American continent from Mexico to Patagonia (around 16 million square kilometres at the time). Brazil and its Portuguese culture and language had strong ties with the Spanish side as Portugal had been integrated into the Spanish monarchy between 1580 and 1640. In this territory "equatorial jungle and desert go hand in hand together with ample savannah plains and closed-in forests not far from the permafrost on a boundless, incommensurable land that reveals an immense living reality, ever changing and diverse; a magic land on which the most diverse ethnic groups can be found under the most dissimilar political structures as well as the most complicated economic systems; a land where social inequalities are widest and where the subtlest of beliefs are to be known, a territory whose cultural roots can only be traced back to the major autochthonous civilisations combined with the Hispanic legacy in all its manifestations" (Vila Vilar, 2012: 27).

This Atlantic space has recently been placed under a constant, major revision. If there is something that really strikes the mind of the present reviewer, it is the fact that British and North American historiography, despite being the intellectual initiator of Atlantic history, did overlook the importance of the Iberian empires, that is, Portugal and Spain, on the integration of an Atlantic world that comprised, and connected, the whole of the American continent with Europe and Africa. Most surprisingly, a wealth of literature available in Portuguese and Spanish has been almost completely ignored; here, the inability to communicate in these languages must have been a major contributing factor. But there are winds of change as many non-Iberian scholars are now stressing the importance of the Iberian colonization and its subsequent influence (Pagden, Elliot, Brading, Armitage, 2004: 12–13). Fortunately, Africa and Asia are now also included in the analyses of integration, and further methodological perspectives are now available which open new possibilities for a stimulating comparative analysis with other colonial empires (Carmagnani, 2005). Also, studies are turning to a deepening analysis of the dynamics of the global systems of integration between 1400 and 1800; this period is already known as the "First Global Age" or "First Mundialization" due to the spectacular increase in exchange of people, products and culture seen during this long period. The 15th century, especially the last two decades, is regarded as the starting point in this first globalization stage as many crucial transformations in the integration of many regions on the planet happened at the time as a result of human action in a global system (Ringrose, 2000; McNeill & McNeill, 2004).

Defining this historic space, the Hispanic Atlantic, the "First America", from a social, political and economic standpoint was complex and ambiguous then, although there are many convincing testimonies that show the image of an aggressive and authoritarian empire which the Spanish Monarchy or Spanish Empire was eager to project to its competitors in the race to hegemony. The origins of the spatial configuration of this empire can be traced back to the Capitulations of Santa Fe in April 1492; this agreement enabled Columbus to organize his expedition across the sea to the West, and it marked the end of the last stage of the "Reconquista" as conducted by the Catholic kings within the Iberian peninsula, and the beginning of a new era of conquest. A maritime frontier was blown open followed by one on land as new territories were discovered in America; while the crown of Castile shaped an imperial image related to the fights against political oppression, ethnic domination, the establishment of a power-race pairing, and socioeconomic exploitation. This was the beginning of a spatialization of power in the Atlantic world along geographic lines of superiority and subordination (Colas, 2007: 21).

As for cartography, there was a convention in the second decade of the 16th century by which a continent was always depicted as a single mass of land (Fernández-Armesto, 2004: 21). From a socio-political and territorial perspective, Spain, despite being relatively poor, sparsely populated and peripheral, was able to engulf the old Peruvian and Mesoamerican cultures by befriending existing oligarchies and taking advantage of the often extremely violent existing rivalries between the local peoples, and by securing the cooperation of those peoples by peaceful means, or forceful ones in most other cases. Despite certain sectors of society denouncing this state of things since early into the Spanish colonization (something that was unheard of in other colonial empires in the 16th century), the idea of converting native Americans to Catholicism was always present in this

Hispano-American endeavour. Let us not forget that Bartolomé de las Casas, one of the staunchest advocates for the welfare of the natives, expressed his ideas in favour of the Catholic faith as a means of safeguarding the barbarian population in America in his 1531 treatise entitled "On how to attract all peoples to the true religion", which was written after his "Memorandum for the Reform of the Indies" in 1516, and before his well-known "Brevísima Relación".[3] However, it is essential to take into account the dilemma between evangelization and commercial business in order to understand the image of imperialism the Spanish monarchy had; this image was instigated even by the chroniclers at the time, such as Antonio de Herrera. He was intent on blaming the Castilian kings' representatives for the atrocities that were being committed, while the objective of converting to Christianity all peoples in the America already discovered as well as in that yet to discover was present in the colonization ventures as it had been decreed in a series of papal bulls. And it has been recently discovered that there was also a great deal of popular imaginary in which "conquistadores" and colonists were depicted either as cruel, blood-thirsty killers or as altruistic, pacific human beings. But in practice, the ideological pretext of evangelization did hide the actual political and economic motivation behind the building of the Spanish empire during the following three centuries (Aram, 2008: 61–85). By 1750, the configuration of the West Indies or Hispanic America was almost complete in its essentials, and its specific structure remained unchanged until the new period of territorial expansion began in the 18th century. Both of the pre-existing, major cultures in Mexico and Peru had been subdued, but large areas had not yet been colonized; the Amazonian forest, the North of New Spain, and the farthest areas of the South Cone, and a number of small islands in the Antilles, although the latter were quickly occupied by other European powers. In the 17th century, territorial expansion had come to a halt while Spanish migration to America dwindled and "creolization" grew considerably; between 1630 and 1680, European powers were encouraged to settle in the unoccupied territories as the colonization process of the Spaniards was discontinued.

Within a context of European wars, the Spanish monarchy presented an unambiguous image to its rivals and competitors in the imperial race. This Spanish and Catholic monarchy was ruled by a Flemish-German family, the Habsburgs; they had waged several religious wars across the European continent, and had gone from a hegemonic position to the failure of the ideals of "universal monarchy" they had attempted to impose. In Munster

[3] I have used an edition of 1983: Bartolomé de las Casas, *Brevísima Relación de las Destrucción de las Indias*, Anjana Ediciones, 1983.

and Westphalia, the monarchy lost its control over large territories in Europe, and had to accept the independence of the Dutch Republic (Herrero Sánchez, 2000). This hegemonic shift in Europe led to a period of transition between empires which has not been fully studied yet; this period would see the emergence of new ideological concepts and various differing ways of understanding an Atlantic empire. However, the American empire was still held by the Spaniards and its competitors were aware of it even in the 18th century as the hegemony in Europe was no longer in the hands of the Spanish monarchy and the American empire had become too costly for the crown of Castile. The image of a powerful and fearsome empire was present in the political and economic literature at the time; Jacques Accarias de Sérionne (1709–1792), philosopher, wrote: "We see part of the wealth of the European nations in that of Spain, Mexico, Peru and the rest of the regions in the New World, in which this monarchy extends its domination. No other nation in the world possesses such valuable assets to such a large extent and that can be so easily exploited; and no nation would surpass it if the Spanish population and their industriousness were of such high value as those assets" (Sérionne, 1772: T. I. Ch. IV: 144). This idea was also in the mind of merchants and mariners; in 1679, a sailor on board a ship sailing as part of a fleet that departed from the port of Hamburg heading for Cadiz expressed the idea that the area along the course west-south-west was the "Spanish Sea": an area exceptionally dangerous (Martens, 1925 quoted in Pietschmann, 2010: 345–360). Besides, in the historical context at the time it was very common to associate power with an indiscriminate use of violent means, which was normal in pre-industrial societies (Pinker, 2011).

The Spanish Atlantic was a very heterogeneous space. It was described as an empire as it boasted several distinct features that implied a relative lack of permanent and exclusive borders. From its inception it strongly resembled the ideals of the Roman Empire as the Roman occupation of Hispania was taken as a model. Historian Edward Gibbon stated that "Hispania had had the singular fate of being the Peru and Mexico of the Old World" (Gibbon, 1993: vol. 1, chapter VI, 178). Somehow the Spanish colonization of the Indies served as a model too for the Atlantic empires subsequently established by other nations, especially since early in the 17th century. A constant expansion of a nation's borders is nowadays regarded as "imperialism" combined with a capital city or geographic centre in which all institutions of power and wealth are concentrated and, above all, with a definite attempt to subdue other peoples and nations. Despite all this a peripheral status emerged and was consolidated, leading to a fragmentation of wealth and power through a process that made local authorities become centrifugal elites and resulted in a polycentric kingdom in which, conversely, there was no scope for independent relations between the various peripheral entities (Cardim, Herzog, Ruiz Ibáñez & Sabatini, 2012). According to J. H. Parry, the Spanish empire was feared as it maintained a "maritime life-line" that became one of the oceanic commercial systems having the greatest impact on global processes (Parry, 1990: 117–122). It was an overlapping, multi-layer system of interactions. Parry described the two most important transoceanic systems in place in the 16th century: the Portuguese one with a foot in Lisbon and another in Goa; and the Spanish one between Seville – and the other leading port cities in Andalusia – and the maritime enclaves in the Caribbean and the Gulf. Along this route a two-sided trade system between Spain and its colonies was consolidated as a result of the implementation of a hybrid monopolistic system that remained in operation from the 16th century to the passing of the Spanish Free Trade laws with the Indies in 1776. The Hispanic expansion was devised around a licence-based trading system that was overseen and inspected by the crown of Castile although privately run, and which delimited a number of production areas and markets in Spain as well as in Spanish America.

But that so-called "monopoly" was in fact an institutional flowchart with two objectives in mind; a) to impose on the colonies a centre of wealth and power, and b) to regulate, arrange and levy the fiscal dues on private trade between Spain and America. Several layers of the social spectrum were involved in this trade ranging from aristocrats and noblemen to foreign merchants. A great number of people did participate as transient and non-resident agents, craftsmen and labourers, and they grouped together in merchant colonies or "nations" established in port cities along those routes where they arrived on board merchant vessels. After some time, certain foreign merchants seemed to "fit in" the urban societies in which they integrated and became an elite acting as a powerful and influential lobby for their own trading interests (Crespo Solana, 2010). This commercial system featured strong social foundations as wide as the social spectrum of its human resources and social capital. It is not surprising then that it has been dubbed "the monopoly that never existed" (Oliva Melgar, 2004). In 1580 the Iberian Union was created, and this provided Spain with an advantage until Portugal won its independence in 1640. Despite this temporary unification the two empires operated separately; the trade between Brazil and Portugal was based on relationships between merchants and skippers rather than on a monopolistic system and therefore grew considerably.[4] Trade as well as

[4] See Chapters 5 and 6 in this book.

cultural and economic exchanges took place in combination with other internal – both local and regional – processes that happened in well connected areas such as Mexico, the Caribbean and the Guianas, and Brazil, as these areas were more open to continuous human and economic migration. The other basic pillar of this system was created by the Portuguese on the coast of West Africa as a very active structure (especially after 1550) for the extraction of slaves and their transportation to America. The Slave Trade is central to understanding the involvement (both at private level and as monopolistic companies) of the Portuguese, the Dutch, the French and the English in this so-called "triangular trade"; a "third layer" of interactions was raised in relation to this trade, due to the expansion process undertaken by Holland, France, and England which led to their integration in the American markets, especially after 1620.

The historiography of trade, merchant networks and international maritime routes between the 16th and 19th centuries has shown the most important factors in the evolution and consolidation of the so-called Atlantic World. Atlantic history and global history are two concepts which are, to a certain degree, complementary in the light of new developments in the study of European expansion, its interaction with other continents and the various interconnected systems and sub-systems for trade and social and cultural exchange resulting from these connections over the centuries. However, it is a line of research which is still being reviewed and new contributions have also been made from a socio-economic and political and institutional point of view ((Baillyn, 2005; Canny, 2002; Armitage and Braddick, 2009). For certain specialists in the Modern age and colonial expansion, this interaction began with European expansion in the 16th and 17th centuries, in an era dubbed Mercantilism. According to sociologist I. Wallerstein, who created the concept of a *Modern World System*, the economic interdependence between different worlds began in the 16th century (Wallerstein, 1984). Later, other historians, perceiving a certain economic interdependence on a worldwide scale, fine-tuned the description of the creation of a global silver market and the interconnection of global routes linking various trade areas. Historians also perceived certain economic interdependence at a planetary scale in the early modern age, at least in so far as the world silver market was concerned (O'Flynn and Giráldez, 1995). Further recent studies on history and globalization depart from the economic aspect and enter cultural and anthropological fields in order to explain the emergence of cultural systems at regional level. Studies on the Atlantic system in the Modern age were intent on finding answers to more complex questions as the perspective of possibilities broadened in the comparative history on societies that have, or had, the Atlantic as the centre of their exchanges not only at economic, but also at ideological, human and cultural levels. Specialists on Iberian, British, Dutch, or French expansion have highlighted the fundamental role of the Atlantic exchanges in the regional integration of Europe, America, Africa (due to the traffic of black slaves), and Asia. Top specialists in Europe are approaching the topic from the perspective of constructive criticism and based on well known studies of European expansion between the 16th and 18th centuries. This is a key period to understanding some of the current consequences of this expansion process. Historians such as John Elliot, Horst Pietschmann and Pieter Emmer (to name but a few) have been able to call on the debate of the experts in Early Modern European history. There may still be an outstanding subject: analysing the influence of the Spanish commercial systems in how the other empires' spaces were articulated from a perspective that is broad and comparative and takes into account cross influences.

It is obvious that the Atlantic world could not be regarded as a separate system. In addition to this, by the 18th century only two facts would appear to contradict either view. First, during the 18th century there was a sharp increase in the market availability for certain Atlantic products (for example, furs) in the Indian Ocean basin (and probably in East Asia as well). Second, the discovery of gold in Brazil in 1689 lowered the "world" price of gold, undercutting demand for gold from Mozambique and increasing the export of African slaves from Mozambique to the Atlantic destinations. Two key issues have to be taken into account in order to understand how the Atlantic became an integrated area and what types of interrelations occurred between the port cities/ports of scale along these integrated routes. Those two key matters are: first, how the Spanish fleets and galleons operated; and second, how this vast territorial and oceanic space for exchange was articulated as one of the main linking routes. In this paper I will devote my analysis specifically to the description of the Caribbean area and the Mexican Gulf as these were crucial in the integration of the New Continent in this emerging socio-political, economic and cultural stage that formed in the Spanish America and where syncretism was a key feature.

2. Theoretical Framework

There are methodological obstacles to the idea of studying the spaces created by Iberian expansion as a unit which can be overcome by an approach based on spatial analysis. The first voyages of westward expansion crossed the ocean with the aim of reaching the Indies. These rich kingdoms of Asia had been known for centuries and European merchant states such as Venice and Genoa maintained trade with them. However, when these westward voyages began, there was very little understanding of the Atlantic. Until the 1493 letter from Martín Alonso Pinzón and the map made by Juan de La Cosa (circa 1500), most cartography schools focused on the Mediterranean and on the collective imagination or fantasy about what might be found when sailing west, away from the coastline of Africa, which had been explored using coastal navigation techniques as early as the Phoenicians. Juan de La Cosa's map is without a doubt the earliest known representation of the American continents and it contained previously unknown information. Shortly afterwards, in 1507, Martin Waldseemüller produced the map that gave America its name and depicted a world very distant from the geography of Ptolemy, made by Juan de La Cosa in El Puerto de Santa María (Cádiz), circa 1500 (Álvarez, 2003: 1–15; Fernández-Armesto, 2007: 86, 168).[5] Like him, many other cartographers of the period tried to express the idea of a space in movement, where there were no clear boundaries and in which territorial principles, ideas and frameworks took shape. Historical cartography thus became established as "the science that studies human events through their representation in the theatre of operations in which they occurred, or how they are expressed as comprehensible schemata, as in maps, literary portrayals, paintings or diagrams, which use images or symbols to describe a successive reality" (Varela Marcos, 2012: 17). This idea of relating space in movement to cartography would be present throughout the formation of the Hispanic Atlantic world that took shape around the so-called *Carrera de Indias*, from the first enterprises seeking to colonize and occupy the territory. This is especially true after the 1513 voyage of the armada headed by Pedro Arias Dávila, known as Pedrarias, to what was later called *Castilla del Oro*, an area currently occupied by the countries of Nicaragua, Panama, Costa Rica, and northern Colombia. This expedition in effect meant the occupation of the geographic space which the Spanish crown would designate as the West Indies. It was also an attempt to open up a new route, "a short and easy passage", to the East Indies and the Cathay described by Marco Polo, an idea older than the discovery of the New World itself (Mena García, 1998).[6] From that time forward, the idea of the *Carrera de Indias* meant a maritime-terrestrial

[5] The original map is in the Museo Naval, Madrid (MNM).

[6] AGI, Contratación, 3253: "Cuentas del Gasto de la Armada que fue a Castilla del oro a cargo de su gobernador Pedrarias Dávila, dado por el tesorero de la Casa de la Contratación Don Sancho de Matienzo". Libros de Armadas, 1513-1515.

landscape that would be used officially to carry the wealth of the Americas to the Spanish ports authorized to put together fleets for the New World (Seville and Cádiz). Historically, the *Carrera de Indias* can be defined as "maritime activity between the Iberian Peninsula and the American colonies as well as every business and other endeavours related to that activity. When a trader engaged in American trade by loading his merchandise onto the fleets and galleons, it was said that he was involved in the *Carrera de Indias*. After all, this term defined a historical category that entailed the development of a definitive way of life, which was strongly linked or even subjected to the evolution of a specific, but not limited, mercantile-geographical system, for this system was connected to other trading areas that did not belong to the Spanish empire but were intrinsically linked to it" (Crespo Solana, 2011: 19).

However, beyond purely geographical frameworks that involve thinking about the Atlantic Ocean and the world of relationships and interactions that began to establish themselves in all geographic areas connected to one side or the other, there arose a historical "system" which has been constructed as part of an intellectual tradition that has defined the "Spanish Atlantic system" as a world of interactions rather than a closed world. Carlos Martínez Shaw described the actual creation of this system as the sum of a dense network of economic, social, cultural, biological and political relationships taking place over more than three centuries (Martínez Shaw, 2005: 12). This was an open system where the influences of other nations interacted, but which was primarily dominated by the formation of a model which in turn served as a model for the other "Atlantic systems", as Horst Pietschmann has also indicated (Pietschmann, 1999).

In this context, geographically integrated history offers new possibilities for analysing global exchanges. Within this framework, the Hispanic Atlantic space serves as an essential benchmark for comparative studies. The basic premise would be cartographic representation and data visualization. This explanation implies a map of complex relationships which is difficult to describe with a simple historical narrative using a teleological approach. As Jack Owens has stated, like other disciplines, the historical sciences do not have clear concepts or offer methodologies and tools which facilitate analysis of the complex systems of history (Owens, 2007). Historical systems in fact comprised complex systems which can be analysed from a number of perspectives. It is also necessary to take different factors into account in order to analyse the dynamic transformations that occur in such systems, how social networks operate within them, and how human activity affects the evolution of these systems. Geographically integrated history offers a suitable theoretical framework for analysing these complex systems. This approach takes the form of a historiographical paradigm which proposes three basic issues: a) The history of any geographic area must be understood from the perspective of its connection to other places. b) Historical periods are complex, dynamic, non-linear systems. This defines them as spatially large and, especially since the 15th century, heavily interconnected and involved in continuous global expansion. For this reason, one of the most important characteristics of historical periods is that they are unstable and often go through major transition phases, with the arrangement of their structures changing. And c) within these systems, social agents and places are connected primarily by self-organizing networks, defined as complex and flexible centres that form integrated evolutionary networks within the historical framework corresponding to the centuries of the early modern period. These agents, traders, producers, lessees, civil servants and merchant aristocracy were the chief actors involved in innovations and changes in the systems, more so than a good portion of history has traditionally indicated.

Thus, the Atlantic world of the first global age was made up of various systems that operated and interacted within it as a result of co-operation, exchange and the continuous movement of social agents among the different models for political, economic, and social behaviour imposed by the colonial empires. The first global age was a complex, dynamic, non-linear system deriving from the ongoing process of human interaction. It was characterized by its predictability and the emergence of new developments arising from the significant amount of spatial integration which took place beginning in the 15th century on almost every continent of the planet. New social and cultural environments took shape, with significant consequences for human action. From this perspective, key problems or processes for understanding how this system worked are already the subject of analysis. These include the evolution of long-distance trade, the consequences of migration and the formation of localized minorities, racial interbreeding, slavery, and the impact of colonialism. In fact, what gives shape to this heterogeneous geographic space known as the Atlantic world, fed by four continents (North and South America, Europe, and Africa), are these global processes of migration, colonialism, trade, and cultural and intellectual exchange that make the social space of the Atlantic a finite and definable world. Studies of these processes which take a spatial approach seek to move away from traditional jurisdiction-based ideas. Originally championed by French historians, these have become pervasive in historical studies that have primarily emphasized the study of sovereignty and the formation of the modern state, while a spatial and global perspective extends beyond national boundaries.

Positioning the analysis of the formation of the Hispanic Atlantic world with the study of the ocean routes which developed and how the spaces connected to the *Carrera de Indias* took shape is a challenge for spatial history, as it considers this process from the point of view of geographically integrated history. One of the approaches used in this method of analysis is to put forward new views on how interactions evolved, from local and regional perspectives. The majority of global regional integration since the 15th century has been created by extensive trade and sea voyages. In this context, small causes have large and complex effects which have taken centuries to change. First, foreign trade, especially colonial trade, defined European economies. Secondly, it is not yet clear to what extent internal and external forces converged in the formation of colonial policies, institutions, and the social, cultural and economic issues inherent to the societies of the different regions (Emmer, Pitri-Grenouilleau, Roitman, 2006).

The new challenge is to initiate research devoted to studying, in a parallel, comparative way, the role of the trade networks as actors in the processes of spatial (geographic) integration, in the social and economic relation of the interconnected areas, their evolution and function in each specific spatio-temporal context. J. B. Owens argued (2007) that as an answer to these conceptual and methodological challenges the study of world connected history requires the use of new communication technologies. This approach focuses on the study of problems deriving from spatio-temporal representation and the analysis of networks and routes of great geographical and historic cartographical importance, as previously claimed by Fernand Braudel, but it would not be possible to carry this out without the help of new technologies which will doubtlessly complement and enrich the work of specialists in European expansion and global trade between the 16th and 19th centuries. One of the reasons encouraging us to utilize information technologies (such as GIS) is the fact that those technologies can be of great assistance to historical analysis as visualization is one of the best methods for the empirical observation of phenomena. In view of this consideration, this line of research requires the creation of new conceptual perspectives which look at the spatial turn (Middell and Naumann, 2010), which is considered to be a constitutive element of historical action which also involves the drawing by historians of certain "mental maps" and new research agendas with focus on historical geography (Krugman, 1999; Pieper and Schmidt, 2002). Global trade between the 16th and 18th centuries would not have been possible without the emergence of new forms of human behaviour (co-operation, competition), the adaptation of which was very much dependent on the complex dynamics from which this evolution developed. Furthermore, this perspective has caused an increase in the analysis of historical data on networks and integration of areas using GIS, with a view to the historian building new narratives of knowledge. In this context, the comparative study of trade activity carried out by various trading companies and the connections between port cities and economic areas becomes an essential topic when it comes to set out new investigations into the commercial systems of the mercantile nations; the Spanish case is a very interesting model for analysis.

Many of the existing studies of the Spanish commercial system (in the Atlantic and, to a lesser extent, in the Asia-Pacific area of influence, due to the trade with the Philippines) make reference to how the trade networks operated in either area. They describe the emergence of new forms of co-operation and competition among economic agents and how merchants developed mechanisms of trading and cooperation. In this respect, a finished project with very interesting results[7] has shown that commerce in the first global age was characterized by high rates of smuggling and this was not possible without co-operation and a close relationship between agents who, on most occasions, lived far apart from each other and had never met one another. In most commercial port cities these agents formed an oligopoly, almost always having previously had ties of kinship through marriage or godparenting by means of various mechanisms of symbiosis and integration. Even when formal commercial agreements could be ratified before a notary, these were frequently hard to enforce and their validity depended heavily on the willingness of the parties to co-operate with each other. For this reason, trust and reputation were crucial factors to understanding merchant behaviour in social environments. Therefore, the behaviour of merchants, financiers, and others in the development of commercial networks will provide evidence of cooperation in trading activity, which has not been considered in other studies. Global trade between the 16th and 18th centuries would not have been possible without the emergence of new forms of human behaviour, co-operation, defection or competition, depending on the complex dynamics of this evolution. Furthermore, this new, proposed, perspective may encourage the analysis of integrating evolutionary social networks into dynamic GIS to form a complex narrative system. According to recent historical approaches to the study of the evolution of social networks, where co-operation is dealt with

[7] See http://www.scientificpapers.org/special-issue-june-2012/.

in the social sciences, especially in evolutionary and historical anthropology, economics, political science, and sociology, the dominant perspective states that greater human co-operation in trading activities became possible with the increasing effectiveness of the state, which was accompanied by the rise of capitalism and individualism characteristic of modernity. This paradigm appears largely based on the standard periodization scheme of the discipline of history, which has viewed the year 1500 as the main turning point after which the metanarrative has been based on the rise, primarily in Europe, of the state, capitalism, and the modern individual, concepts that are frequently reified into actors. In this context, the comparative study of the mercantile groups between different cities and the interests that they represent, as well as their influence in the evolution of politicians of different states, the regional integration analysed at levels ranging from local to global, the influence of networks and integration in economic transformation and their role in the modelling and stabilization of intercontinental maritime routes and the hierarchical structuring of markets and productive areas, are key to understanding the emergence and development of the so-called Atlantic System and/or the First Global economy. Most importantly, this type of study can become the foundation for future comparative studies between empires. This proposal aims at going beyond the previous studies of trade networks and communities conducted so far which have provided us with a wealth of information, in order to set the challenge of relating merchant behaviours to economic spatial analyses.

At the same time, the application of GIS technology to the study of world connected history is a recent line of research which is currently drawing the attention of the historian. It is a topic which has given birth to projects which really are on the frontier of knowledge. This line of research offers new analytical perspectives on databases and historical conceptual models designed for GIS technology, but these still require systematic testing with real data. In Andrea Kunz' opinion (2006), the results from historic GIS have two aspects. On the one hand they are related to the history of the evolution of a specific phenomenon such as transport, the administration of a territory, or the development of human relations; by working on simple maps through to related series of subjects the outline of a multimedia atlas becomes apparent. On the other hand it is obvious that the cartographic presentation of historic data not only complements the traditional function of a map by enlarging the description but it may also become a tool in itself to the economic historian as he will be able to observe new connections between the spatial dimensions of historical processes (Gregory, 2003; Knowles, 2008).

In fact, one of the most highly valued perspectives is geography, especially the economic geography of Krugman (1999), who asserts that geography and economics must be included when creating models which explain the historical evolution of regions and countries. Krugman proposes a model for spatial representation which has been used primarily by geographers and economists (Ramírez, 2001: 25–38). The idea is to include geography as a variable for understanding macro- and microeconomic data, as well as using mathematical models and the critical approach found in neoclassical theory. The most important part would be spatial representation in order to understand the geographic location of economic activities. Here, this would not be understood as the study of the location of the factors of production – land, capital and labour – in space. Instead it would be a process of empirical data-based knowledge which allows historians to explain where the economic activity took place and why, what factors affected the evolution of these locations and how they acted internally and externally in the medium and long term, with the aim of relating the theories of global history to the search for the background to the world's current problems (Northup, 2006: 249–267).

Processing and analysing large repositories of data on a historic system and their relationship to the spatial distribution of processes and their evolution have developed significantly thanks to spatial history. In the words of Philip J. Ethington, the concept of "spatial history" brings to mind "the problematic way we tend to view history and geography as disparate subjects: as though study of the past does not out of necessity examine the intersection of time and place. That is, both historians and geographers concern themselves with change, measurable in space with the passage of time. All history has a spatial component". He went on to say that: "All human action takes and makes place. The past is the set of places made by human action. History is a map of these places. The past thus exists not in time but in space" (Ethington, 2007).

Spatial history is a historiographical turn that proposes an analytical schema which involves interpreting the dynamics of historical events based on their spatial presence. It is not entirely new, as the theoretical framework owes much to Fernand Braudel, Jules Michelet, von Ranke, Henri Pirenne and Joseph Lefevre. If I had to come up with a phase that is common to the beginning of all of their works, I would summarize it as: "modern history begins with a landscape". For the subject of study, although not the object, was a landscape. This work already forms part of the tradition and it is in new research that further progress can be made, from both a methodological and technical point of view, as well as in relation to the complex issue of concepts. In his book *The Production of*

Space (1991), Henri Lefebvre stressed that space does not simply refer to natural geography, nor is it an empty container filled by history. Rather, space includes the importance of spatial relationships. Space and human relations change over the course of history and each influences the other. Lefebvre spoke of "social space" when considering the idea that space and humankind are mutually constituent elements and not separate categories in an abstract model. Relationships are important in the geographic and social construction of history. The work of the historian is in fact dedicated to highlighting this important phenomenon: the relationship between space and mankind. Relationships are categorized and analysed by the movement of human beings through space and this movement is dynamic. The new approach since the "spatial turn" clarifies new concepts related to space. Its importance stems from two issues of significant value for historiography today, as it champions a reconsideration of "space" as a historical and analytical category. Aside from notable exceptions, the study of space as an object was already starting to be found lacking in the majority of previous historical studies, as analyses that involved the configuration of time were traditionally given greater consideration. However, space as an object of study was also subliminally present in studies of Atlantic history, as well as the intellectual tradition of historiography on European expansion and global history. Nonetheless, these schools considered the oceans and associated territories around them as cultural and historical constructions overlaid on heterogeneous geography, rather than applying truly scientific economic and social criteria. This idea has been broken open with the publication of a group study on the importance of oceans as means of communication and global union, while at the same time considering the waterscapes or marine landscapes where human social action takes place not as empty spaces, but as key for analyzing global integration and its evolution (Mukherjee, 2013: 3). These studies must be interdisciplinary, as they analyze social, economic and political issues with the help of cartographic visualization.

But what is truly novel about spatial history is that it uses computer technology, particularly all those tools that facilitate unification, processing, analysis, and visualization. Spatial history is becoming virtually synonymous with digital humanities, as described in many works written since the late 1990s. Based on this work, a methodological model for spatial economic analysis has been established, incorporating four main aspects. First, group research is capable of achieving a scale suitable for this type of multi-disciplinary study, in which the results must be visualized alongside the discourse of the narrative, because the aim is in fact to visualize historical narratives. Secondly, visualization of the results is strictly dependent on digitizing the results, on geographic information sciences (geographic information systems or GIS) and their various related analytical tools. Third, Digital history makes it possible to exploit many types of information and sources using digital repositories and databases (White, 2010).[8] This interdisciplinary co-operation uses not only texts and charts, but also images and maps, to visualize processes in order to see changes in spatial relationships, making it possible to analyze changes in space-time, that is, in space-movement. This term goes beyond economic space and can be defined as the network of cartographic points that underpins the network of social agents that appear grouped between the geoentities and cartographic areas where the polygons which are the object of study can be defined, and from which it is possible to extract variables for analysis (Pinto Gomes, 2013), and the nodal points or key sites for understanding the flow of historical processes (Dolman and Ettinger, 1992). Spatial history projects are producing a number of repositories with maps for use in GIS, but they still lack proper conceptual and theoretical contextualization regarding whether this is a line of historical research or a new interdisciplinary branch of the humanities. Generally speaking, the digital humanities require a suitable theoretical and methodological context (Siebert, 2000: 537–574; Carter and Malouf, 1989: 173–183). Spatial history may be better understood as a sub-field of history which offers the possibility of accepting the spatial turn, so that it "represents an acceptance of interdisciplinary methods, whether historians realize it or not. Like text analysis, which often brings literary methods into historical analysis, spatial history, especially spatial analysis with digital tools, brings geographic methods into historical analysis". Thus, the most important aspects are the digital applications and tools that make it possible to visualize historical processes. Lastly, it seems that these projects are open-ended, a "never-ending story" that remains open for ongoing use and to check historical data.

Within this historiographical framework and with the help of spatial representations, it is possible to discuss space from a historical and conceptual point of view. Spatial issues have often been considered the sphere of geographers and town planners, but it has been the attempts made by specialists in geography and even economics that have made the concept of space accessible to historians. These studies and the use of digital tools have provided an adequate methodological apparatus to follow this spatial turn which historians dedicated to studying global connections now champion (Schmidt, Wardf and Arias, 2009). It involves

[8] Richard White, "What is Spatial History?" Spatial History Lab, 1 February 2010. Stanford University. www.stanford.edu/group/spatialhistory/cgi-bin/site/index.php.

the need to construct mental maps in order to analyse the evolution of time and space within socio-historical processes. In fact, the spatial turn is now found in studies of the global expansion which took place during the centuries of the early modern period. Although from a conceptual point of view, the idea of space still needs to be clarified, the concept can be approached from different perspectives. As Stuart Elden argued in a doctoral thesis on the philosophical and metaphysical foundations of the historical concept of space according to the ideas of Heidegger and Foucalt, it is necessary to historicize space and spatialize history (Elden, 2002). In other words, rather than providing an analysis of how the meaning and use of the word "space" have changed over time (undoubtedly a useful analysis for geographers and cartographers), historians must recognize that space, places, and locations are crucial determining factors in any historical study.

Furthermore, spatial history has an interesting philosophical background related to the metaphysics of place. Stuart Elden took an interesting approach to the metaphysical concept of space in the work of Foucault and the philosopher Heidegger, suggesting that historical space should be reconceived as spatial history rather than the history of space. This means that space should not be just an object of analysis, but also a tool for analysis. As a result, it has been possible to analyze how external and internal influences converge in local configurations.[9] An excellent example is the work of Jonathan Amith which "explores the history, political economy, and culture of space in central Guerrero, Mexico, during the colonial period". It analyzes the processes which affect the colonization of non-Western societies, while also offering a theoretical and methodological framework for understanding how a territory is converted under colonialism (Amith, 2005). We can ask ourselves this question: When incorporating spatial history or any other aspect of what we now call digital humanities, it's vital to remain faithful to the narrative dimension, to unpack the meaning of each visual representation. What is this visualization actually telling me? What does it add to the information it contains?

The concept of Atlantic space has been outlined in other ways. Therefore, spatial history represents a challenge for the history of the Atlantic Ocean and its societies, as it forces us to delineate territories which are sometimes only ambiguously related and which have significant environmental, geopolitical, cultural, ethnic, climatological and geographic differences. David Armitage argued that: "The attraction of

---

[9] http://marbl.library.emory.edu/remapping and blog about "Mapping and Spatial History": hastac.org/blogs/bendavidweber/2012/10/21/mapping-and-spatial-history.

Atlantic history lies, in part, in nature: after all, is not the ocean a natural fact?" He later went on to state that the Atlantic is a European invention, "the product of successive waves of navigation, exploration, colonization, administration, and imagination" (Armitage, 2004: 9). However, it is not possible to overlook the fact that Atlantic history was originally founded on certain assumptions with a strong ideological bent which defended the construction of civilizations based on ideas and customs. Therefore, this Atlantic space is defined as a sort of boundary between the Old World (with Judeo-Christian and Greco-Roman foundations) and the New World, where the old combines with new challenges imposed by geographical and historical constraints (Kraus, 1949). Wallerstein even stated that there was a cultural relationship that made reference to Western Europe and North America (2000). The fact is that the Atlantic is not only a physical reality, but also a cultural one. The aim is to affirm its geographic nature as a junction for exchange, rather than imposing ideas deriving from centuries of colonization and empire. It is a physical reality that also includes inland seas such as the Baltic, the Mediterranean, and the Caribbean. It also contains some territories which have been difficult to navigate, this being the reason why they have sometimes been excluded from the economic framework or have been less attractive destinations for migration. Geography in itself does not seem to benefit much from homogenization into one system and in this the history of the Atlantic worlds tends toward a global scale of consideration (Morienau, 2006: 301–316). It also seems that the difficulties of geographic constraints have in part been concealed by global history and the critical approach to the civilization building model found in the work of André Gunder Frank (2008).

With regard to studies of the Spanish colonial empire and system of trade, there has already been some research focusing on detailed analysis of focal points where the activities that made up the networks surrounding trade were situated, locating the places of action where business activities took place and studying the attributes and functions of the different locations. In fact, they are framed within a historiographical context in which the spatial perspective is pervasive in the majority of recent studies on global expansion, globalization and the consequences of colonization and global trade. This had already achieved considerable scale in the work of Guillermo Céspedes del Castillo (1991). Now, there is a great deal of work relating to how human societies have created an interconnected world through commerce and the exchange of ideas and products. This is also being done for other models of colonial empires (Haggerty, 2006). In these empires, there was a struggle to overcome the obstacles imposed by space and distance and maintain communication and order. In many cases, as John Elliot argues for the British and Spanish empires, it was precisely this that led to various mechanisms which gave rise to the fragmentation of the imperial order, as well as the creation of powers on the periphery (Elliot, 2006).

Spatial analysis of this Atlantic world offers many possibilities. Within a line of research related to the history of exploration and cartography, it would also be important to analyse the "significance of imaginative projections" of the different territories that made up administrative territorial bodies, how nations took shape, and the territorial and spatial make-up of the empires. To this end, it is useful to produce cartographic projections and see how they evolved over time in order to study the interrelationships among the various conceptions of territory, colony, nation, or continent. These types of analysis have been done in the case of the New World. This includes analysis of representations of space and place according to different peoples, societies, and civilizations; the iconographic ideas of the Americas, or Europe or Asia; and the associations between maps and policy; as well as how these ideas evolved over the centuries. It is worth noting the important role played by space in two specific types of analysis. Firstly, there is how mercantile networks were structured, the role of mercantile networks and communities seen and analysed from a global, comparative and transnational perspective. In this context, one of the most important phenomena to occur is extensive co-operation between mercantile networks and communities. In relation to this, it is important to take into account the conception and perception of the different routes. One example of this analysis studies routes in Asia (Bose, 2006; Picazo Muntaner, 2012). These analyses of networks offer, for example, the possibility of using certain tools (such as Gephi or Cytoscape) to visualize the connections between historical actors and the broader world from the perspective of a port city, for instance. Secondly, it is necessary to study the interaction between social agents and geography. This could include the study of the specific weight of the different cities on maps, especially in relation to the administrative and institutional schemata of the empires. This perspective could be expanded to study how geographic, racial, ethnic, religious or cultural differences had an impact on the understanding of the constructions of human society, such as the public, institutions and the various forms of human associations and mutual societies. We could also examine the practices and theories behind how different peoples engaged with the land, and the forms and consequences of land distribution in the conquered territories.

One way of defining space may derive from analysis of tangible and intangible (visible or invisible) networks that affected the interdependencies of economic and social life. One problem that appears is whether or not the creation and densification of these networks lessened the problems associated with distance. The types of interaction that occurred influenced the globalization of markets, as well as social and cultural life in general. In the long term, they substantially changed behaviours related to the condition of distance and the idea of spatial interdependencies, something which clearly occurred in the case of the Spanish empire (Echevarria Bacigalupe, 2013). With this, it would be possible to come up with a theory in order to examine the hierarchies which took shape in spatial relations – the position of the territories in terms of the political and institutional administration of the period – as well as how that territory was integrated into the global map of routes and markets. In the long term, this also created a new globalized world characterized by new economic networks and new ways of organizing the market (Andersson, Anderson and Johansson, 2003). There is still an important question which remains to be answered: how can this affect our current knowledge of global history, and in this particular case, our knowledge of the impact of the evolution of the Hispanic Atlantic empire on the make-up of today's world, of the world that has prevailed since this empire?

The aim of the following chapters is to put forward certain theories and present methodological advances which may be of interest for studying the shape taken by the social space that is created around the relationships between different agents (in this case merchants), taking the system of trade in the Hispanic Atlantic as a case study. The goal, in the words of Stuart Elden, is to spatialize this history.

3. The Hispanic Commercial Model of Spatial Integration: An Historical Narrative

The Hispanic Atlantic expansion was a dynamic spatial model which evolved into self-sustaining locational structures. And this spatial structure thus created was, perhaps, what mostly influenced the logistics that the other mercantile empires established when it came to developing their own respective expansions. For over two centuries, and unlike other expanding nations, such as France, or the maritime powers at the time (England and Holland) in which private domestic capital gives them the opportunity to exert strong financial leverage, the complex financial system of the Hispanic monarchy is made up of an intricate network of transnational interests.

Between 1516 and 1700 the Hispanic crown is a political entity formed by a group of scattered territories with citizens on several continents.

Although, as we know, they have a federal political tradition, this conglomerate of territories is called the Spanish Empire. From 1492, when Christopher Columbus made his first trip to America, the crown of Castile created various institutions aimed at governing, administering, and fiscally ruling the vast territories in the New World. The *Casa de la Contratación* was created in Seville in 1503, and the *Consulado de Cargadores a Indias*, also based in Seville, in 1543. While the wealth of resources, especially mining, in the vast American territories are the driving power of the American economy, commerce works as the means of obtaining profits by private businessmen – including the increasing number of foreign merchants – as well as by the crown, as the latter is an active partner in these commercial ventures with America. In order to safeguard this established order a number of mechanisms are put in place – official supervision, private collaboration, a single-port policy, and escorted shipping. The Sevillian *Consulado de Cargadores*, a guild of merchants, (so called *cargadores*) in the Indies Trade, has the exclusive right to trade with America. By doing this, the crown is effectively transferring the royal monopoly to this small but powerful group of subjects. Some years later similar institutions are created in Mexico (1592) and Lima (1613) with identical functions. Throughout the 17th century, the *Casa de la Contratación* begins to cease taking the initiative in matters relating to the Indies Trade, while the *Consulado* becomes the true ruling body of the *Carrera de Indias* (García Fuentes, 1980). Commerce is centralized in a single port city with the object of facilitating traffic control and tax collection. This town is Seville, although from the early 17th century Cádiz became an important port of arrival owing to the difficulty involved in sailing up the River Guadalquivir. Generally speaking, this single-port policy was loosened after 1529, with other ports such as La Coruña, Bayona, Avilés, Laredo, Bilbao, San Sebastián, Cartagena, and Málaga being allowed to trade with America. However, all ships had to arrive in Seville in order to allow for the *Casa de la Contratación* to carry out its fiscal operations. In 1573 Philip II reinstated Seville as the monopoly's single port; it remained so until 1668, when fleets began to load and unload in Sanlúcar and Cádiz.

The government of the Hispanic monarchy could barely exert proper control over all of the territories that it incorporated into its complex political system as kingdoms. With the aim of governing, administering, and taxing the territories which the Crown considered most important, it created the two viceroyalties of New Spain and Peru. For a long time, these were the two most important manifestations of Spanish territorial control in the Americas. The Castilian monarchy installed a viceroy at the head of each and appointed governors with governing, military, financial, and judicial authority at a time when there was no separation of powers. Each of these viceroyalties united large areas which had not been fully explored within vague boundaries. The Viceroyalty of New Spain included a large part of Central America (Guatemala, El Salvador, Nicaragua, Honduras, and Costa Rica), except for the government of Castilla del Oro (which included Panama and the isthmus area). It extended east to the Gulf of Mexico and the West Indies, although the islands (including Cuba, Santo Domingo, and Puerto Rico) were governed independently. As it spread further into North America, New Spain came to include today's American states of California, Texas, New Mexico, Arizona, Utah, Nevada, and part of Colorado. In the Pacific, New Spain was put in charge of administration of the Philippines, conquered by López de Legazpi in 1564. The Viceroyalty of Peru was a very troubled territory until 1555, due to opposition from the *encomenderos* (individuals to whom the Spanish crown had granted control of a certain number of indigenous people) to the new laws established for the territory. These laws had been enacted by the crown to halt the abuses of the early colonists. It was a vast territory that encompassed virtually all of South America, except for Brazil, the Guianas, and the coast of Venezuela in the Caribbean. It also included the governments of Rio de la Plata, Chile, and Panama. On this vast map, there were hot spots over which the Spanish crown sought to gain economic control by means of a shipping lane across the Atlantic. This shipping lane was extended with ports of call on the Atlantic islands between Andalusia and the Caribbean and Gulf of Mexico. It linked the Spanish metropolis with the mining areas (Zacatecas and Potosí being the most important) from which the gold and silver to be shipped to Spain were obtained, and the cities where fairs, a meeting place for traders, were held (including Portobelo between 1606 and 1739, and Veracruz, a Mexican port). The fairs constituted a space which gave rise to markets based on the networks. It could even be said that the markets as such would not have existed without the networks of merchants (Álvarez Nogal, 2011: 53–87). The voyages of the fleets to the Indies were organized around these activities.

The organization of the Indies fleets has been studied at length, and there is an extensive bibliography on the subject. Essentially, a spatial system took shape around a network of ports, primarily between Spain and the Americas. However, other cities on the European Atlantic soon joined the network, establishing themselves as commercial centres for warehousing and redistribution. The system of fleets was eventually regulated in 1564; it consisted of two separate convoys: a) the fleet to New

Spain, setting sail in April bound for Veracruz and calling at Santo Domingo, escorted by two warships, the *Capitana* and the *Almiranta*; b) the "Galleons to Tierra Firme", setting off in August bound for Cartagena de Indias and then to Nombre de Dios in the Panama isthmus (in 1598 this final destination changed to Portobelo). These galleons bound for the American continental area, the mainland (the so-called Tierra Firme), totalled six or more; their main mission was to carry the silver from Peru on their journey back. Once the various fairs at the destination ports had been attended to, both fleets hibernated and then gathered at La Habana in March or April, from where they carried silver and American produce. The main fleet, the Fleet to New Spain, sailed from Seville and Cádiz, connecting Havana and Mexico. It also made the return voyage in the opposite direction. There were certain ports in Spain which the Spanish crown considered to be the official ports of entry for the fleets sailing this route, two to be specific: Seville and Cádiz, apart from their chain of outer harbours. But in reality, there were other connections. English and French commerce in Spain had been very active since the Middle Ages. As a result, the ports of the Bay of Biscay and even the Levant region took part in American trade, either directly or as part of a system of ports of call, what the Catalans and Dutch called "port to port". This consisted of exchanging products between various secondary ports on the Iberian Peninsula in order to connect with the departures of the fleets or galleons from Seville and Cádiz. This route generally carried manufactured goods bound for the Americas and silver on the return voyages. They brought gold to Spain from the mines of Potosí and Huancavelica (in Peru), unloaded products for distribution in the Americas and made connections with domestic fairs, especially the one at Portobelo (Lang, 1998). This chain was vital for financing the wars of the Hispanic monarchy. However, by the 17th century, Spanish navigation of the Atlantic was sharing spaces with the expansion of other nations, with the Caribbean and Lesser Antilles as the Achilles' heel of the empire. Because of its desire to obtain gold and silver, Spanish colonization focused too heavily on the continent, leaving behind open flanks where powerful influences gained access, soon turning the West Indies into a favourite area for smuggling and piracy. The increasingly common presence of foreign fleets in regions near the colonies led ships to sail to certain areas excluded from the fleet system. In 1627, registered ships began sailing to Guyana[10] and something similar happened on the coast of Venezuela.

In economic terms, the fleets can be said to have crossed the system of triangular trade that had emerged in the Atlantic world as a result of the

---

[10] AGI, Contratación, 1694.

convergence of systems of trade which developed between Europe, Africa and the Americas. However, apart from these two main fleets, there were other secondary fleets that gradually interconnected different spaces and integrated them into the principal polygon of ports linked by the *Fleet to New Spain* and *Galleons to Tierra Firme*. Small flotillas linked to other routes which linked marginal spaces where active interregional trade had developed. In the long term, they represented strong competition for the main economic route of the fleets and galleons. The secondary fleets connected places of production and mining in the Americas with the nerve centres of the fleets and galleons in their mission to return to Spain with the resources of the different regions of the Americas. There were in fact three secondary fleets: one in the Caribbean and one each in the southern and northern Pacific. The latter two moved along trade routes that ranged from Río de la Plata to the Philippines, sailing far into the Pacific Ocean. In theory, the secondary fleets would deliver their goods to the primary fleets (those which crossed the Atlantic) at the time the latter reached the metropolis, but in reality this almost never happened. This was due to a number of factors, including the technological obstacles of the time; the enormous distances they had to traverse, sometimes through unfavourable winds or storms; and the delays in finalizing business agreements due to complex dealings between the merchants. The bases functioned as the main points of terminus for the Carrera de Indias in the Americas: Cartagena, Portobelo and Veracruz (sometimes Havana) for business with the Caribbean, and Acapulco and Panama for trade with the Pacific and the Orient. Veracruz and Acapulco were linked by a road that crossed Mexico from coast to coast. This was the route also followed by Mexican silver from the north travelling from Mexico to the coast. Panama and Portobelo, with a distance of some 60 miles between them, were connected by a small road called De Cruces, which led to the upper stretches of the River Chagres, at that time used to reach Portobelo. The route crossed the isthmus, parallel to what is today the Panama Canal. It was a real bottleneck for trade along which the American silver travelled, making it the object of pirate attacks and very dangerous territory (Serrera Contreras, 1992: 14, 84–92). The American fleet in the Caribbean was enormous and primarily made up of small vessels that connected ports in Cuba (Havana, Santiago, Matanzas), Puerto Rico (San Juan, Ponce), Santo Domingo, Venezuela (Cumaná, La Guaira, Puerto Cabello, Coro, Maracaibo) and the New Kingdom of Granada (Riohacha, Santa Marta, Cartagena), as well as Central American and Mexican ports. On this map, Cuba appears as an important island in the Atlantic-American connection, as it became the location for trade brokerage for the Spanish fleets and a gathering place for other local and Northern European commercial interests. In addition, following the establishment of an urban centre on the coast, it was a preferred location for boat building. In 1629, Philip IV ordered "that the residents of the city of San Cristóbal de La Habana may enjoy and do enjoy, a corps of builders of these kingdoms, and that their ships be admitted into any fleet to the Indies at the location where they may be found, according to their order of precedence, based on the date of their arrival in the ports of these kingdoms"[11].

In contrast to what has sometimes been stated, the transoceanic voyages of the fleets and galleons occurred at regular intervals, although there were some ups and downs in certain circumstances, for example between 1596 and 1622 (Lynch, 1984: 227) and after 1654, if we can rely on the testimony of the sea and military captain Antonio Garrote, a native of Seville.[12] The fleets usually departed every two years, according to monopoly regulations. In 1621, the armada under Thomas de Mazpuru departed with the fleet headed by Juan Benavides and the Tierra Firme galleons with Juan de Florez Ravanal. In 1622, another fleet was readied and departed under the charge of Fernando de Sousa, with the armada under the Marquis of Cadereitas and the galleons under the charge of Juan de Camorán, although it returned to Spain with treasure in 1623. There were some minor irregularities, as in the case of the 1628 armada under Thomas de Mazpuru, "which due to passing the winter, did not arrive that year, but arrived in 1629, Saturday, 7 April"; or the 1636 armada under Carlos de Ibarra, which accompanied two fleets to New Spain under the charge of Martín de Vallecillas and Juan de Vega Bazán, as "due to the death of the first, both fleets arrived under the charge of the second", together with the galleons under the command of Luis Fernández de Córdoba. In 1637, the armada under Carlos de Ibarra, Marquis of Caracena, had problems on the return voyage "due to a fight with the Dutch enemy within sight of Havana, and not being able to hurry, it spent the winter in New Spain and could not reach these kingdoms in the year 1638 and came from said provinces of New Spain in the year 1639 with its fleet, under General Martín de Orbea in the month of July". The voyages of the fleets and galleons began to be organized at greater intervals after 1656, although fleets departed in the years 1656, 1658, 1660, 1662, 1665,

---

[11] Quoted in: *Recopilación de Leyes de los Reinos de Indias*. Madrid, 1841, Ley XI, p. 41.

[12] BNE, "Manifiesto al Rey y al Consejo de Indias del Capitán de mar y guerra, Bartolomé Antonio Garrote, en que demuestra que las Armadas y flotas de Nueva España y Tierra Firme han salido de estos reinos para América todos los años desde 1580 a 1699", Mss. 12633, fols. 192–214.

and 1666. After that, the preparations continued, although the number of merchant ships decreased, a problem which would later result in the inclusion of foreign ships. From 1645 to the end of the reign, silver continued to be shipped on board the fleets, although to a significantly lesser extent. This did not mean that the fleets halted their voyages, and they continued returning to Spain except when special circumstances forced them to spend the winter in the Americas. However, they carried less silver. The other legs of the trade route remained stable, although with some changes in the types of goods shipped, which included increasing amounts of European manufactures.[13] During the War of Spanish Succession, Spanish trade with American ports briefly suffered, but resumed in great force with the fleet of 1711, organized by Minister José Patiño in Cádiz. It set sail that year under the command of Andrés de Arriola, and in addition to 1,596.85 tonnes of merchandise, it carried the Duke of Linares, newly appointed as viceroy of New Spain. On its return, several ships were separated from the others due to the storms that lashed the fleet as it left Havana in 1713. The ships reached Cádiz and Faro, Portugal, under the command of Pedro de Ribera, due to the death of Arriola in Veracruz.[14] During the 18th century, the fleets returned to normal, although the voyages of the galleons gradually decreased until they disappeared. This system was reformed and transformed, with the most important changes being introduced by a law which permitted the inclusion of "miscellaneous ships", recognized in the 1720 Royal Fleets and Galleons Plan, although it functioned poorly until 1738. It was between 1739 and 1754 that the War of Jenkins' Ear, with the Austrian succession as a backdrop, led to the actual establishment of the registration system, also creating companies which were given trade privileges with the Americas (García-Baquero González, 1976).

In this complex setting, stories of piracy abounded. Pirate activity played an active role in the process of imperial transition and was common between 1620 and 1648, when attacks on the Spanish fleets increased. Piracy and smuggling and the trade war, with which other nations (England, the Netherlands, and France) mounted a counter-attack against the Spanish Atlantic offensive, were forever linked for the Spanish crown. But the societies where pirates lived grew up on their own and soon went from being a weapon in the war against Spain to a general problem in this vast frontier space between the power of the European colonial institutions and the social and economic reality that gradually took shape. Between 1650 and 1720, piracy reached its height, affecting all trading nations

[13] AGI, Contaduría, 570.
[14] AGI, Contratación, 1272, 1273.

equally. In England in the 1620s, it began to be considered an evil that had to be eradicated but in reality it was a problem in the organization of a society at the limits of the regulations of the absolutist states and introduced enormous instability into the world order (Earle, 2004: 80). With regard to the issue of piracy, there were essentially two activities which most brought the Hispanic monarchy into conflict with other powers. The first was confrontations between corsairs who threatened the Spanish fleets carrying treasure to Spain. Secondly, there was the emergence of pirate societies on the margins of the empire that made their living from smuggling and illegal trade, generally aided by the co-operation of royal civil servants, the majority of whom were *criollos* (people born in Latin America to European parents). This situation is difficult to separate from the occupation of some of the islands in the Lesser Antilles, making them a nest of outcasts.

In order to defend the coasts of New Spain, a special fleet was created under the name of the Armada of Barlovento. It was organized in 1635 by order of the king, with the involvement of the Count-Duke of Olivares, for the safety and defence of the ports of the Indies and to contain the Dutch enemies who attacked in 1645 and 1643. The Armada of Barlovento was kept on to serve as a coast guard fleet throughout the entire section of the Caribbean which suffered the most attacks by pirates and corsairs from enemy nations. This area was also undergoing an accelerated process of constructing a new social and economic reality outside the Spanish empire. From what we can gather from statistical data this system was safe and efficient; in fact, fleets were successfully assaulted on rare occasions. The whole treasure from New Spain was successfully captured only once, by the Dutch admiral Piet Heyn in the Cuban Bay of Matanzas in 1628. For safety purposes a number of armadas were created, such as the Windward one based in Veracruz, or the *Armada del Sur*. Money was levied from the merchants themselves through the *impuesto de avería* to cover the military expenses for these escort operations. Other than safety, a number of factors hindered the system of fleets: the inflationary situation inherent to this system, the slowness of the convoys (especially on the return trip), and an inflexible trip schedule, combined with the on-going state of warfare in which the Hispanic Monarchy was involved from the 1620s.

There was another route between West Africa and the Caribbean, which primarily carried slaves. Spain played a special role in this trade, especially during the period when Portugal was part of the monarchy. Later, significant *asientos* (contracts to provide slaves) were awarded to Dutch merchants, which also allowed the Irish and Genoese to co-operate in the slave trade (Herrero Sánchez and Pérez Tostado, 2010: 307–323). There were two basic routes for the slave trade which affected Spain: the route from Africa to the plantations of the Caribbean, which would later come under Dutch and English control; and the route that went from West Africa to the colonies of South America, especially Brazil, run by the Portuguese. There was also an important supra-Atlantic route, travelled by the galleon to Manila, the connection between Europe and Asia to the west, via Mexico. There was also other commerce and other routes connected to these, just as in Europe the Baltic and the Mediterranean were incorporated into the Atlantic economy. Over the course of the 17th century, the commerce of other nations and that which took place in the areas excluded from the system of fleets and galleons soon constituted one of the most important foundations for the development of the Spanish colonies in the Americas (along with other colonies). In this regard, noteworthy examples include trade between the Caribbean and the North American colonies or within the Caribbean itself. Transport, financial services, and slave trading featured heavily in this commerce, and it was carried out in collusion with other nations, especially the Dutch, French, and English. This system, which to a certain extent had been established and opened up by the Hispanic monarchy, was fed by transnational co-operation in economic matters. And, as André Gunder Frank argues, the profits it provided were especially useful for allowing the American colonists to cover the deficit in the balance of trade with Europe and to accumulate their own capital (Frank, 1998: 102). In fact, the slave trade took place on the fringes of the fleet and galleon system. Except for the early years after the conquest, the Spanish crown never had a direct monopoly on this trade, which was carried out by the Portuguese and the Dutch, and after 1711 by the English. The exchanges and business deals surrounding the slave trade made cross-national business between traders of different nationalities possible. Transporting slaves always went hand in hand with other business dealings, such as sugar, which made bilateral trade between all the nations necessary, despite the rivalries among them. And so this triangular trade became the spatial framework for multilateral trade, although the structure of the markets between Africa, America, and Europe remained in place. However, there was also direct trade between different European and American port cities. Therefore, a new polygon of exchanges was established, superimposed on that which surrounded the triangular trade system.

The conversion of this space into the setting for multiple exchanges took more solid shape between 1650 and 1690, when competition between the trading nations and France became stiffer. The routes sailed by the

Spanish fleets travelled right through these two trade polygons created by other nations (that formed by the triangular trade system and that formed by direct trade in goods). This reality seems to have been reinforced beginning in the 1630s, when the system of commercial networks began to consolidate around Atlantic trade. This decade also marked the start of a decrease in the amount of American silver reaching Spain (Lynch, 2007: 184). In this period Spain's granting of *asientos* coincided with more widespread use of permits or licences to award trade privileges in other countries, especially England (in the specific case of England, they began to increase between 1630 and 1660). As occurred with the *asientos*, this created a trade model based on granting monopolies, halfway between royal prerogative and modern, liberal concessions. In the long term, this would establish a new system of trade, economic liberalism. The 1630 Treaty of Madrid represented an attempt at an alliance between England and Spain to co-operate in the wars in Flanders. However, for the English, it was merely intended to gain trade privileges in Spain, on condition that they would not act as trade intermediaries with the enemy (Girard, 2006: 121). The changes in the "Hispanic Sea" began to be more obvious from this point on, when peace negotiations with the United Provinces became a necessity due to the difficulty of continuing with the war. In the second half of the 17th century, co-operation between Amsterdam merchants and Portuguese Jews, many of them residing in Hamburg, Spanish ports, or Portugal, grew stronger in response to the problems arising from the Atlantic war, which seemed to have a negative effect on retaining the sums of money invested in travel ventures and on payments on insurance policies, which had increased due to the conflicts. This problem is mentioned repeatedly in commercial correspondence.[15] During this period, Spain lost control of some of the West Indies, including Curaçao and Sint Eustatius, which became major areas of trade and smuggling after 1630. Jamaica, taken 1655, came under English rule. France and England occupied other areas, including the Bahamas (Spanish until 1670), Antigua and Barbuda (1632), Jamaica (1655), Barbados (1524) and Santa Lucia (1654). France consolidated its occupation of the northern part of the island of Hispaniola, which had been abandoned by the Spanish crown since 1603, when the king of Spain ordered all of the island's inhabitants to move to Santo Domingo to prevent them from interacting with pirates. This area was officially recognized by Louis XIV in 1665 (Sevilla Soler, 1991: 163–185). The non-Spanish invasion of the Caribbean was soon joined by the problems resulting from the Dutch occupation of Brazil,

[15] Case of Lucas van Peenen, servant of Nicholaas van Hulten, merchant in Amsterdam, 22 April 1630. GAA, 664 (17) 119v.

which began in 1624. This conflict would overlap with the crisis caused by the revolt of Portugal beginning in 1630. Additionally, the Spanish viceroyalties of New Spain and Peru had authority over a large number of royal courts known as *audiencias* scattered throughout the vast territory. This led to a significant dispersion of power, which was diverted into the hands of Spaniards and *criollos* (Sánchez-Albornoz, Lockhart and Bowser, 1990). This structure was retained with few changes and many internal conflicts throughout virtually the entire 17th century. The main ports of the viceroyalties of New Spain and Peru were the areas under the greatest control. Veracruz, in New Spain; Nombre de Dios, Portobelo-Panama-Cartagena de Indias in connection with the hazardous isthmus region; and El Callao (Pacific port) and Havana in Cuba, were the key port cities for the institutional monopoly. The extent of their connection to the route of the fleets and galleons led to the development of colonial institutions in neighbouring areas.

Map 1. Spanish Routes in the Caribbean

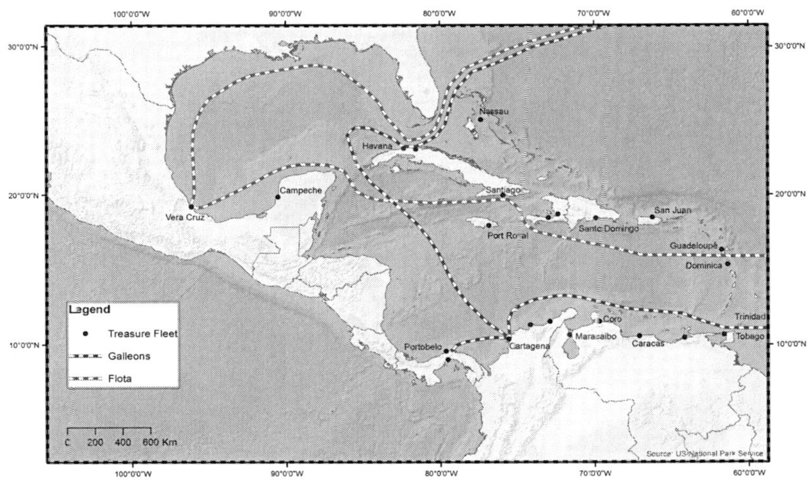

For example, a 1549 royal decree issued by Emperor Charles V created the court known as the Audiencia of Santa Fe. This was the result of the economic boom in the newly founded Kingdom of New Granada and its size in comparison with the difficulties and lack of communications in the region. To the north, the Magdalena River was the only major communications route, while to the west, towards the Pacific Ocean, the challenges for traffic were almost insurmountable. This new *audiencia*

encompassed the territories of Santa Marta, Rio de San Juan, Popayán, Cartagena de Indias and Guayana or "El Dorado", and even some parts of Popayán (divided between Santa Fe and the Audiencia of Quito, created in 1563) and Maracaibo (divided between the jurisdictions of the *audiencias* of Santo Domingo, Santa Fe and Caracas).[16] The boundaries of this territory were vague, even though it was governed by a governor and a captain general, five civil judges (*oidores*) or criminal judges (*alcaldes del crimen*), one crown prosecutor (*fiscal*) and other ministers. Territorial control over these areas was incredibly complex and very ambiguous. For this reason, today it is possible to locate documents for several different areas in various territorial sections of the Archivo General de Indias (General Archive of the Indies) in Seville.

Beginning in 1717, two new viceroyalties were created, again producing a certain fragmentation of power in the Americas. This helped consolidate the rise of the *criollos* to spaces of power, their discontent with the discrimination which they supposedly experienced when seeking public positions and complaints regarding the metropolitan obstacles to trade and industry (Mires, 2001: 155). The Viceroyalty of New Granada was created in 1717 and its jurisdiction covered all of the territory of the Audiencia of Santa Fe, Maracaibo, Caracas, Guayana and that of the *audiencias* of Panamá and Quito, which were eliminated as such. It was abolished in 1723 and then definitively re-established in 1740. Sebastián de Eslava was appointed viceroy, captain general and president of the Audiencia of Santa Fe by the Royal Decree of 20 August 1739. The instructions to the viceroy stipulated that he would be president of the Real Audiencia of Santa Fe and governor and captain general of the jurisdiction of the New Kingdom of Granada and attached provinces, which would include "Chocó, Popayán, Kingdom of Quito and Guayaquil, provinces of Antioquia, Cartagena, Rio del Hacha, Maracaibo, Caracas, Cumana, Guyana, islands of Trinidad and Margarita, Río Orinoco and provinces of Panamá, Portobelo, Veragua and Darien".[17] The *audiencias* of Panama (abolished in 1751) and Quito would still survive, subordinate to the viceroy, who appointed three commanding generals, one for Cartagena with command over Santa Marta and Rio de la Hacha; one for Panama with command over Portobelo, Darién, Veragua and Guayaquil; and one for Caracas with command over Maracaibo, Cumaná, Guayana, Río Orinoco, Trinidad and Margarita. Following the Royal Decree of 12 February 1742, the Government of Venezuela no longer reported to the Viceroyalty of Santa Fe. When the Captaincy General of Caracas was

[16] AGI, Santa Fe, 540, 1, fol, 141–143.
[17] AGI, Santa Fe, 541, 1, fol. 15.

established in 1777, by Royal Decree of 8 September 1777, the provinces of Maracaibo, Cumaná, Trinidad and Guayana were separated from the Viceroyalty of New Granada. The newly created Alto and Bajo Orinoco and Rio Negro had been added to the last of these provinces, set up as a General Command, in May 1768.[18] The captain general of Venezuela reported to the Audiencia of Santo Domingo until 1786, when the Audiencia of Caracas was created. This "nationalization" of territories previously subordinate to bodies or viceroyalties that were more centralized, but scattered and weak in political and military terms, went on over the entire 18th century, when European trade with Spanish America was becoming established and trade routes had multiplied multilaterally in relation to the economic interests of the different nations.

Spain's social and economic circumstances caused a situation very different to what the crown's laws sought. Spanish trade with its American colonies was based on a rigid institutional system but it degenerated from the end of the 16th century due to several factors: limited shipping activity of the fleets and galleons due to the lack of an effectively armed navy; a period of crisis and the outbreak of the Atlantic wars between 1629 and 1699; smuggling undertaken by all European nations in the Hispanic America, and especially, a widespread corruption of the officials of the monopoly both in Spain and America. One factor in Spain's singular role throughout the 16th century is the fact that Spain had become the main centre for silver and other precious metals destined for the various international circles of commerce (Chaunu, 1955; O'Flynn and Giraldez, 2002). As a consequence, several Spanish cities became trade centres engaged in re-exporting colonial products along various maritime routes, which also attracted merchants from different European and Mediterranean countries arriving and settling in those cities, thereby forming merchant colonies where they, in turn, directed their economies to the Atlantic. After 1640, Cádiz became the main port for the Carrera de Indias. The city was home to many foreign merchants who had settled there in order to trade with Spanish America. This new capital of Spanish colonial trade, replacing Seville, which had declined over the second half of the 17th century, received official support from the Spanish crown in 1717, when the Casa de la Contratación (Board of Trade) was transferred there (Crespo Solana, 1996; Bustos Rodríguez, 2005: 47–52). The activity of foreign merchants in the city had much to do with this decision, due to their control over certain resources, such as salt, and their ability to form relationships with the local élite.

---

[18] AGI, Santa Fe, 542.

4. Space and Perception

The ideal method of analyzing and understanding the spatial system of the Hispanic Atlantic world might be to build databases relating to two main issues: First, a group of digitized thematic maps showing the territorial structure of the areas of the Americas linked to administration of the empire, making it possible to see their spatial-temporal evolution. Secondly, several of the existing databases should be linked in a standardized data model combining data on quantitative and qualitative indicators relating to economic and social issues, with special reference to the role of social agents in the design and behaviour of the system. This would be important for developing true thematic mapping in order to better understand the logistics of the system described in the previous chapter. As it is the job of the historian to analyse changes in processes over time and space, I would emphasize the importance of illustrating this model for other comparative analyses, both diachronic and in other later periods for which we are not yet sure whether these behaviours were altered due to the internal dynamics of the networks themselves or as a result of various external political, cultural or economic determining factors. Thus, this type of analysis would be of interest for later comparative and evolutionary studies on the functioning of trade networks when integrating such diverse, geographically distant, and heterogeneous spaces. It would even be possible to, for example, establish a behavioural model related to situational causes or determining factors for a period.[19]

The difficulty of maintaining such a complex system of trade was mentioned in narratives of the time. Study and discussion of some of these sources could facilitate the work of constructing local historical narratives connected to the overall history. Some of these sources discuss the route of a voyage or an itinerary, which is explained by means of a chronology that includes a description of events. However, *mirabilia* (marvels) also form part of the discourse that gives way to construction of the space. These characteristics are also found in accounts of trips (López de Mariscal, 2004: 27). Some of this evidence about space appears in descriptions of trips and pilot books. These sources offer information about the perception of space and indicate that one problem was physical access to the different geographical enclaves. Even so, emphasis is placed on the efforts of the sailors to overcome the "servitude of space" (Echevarría Bacigalupe, 2013: 136) and complete their political and commercial missions. Narratives of trips are rich in information about the perception of space

---

[19] See map in Figure 3, Chapter 4: Georeferenced West Indies route showing the direction of travel of ships.

and time. In 1534, a German Landsknecht from Straubing called Ulrich Schmidl recounted the story of a journey of colonization on which the city of Buenos Aires was first established. He spent years travelling through what he called the "paradise of the jungles of Paraguay and El Chaco", but first offered a unique source of space-time narrative: "departing from Antwerp on the second of August, I reached Hispania by sea and later the Indies, by the will of God".[20] In fact, this area did not become important to the Spanish crown until the 18th century, although there were earlier expeditions. This region was part of the Viceroyalty of Peru, the destination for trade on the route of the galleons that called at Portobelo. The enormous size of the area encompassed by this territory, virtually all of South America except for a territory ambiguously bordered by the Amazon (Portuguese Brazil), led to the creation of the Viceroyalty of the Río de La Plata or Buenos Aires in the 18th century (1776–1777). The intention was to protect a region that was the object of active illegal trade outside the control of the Spanish crown (Moutokias, 1989). However, there were some expeditions to this territory beginning in the early 16th century. Schmidl's account of his journey with the fleet under Pedro de Mendoza describes the travel times for the voyage across the mid-Atlantic and his stop in Cádiz, having recently arrived from Antwerp. There he saw "lying on the coast before the city a whale or *walfisch* with a length of thirty-five paces and from which they have removed some thirty barrels like those used for herring filled with blubber". His journey from Antwerp took fourteen days "as the distance by sea is calculated to be four hundred leagues from Antwerp to the above-mentioned city of Cádiz, where fourteen large ships were prepared and well equipped with all the necessary munitions and supplies". From Cádiz he describes the passage to the Canary Islands, a mandatory port of call in the "Hispanic Sea", on ships which he said belonged to Nuremberg merchants Sebastian Neidhart and Jacob Welser, and accompanied by "other High Germans and Dutchmen, up to eighty men, well equipped with our firearms and other weapons", departing from Seville. "On St Bartholomew's Day, we reached a Spanish city called San Lucas [Sanlúcar de Barrameda] twenty leagues along the route from Seville. There we remained anchored due to the force of the wind until the first day of September of that year. Then we set sail from the aforementioned city of San Lucas and from there we reached three islands that are near to each other. The first is called Tenerife, the second is called Gomera, the third Palma; and from the city of San Lucas to these islands it is more or less two hundred leagues. There the ships

[20] Expedition of Pedro de Mendoza in: http://www.biblioteca.org.ar/libros/10069.pdf.

were divided amongst the islands. These islands belong to his Imperial Majesty and the inhabitants are pure Spaniards with their wives and children and they make sugar. We also went with three ships to La Palma and remained anchored there for four weeks and again took on supplies and repaired the ships" (Mitre and Lafone Quevedo, 1903).

From the earliest Atlantic voyages, the crews of the ships that made up the expeditions were comprised of people of different nationalities. There were not only Spaniards and Portuguese, but adventurers from the feudal lands of the king of Spain (especially the Low Countries and the territories of the Hapsburg Empire) would also make the journey in search of wealth and opportunities. It is important to highlight that, as the author of this account states, many preferred to board the fleets going to "New Spain in Mexico", rather than embarking on the enterprises destined for frontier areas in the viceroyalties of New Spain or Peru, voyages on which overcrowding and hazards were common, or where the commander had the power of life and death over disobedient travellers. Most of these voyages were undertakings involving occupation and trade, especially the voyages of the fleets and galleons whose destinations were the focal points for the exchange of raw materials and precious metals for provisions for the colonies. These products were generally manufactures of various kinds which it was believed the colony needed for the use and survival of the newly settled. The cities which hosted fairs were gathering places. These included Portobelo, which was described as an "unhealthy little town because of its location and boggy due to the hot climate" by Pierre Vilars, ambassador of the French court to Spain (Sanz Camañes, 2004: 288).

In addition to the problems resulting from the control over the fairs which was held by local merchants – as was the case at Portobelo, controlled by merchants from Lima and later by the Peruvians – the galleons sailing to Tierra Firme encountered their own problems. This voyage was sometimes more hazardous than that of the fleets to New Spain. A 1654 report describes a departure from the Bay of Cadiz, the flagship leaving on 2 June of that year, with the rest of the ships in the armada departing the next day with five merchant ships. Two galleons and the fleet of New Spain were dispatched, as well as some *pataches* travelling with the galleons to Portobelo, Cuba, and Margarita Island. With various vessels, the expedition sailed along the coast and then between the Leeward Islands, arriving in Cartagena on 12 July.[21] The return crossing was very hazardous. In addition, it was necessary to negotiate any problems resulting from the possible condition of the ships following the

[21] BNE, "Relación de viaje de una flota enviada por el rey Felipe IV desde Cádiz a Portobelo, 1654", Mss. 18719/40.

voyage: "With the ocean-going galleon San Felipe it sailed away the previous night with a fresh wind, and having here settled the fine for the smuggled clothing at 23,000 pesos and leaving the galleons Santiago and Hércules, also ocean-going, the latter being sold for 6000 pesos as it had made the voyage from Spain taking on a great deal of water and could not assure the return voyage without any work, and the Santiago, so that after a great deal of carpentry work and caulking and with a new main mast brought from Spain it could then follow me." When a ship could not be repaired, it might be sold at the destination port and never return to Spain, with its munitions and supplies being used to fit out any other vessel in the fleet that might have lost its equipment and rigging. The anonymous captain in this report, perhaps galleon commander Marquis of Conde Alegre, stated that he reached Portobelo on 2 August. It was normal for the arrival of the armada of Peru to be expected, with prior correspondence between the president of the Casa de la Contratación and the admiral of the armada; so that when the galleons reached Portobelo they could "start bringing down the treasure to be loaded, as had been done since the year '57". Normally there were complaints, like those of other admirals of the period, about the regular delays in all port activities related to guarding the armadas and communication with the authorities, who were not always of a mind to co-operate, given the corruption present in the Indies institutions themselves and the long waits for dispatch boats. These dispatch boats were the only way to keep the Casa de la Contratación informed about any possible problems. The captain tried in vain to persuade the Count of Alba, viceroy of Peru, and president Pedro Carrillo (sic) to have a dispatch sent as soon as possible so that royal officials in Panama would send the treasure he was to take back to Spain. It took Carrillo one month to hand over the register with the reports on "what all those trading in Peru had presented before his person". With this material, the captain began making the register for the return voyage, finishing on 18 September. He took delivery from the masters of the different quantities of silver that were destined for Spain, including the part for the king's treasury, the assets of individuals and the proceeds of sale of certain products (629,676 and 38,400 pesos, respectively), as well as entries for 204,114 pesos, 176,114 reals and 28 thousand pesos in indigo. These sums were, in the captain's judgement, "very little of his majesty's silver and none of individuals", due primarily to the irregular arrivals of the galleons and the fact that the majority of this silver followed other routes. In fact, this was a reflection of the increase in activity by the Peruvians during the reign of Philip IV, which among other things, caused the galleon voyages to be more and more infrequent. This fleet scattered, with the flagship reaching the coast of Cantabria on 12 April 1659. Apart from the delays and the apparent lack of trust, there was the problem of the hazards of the return voyage, which are also described by the anonymous author in this report. In this and other sources, space appears as pure intuition within which impressions are also described. These sources to some extent show the "three spaces of human experience, perceptual, behavioural and physical" according to the reasoning laid out by Monica Wachowicz based on the views of philosopher Ernst Cassirer (1874–1945) (Wachowicz, Orellana, Crespo Solana, Owens & Bernabé Poveda, 2008).[22]

Based on the historical data collected in my database *CrespoDynCoopNet Data collection*, held in the institutional repository at the Consejo Superior de Investigaciones Científicas,[23] it is possible to compile information about trade voyages in order to analyze what type of locations acted as the "nodal points" in this Hispanic Atlantic system. This is a very detailed database on merchants. It contains their relationships of co-operation and their trading activities as well as those activities from large monopolistic companies, which materialized into a series of maritime routes with specific ships and fleets. These databases have been integrated into a data model which has been devised around three basic questions that are common to all these historic databases: (1) socio-economic system, (2) merchant networks and (3) trade circles (Crespo Solana & Alonso García, 2012). The GIS model presented in Graph 1, in UML notation, creates a link between two large information blocks compiled from very heterogeneous sources: on the one hand the life cycle of the ships involved in the trade operations at the time, which includes each trip with the ports of departure, call and destination as well as the commercial transactions that took place and the incidents that occurred, and on the other hand the relationships of cooperation established by the agents on specific geographic locations, be it private traders or large monopolistic companies.

---

[22] "A spatial representation for reasoning about the Atlantic commercial routes of the 18th century", unpublished paper, presented to *2008 Annual Meeting of the Social Science History Association*, Miami.

[23] http://digital.csic.es/handle/10261/28394.

Figure 1. Data model in UML Martin 8.0 notation by DynCoopNet

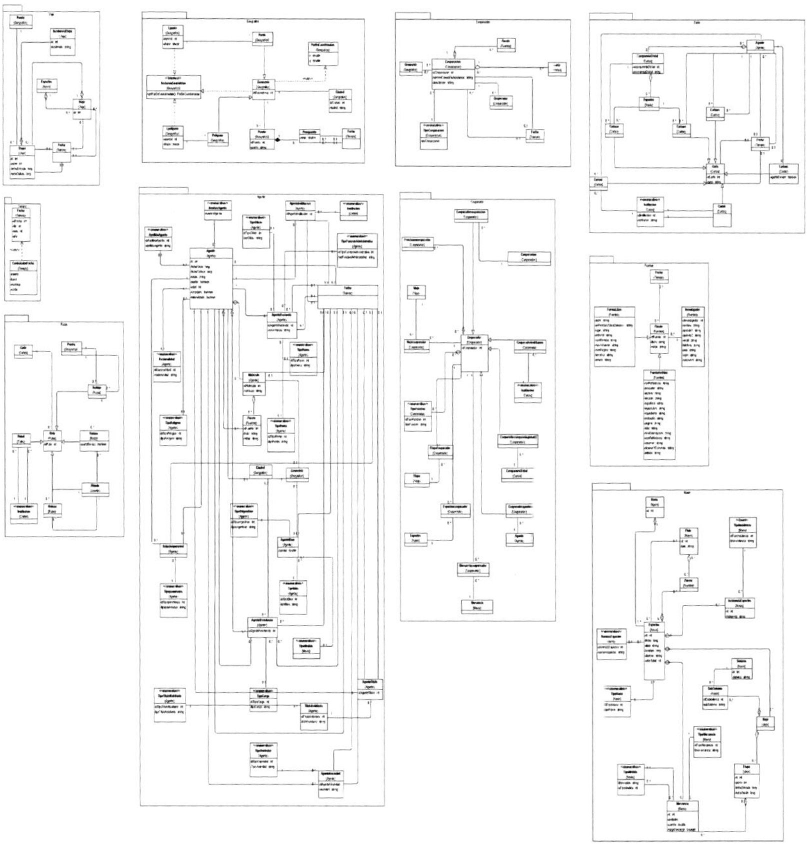

Both blocks are related in the model. This will allow for further analyses that will open up new perspectives in the understanding of the behaviour of the self-organizing networks. Such analyses will address new questions formulated in specific geographic spaces. These 16th–18th century networks involved a great deal of co-operation among individuals from very diversified social, linguistic, and religious backgrounds, and gave rise to a movement of people and goods within a wide open system in an increasingly expanding geographic space. Because of the complexity of the data model, its handling and exploitation is done through a purpose-built API (Application Programming Interface) in order to facilitate the data upload by the investigators involved in the project.[24] One of the two main tables in the database is that for SHIPS. The mercantile history of each one of them is conceptualized in what has been denominated "spectra", which refers to a ship's state in an instant of specific time and in a specific localization (spatio-temporal component of the GIS). This will allow following up on a certain ship on each and every one of the marine-commercial trips it was involved in, as a result of the different co-operation relationships established among diverse agents, whether at individual level or as global monopolistic companies. Among other features, this will enable us to visualize in the GIS over a base cartography the routes followed by the various ships or the fleets those ships were part of; the incidents that occurred, the volume of traded goods, etc. The model's second great table is the AGENTS and the information related to them: family relationships with other agents, religious confession, residences where they developed their commercial activity, positions occupied, institutions they worked in, and even the epistolary relationships maintained with other merchants, etc. The design in this block has been devised to be subsequently capable of describing and representing the co-operation relationships among the different agents in specific geographical localizations. The two previously presented blocks are connected through what has been denominated "co-operation", commercial relations among agents. An AGENT is in fact a merchant that, either one temporarily associated with another agent for a specific commercial dealing in the journey of a ship loaded with certain goods (and with some specific epistolary relationships on occasion), or one belonging to one of the existing large global monopolistic companies in the period studied; either way this agent used to play a certain role in that commercial relationship (factor, agent, master, etc.). This table is the central axis of the data model, establishing the relationship between the two previous large tablesblocks: ships and agents.

With regard to the locations, these may be considered "geopoints" or "geoentities" (Pinto Gomes, 2013). "Geopoints" are an important reference framework within the interface, as their location in geographic coordinates is necessary for them to appear in the interface viewer. One important type of analysis is the perception of space that was had, studying the cartographic weight of the locations (Picazo Muntaner, 2012). This is of great interest in relation to considering ports of trade as "nodal points", understood as key locations for plotting connections from a cis-Atlantic perspective. The aim is to identify the location of these "nodal points" within a fragmented geographic space, although in the Spanish case, these

[24] https://161.111.47.117/dyncoopnet3/login.jsp.

were strategically situated in a regionally integrated space, unlike other trading empires such as the Dutch (Meinig, 1986: 62). A gazetteer has been created with the locations, geopoints, and nodal points that appear in the historical documents. It is made up of 1325 locations, many of which correspond to historical place names.[25] These locations refer to the spatial information found in a total of 6018 trade voyages contained in the database. In this system, a high percentage of port cities act as nodal points between major centres of trade. Between the nodal points and the territories linked to them, there is a sort of unfolding of space which can be observed on the maps. This unfolding of space refers to the existence of numerous small river ports, towns, and settlements which are subsidiary to the main port city where most of the principal economic activity takes place. It is also possible to establish a hierarchical network of centres of trade and finance, distributed geographically, which can be clearly visualized thanks to the graphic device. For example, the cities with greater cartographic weight are:

Map 2. Locations of most important atlantics port cities linked to the Spanish *Carrera de Indias*

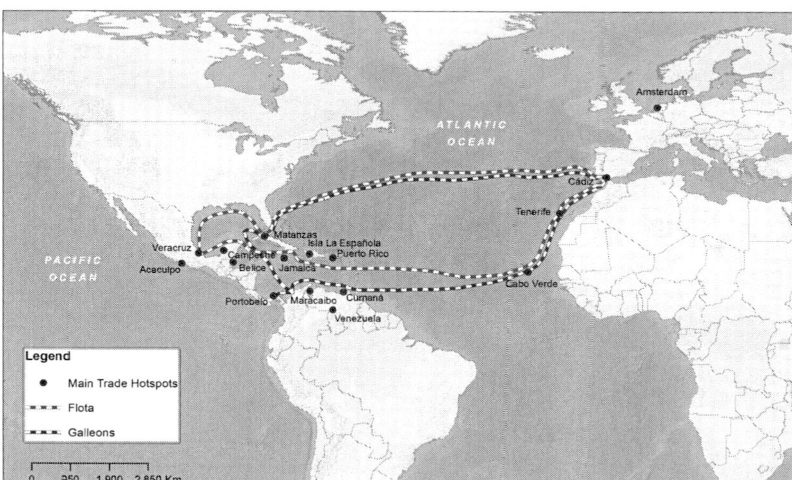

[25] I am grateful to my GIS assistant María José García Rodríguez who has worked on the maps.

Map 3. Locations of Places (detail)

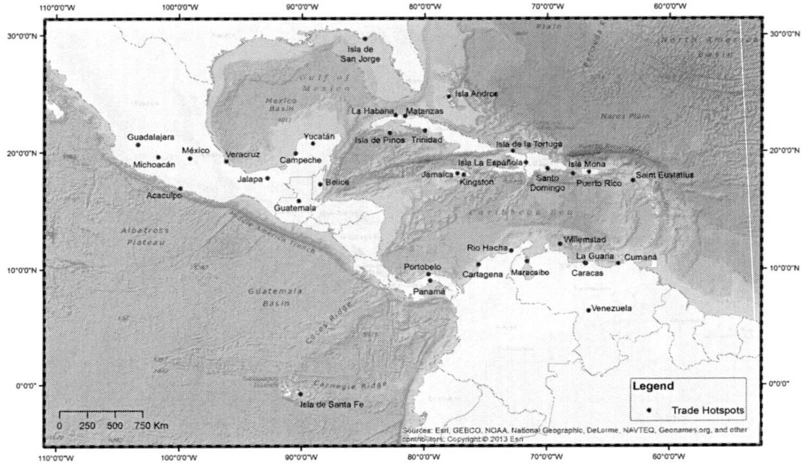

These locations form an economic space, with each of them having a certain function. It is thus possible to create a bibliography of the cities that have the common characteristics of being financial centres, centres of trade, and sea ports. The geography of this system offers the view of an "interesting route" that crossed the Atlantic and which constituted the "Hispanic Sea" or Hispanic Atlantic that was perceived by sailors, seafarers, and traders of the time who travelled this route. The term "interesting route" refers to the spaces with easy access for business ventures and commercial exchange, around which the areas of political-administrative control developed. There is an abundance of thematic maps on the main ports of trade and it is possible to see how these ports began to appear on maps as they became integrated into the commercial system of the European networks. On many occasions, their presence meant the disappearance of other nodes of lesser importance in the collective imagination of the Old World. For example, this is what happened with Nombre de Dios, the earlier destination of the galleons, which was replaced by Portobelo, because the first was a dangerous location on the coast of Panama near the River Chagres with the South Sea (Mar del Sur) and the North Sea (the Atlantic). They began to be included on maps as they penetrated the market and more and much more detailed information was obtained.

Map 4. The Isthmus of Darien and Panama in 1697[26]

The role of distance is also of interest. The absolute cartographic weight of the Hispanic Atlantic is related to two fundamental issues: one is the component of trade penetration and the transfer of information, and the other is the relationship of distance. The connection between markets and distance developed as part of a much wider process. The market is defined not only by economic forms, but also by the physical space in which it is located and by the time factor. Thus, the transnational trade of the Atlantic system is subordinate to three major determining factors. First, traffic decreases with distance. Prices also vary from port to port, with significant differences between them. Sometimes this involved over-speculation, as occurred at major fairs such as Portobelo, and some important port cities such as Veracruz or Cádiz. In Veracruz, local merchants inflated prices for their own profit when goods left the city for distribution to inland markets. In Cádiz, the merchants had become illegal buyers of silver, monopolizing exchange and financing for their benefit and for the benefit of the

[26] A New Voyage Round the World, London, 1697, by Sebastián Wallroth. Source: http://commons.wikimedia.org/wiki/File:Map_of_the_Isthmus_of_Darien_and_Panama.jpg.

European lobbies that invested in Spanish trade with the Americas. Thirdly, geographic barriers reflect the obstacles posed by the many transport costs, tariffs and quotas, as well as the problems involved in finalizing deals. These three variables dictate a comparative model with the system of trade that emerged in the Pacific Ocean, as demonstrated by Antoni Picazo (Picazo Muntaner, 2012).

Despite this heterogeneous Atlantic world, which is problematic when it comes to analysing it as a whole, with many "disease environments" which had catastrophic consequences for native populations as well as new settlers, this perception of space gave rise to political and social issues. The Atlantic is therefore a historical construction as well as an imperial geography. Just what characterized the Atlantic was its heterogeneity; in reality, trade and migration did not interconnect areas due to the profound economic differences which resulted. In fact, "rather than boosting the economy, the 'Atlantic System' stimulated the expansion of European values and norms, such as private property, monogamy, the nuclear family, free labour, and the place of women and children in society" (Emmer, 2008).[27] Indeed, this is related to how we use the word "system" and the fact that the Atlantic, like global integration, is not an economic phenomenon, but rather a cultural one. These may be the general properties of the entire system, in which it is possible to see an evolutionary process related both to the capacity of human beings to give form to hostile space and to unexpected changes deriving from the fragmentation of space.

[27] http://nuevomundo.revues.org/42173.

CHAPTER TWO

TEMPORAL GIS FOR HISTORICAL RESEARCH[1]

MAY YUAN[2]

Historical GIS is now a vibrant area of research that applies mapping and spatial methods (both quantitative and qualitative) to the study of the geographies of the past or the influence of geography on history. A quick search on Google Scholar with keywords "historical GIS" returned over 35,000 results.[3] While historians consider time as the key to organize information, GIS methods use space as the first-order principle for representation, inquiries, and understanding.

The importance of spatial emphasis in GIS pushes a "spatial turn" that leads to geographic approaches in historical research and more broadly humanity scholarships (Finnegan, 2008; Warf and Arias, 2008; Bodenhamer, Corrigan and Harris, 2010). Similarly, GI scientists have recognized the need to make a "temporal run" in GIS since the early 1990s – for example, Langran (1989), Peuquet, (1994), Wachowicz (1999) and Yuan (1999). In addition, many temporal GIS papers with diverse ideas have been published throughout the 2000s. Fundamentally, temporal GIS research investigates conceptual, representational, and analytical challenges to integrate spatial and temporal data in one information system. Much of the spatial dominance in GIS inherits a space-centric or layer view of the world inherited from cartography and remote sensing where line drawings or picture elements (pixels) default to 2D data layers, the space-results in time being considered as an add-on to spatial data. In other words, GIS data modelling first determines where things are and then decides what has changed over time.

---

[1] This material is based upon work supported by the National Science Foundation (NSF) under Grant no OCI0941501. Any opinions, findings, and conclusions or recommendation expressed in this material are those of the author and do not necessarily reflect the views of the National Science Foundation.
[2] University of Oklahoma.
[3] Searched at http://scholar.google.com on 26 April 2013.

This chapter aims to provide an overview of temporal GIS development and its relevance to historical research. Discussions go through the concepts of time, how these temporal concepts are integrated into spatial databases in GIS, and furthermore, what limitations inherited from these temporal concepts consequently impose GIS constraints in the study of spatial history, and what additional temporal concepts will be desirable to overcome the constraints.

1. Temporal GIS with Snapshots, Space-Time Composites, and Spatiotemporal Objects

With layers of spatial objects, time may be associated with a GIS layer to note all spatial objects on the layer to this particular time (e.g. a layer of land use polygons in 2013 to denote the distribution of land use types in Norman, Oklahoma for that year), which is commonly known as a *snapshot* model. In a snapshot, every spatial object is attributed to the same time, which can mean that observations of all spatial objects on the snapshot layer are taken at the same time. Remotely sensed imagery by definition falls in this category. Most studies that analyse and map historical statistics data (e.g. demographics, vital records, disease, etc.) treat temporal information as constant for all spatial objects on the map, and therefore also adopt the snapshot model, which has been the primary GIS application taken by national historical GIS projects (Gregory et al., 2002, McMaster, Lindberg and Van Riper, 2003).

However, snapshots can be artificial. Descriptions or data about the spatial objects on a map might be for different periods of time. In the past, mapmaking used to be a lengthy and tedious process that could take months or years. Giovanni Domenico Cassini started assembling geometers to conduct surveys starting in 1760 to create a map of France at a scale of 1:86400. The map, *Carte de Cassini*, took 30 years to completion with continuing efforts from his son Jacques Cassini, grandson César-François Cassini, and great-grandson Jean Dominique Cassini. Much could have been changed in geographic features or settlements within the 30 years. The snapshot approach ignores the temporal differences in geographic features on a data layer and only captures discrete changes through a series of snapshots over time, which overlooks gradual transitions between the snapshots.

Alternatively, we can associate time information with individual spatial objects to indicate their differences in temporal characteristics. For example, every parcel polygon has the date of the last property transaction. Time-stamped spatial objects enable GIS to support queries about spatial objects in the same time period or according to certain temporal relationships (e.g. before, after, co-occurrences, etc.). Moreover, time-stamped spatial objects allow GIS to analyse spatial diffusion (e.g. spatial distribution of the onset of infectious disease in each county) or landscape of *space-time composites* (e.g. mosaics of wildfire burns). Many early temporal GIS studies on historical cadastral database development took this approach to record dates with individual land parcels (Hunter and Williamson, 1990). With such data, we can search for parcels which were created before or after a certain date. If multiple temporal data are recorded with individual spatial objects, a GIS can support longitudinal analysis of spatial properties at locations. For example, spatially integrated county-based population data from 1990, 2000, and 2010 census surveys will have every county polygon with population counts for the three years, based on which GIS functions can be used to calculate and map decadal population differences to show counties with increased or decreased populations over three decades and moreover to reveal clusters of counties with growing population. A major challenge for such longitudinal studies is the varying administrative boundaries over time (Knowles, 2005). County population is a spatially extensive variable, in which quantity (i.e. population count) is true only of the entire area of measurement (i.e. county). Therefore, changes to county boundaries will result in spatially incomparable population data for longitudinal studies. Spatial interpolation functions in GIS can be used for population estimates when administrative boundaries changed as used in the US National Historical GIS project (McMaster et al., 2003).

2. Temporal GIS with Event, Activity, Process, Change and Movement

Yuan (1996) summarized the key developments in temporal GIS conceptual and database frameworks from Armstrons (1988), Langran and Chrisman (1998), Beller et al. (1991), Worboys (1994), and Peuquet and Duan (1995), up to 1995 when event-oriented GIS data models emerged (Figure 1). Peuquet and Duan (1995), Claramunt and Thériault (1995), Chen and Jiang (2000), Worboys (2005), and many other researchers proposed approaches organizing spatial data based on geographic events. Since then, temporal GIS has expanded into discussions of events and processes. While the definitions of events and processes in GIS vary still, there is a general agreement that events are discrete and processes are continuous. The dichotomy of events and processes in space and time mimics the dichotomy of objects and fields in space. GIS objects capture

Kant's concept of discrete and relative space where GIS only recognizes space where is occupied by objects. In other words, space only exists with occupancy. GIS fields conform to Newton's concept of absolute space which exists in its own right, and hence support the concept of empty space. Likewise, events and processes represent two ends of a space-time continuum along which events mark discrete happenings in time and processes manifest continuous transitions over time. Temporal GIS based on events records only the time stamps when events were recognized, similar to spatial objects in GIS databases. In contrast, process-based temporal GIS shall include a continuous time span of interest whether or not there are processes in operation.

Yuan and Stewart (2008) proposed a conceptual premise that synthesizes the concepts of events, processes, activities, change, movement, and spatiotemporal data (Figure 2). An event denotes that something has happened, a process characterizes how it happens, and an activity is an action carried out by an actor. When events and activities take place, they may trigger a process to initiate or become intensified. An example is Dr Martin Luther King, Jr.'s "I Have a Dream" speech during the March on Washington for Jobs and Freedom, which regrettably triggered the United States Federal Bureau of Investigation (FBI) to expand their COINTELPRO operation and target Dr King as a major enemy of the United States. On the bright side, the event also successfully pushed the Kennedy administration to advance civil rights legislation in Congress. Propitious processes can lead to change or movement, which, when certain thresholds are exceeded, become observables that can be collected as spatiotemporal data (including facts and evidence) and should be further used to project events, processes, and activities that generate these observables in the first place. Moreover, contingency among events, processes, and activities help form narratives that give meanings and interpretations to the embedded causal relationships.

However, the proposed premise in Figure 2 remains mostly in a research phase. Temporal GIS used for historical research is dominated by snapshots, space-time composites, and event-based records. Consequently, support for historical research has been on mapping, visualization, animation, spatial analysis of point or polygon features, or statistical analysis of spatial patterns and arrangements. Admirable efforts in building national historical GIS contributed to digital archives of databases for mapping, georeferencing places, and longitudinal or spatial analysis of statistical data. Consequently, the GIS-enabled spatial turn in history is largely on the rediscovery of the power of map and cartographical representation of the past (Bodenhamer, 2013). Thrift (2008) argued that space is non-representational and concerned with the "geography of what happens" to capture the flow of everyday life. Happenings and events are very much at the centre of Time Geography (Hägerstrand, 1970). "The spatial turn in the humanities must be more than methods: it must understand the role of space in human events" (Bodenhamer, 2013: 7).

Figure 1. Key temporal data models up to 1995

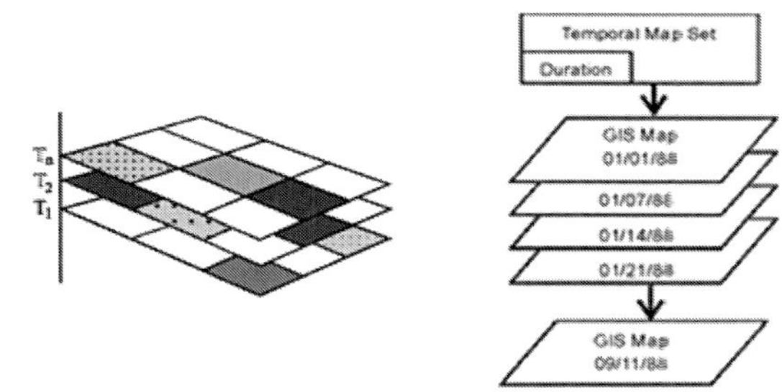

Layers of snapshots by (a) Armstrong (1988) and(b) Beller et al. (1991)

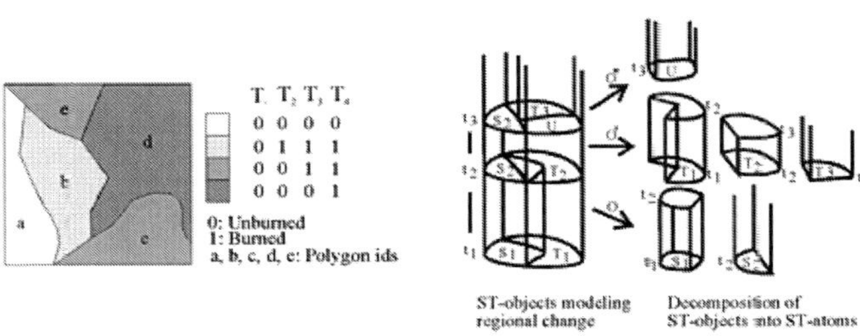

Space-time objects: (a) Space-time composites by Langran and Chrisman (1988) and (b) Spatiotemporal object model by Worboys (1992)

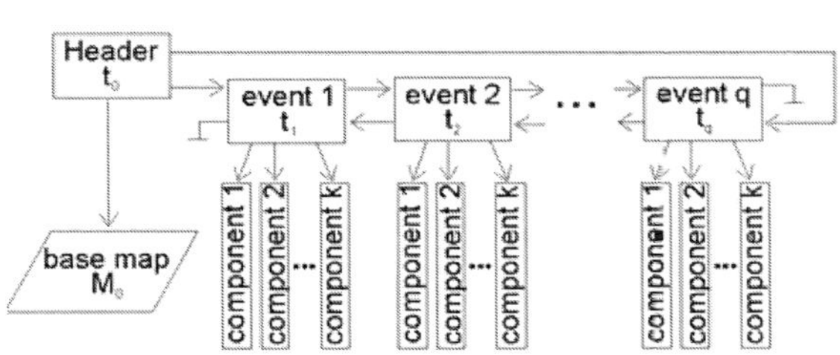

Event-based Spatiotemporal Data Model by Peuquet and Duan (1995)

Figure 2. A premise: events and activities trigger processes to initiate or intensify; processes drive change or movement as observables that can be measured and recorded as spatiotemporal data

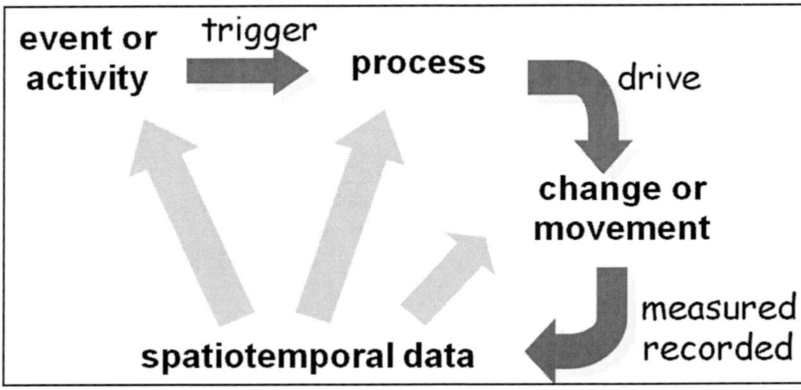

The proposed premise in Figure 2 goes beyond the cartographic constraints of objects and fields in conventional GIS by enabling connections among events (activities), processes, change and movement, and data (facts, evidence) in space and time. By making proper connections, temporal GIS will be able to construct spatial narratives through historical research exercises of selectivity, simultaneity, and shifting of scale in manipulating space and time to interpret the past (Gaddis, 2002). The proposed premise supports the use of time as an organizing principle and the mutually constitutive relationship between space and time as inherently in the definitions of events, activities, processes, change, and movement as well as their contingent connections.

3. From Temporal GIS to Narrative GIS

The development of temporal GIS by and large parallels to the trend in temporal database management systems with an emphasis on how to store temporal data. Historians and humanity scholars need to embrace space and time in ways to contextualize events and behaviours, based on which they can explore causation and answer questions on "why" (and for them, "why" matters more than "who," "what", and "when") (Bodenhamer, 2013). To this end, the proposed premise in Figure 2 may suggest a brand new way to use both space and time with equal importance in organizing geographical and historical information. The new approach transforms temporal GIS to narrative GIS with spatial narratives as the core concept.

While temporal GIS builds upon layers of snapshots or spatiotemporal objects, narrative GIS uses spatial narratives to geographically contextualize events and interactions in a time-orderly fashion to reveal causation, contingency, and human experiences. Points, lines, and polygons are ingredients to build GIS layers or feature classes. Likewise, events, activities, processes, changes, and movements are all elements available to construct spatial narratives. Even though these narrative-building elements are abstract and use of their spatial and temporal representation schemes for visualization is inevitable, narrative GIS privileges happenings over entities and therefore allows for ambiguity and flexibility with a hierarchy of events (such as wars and battles) associated with multiple spatial and temporal footprints. Selection of narrative-building elements may be based on semantic, spatial, or temporal considerations, or any combination of them. A narrative GIS aims to provide the element pools, frameworks and tools to facilitate selecting relevant elements and assembling them to form narratives that represent meaningful flow of happenings from the chosen perspective.

Events, activities, and processes are incorporeal, not bounded to a particular space and time, but they can be actualized in multiple ways and experienced as such. Movement (or motion) and change result in differences between successive events, activities, or processes, and movement and change inherit distinctive durations. Fraser (2006: 130) synthesized Whitehead's thesis that "duration is the field of the event: points and instances, spatial and temporal divisibility and extensiveness, are the properties of a duration". In other words, movement and change actualize events (and activities and processes) as durations with discrete spatial and temporal extents. Table 1 gives simple definitions of these narrative-building elements. Whitehead (1978: 35) posited that "There is a becoming of continuity, but no continuity of becoming": discrete activities and events lead to a becoming of continuous processes. Events and activities take place, and processes take time. Change and movement denote the differences and effects in space and time. Together, they form spatial narratives to express what happens, how it happened, what the consequences are, and the historical and geographical reasons.

Table 1. Definitions of narrative elements

Term	Oxford English Dictionary	Narrative Elements
Activity	The state or quality of being active; exertion of energy; the degree to which a substance, esp. an enzyme, exhibits its characteristic property.	Action taken by actors (e.g. individuals, agencies, or organization) in space and time. The consequence of activities may generate movement of the actor or other objects. Activities can also cause change to the characteristics of someone or some environment.
Event	The fact of anything happening; the occurrence of; an incident.	An occurrence of something with significance that drives noteworthy motion or change at locations over time. The decision on "significance" and "noteworthy" is situational and problem-dependent.
Process	The fact of going on or being carried out, as an action, or a series of actions or events; progress, course.	A gradual transformation that transcends properties, forms, and patterns over time. The determination of "gradualness" is scale-dependent.
Change	Substitution of one thing for another; succession of one thing in place of another; substitution of other conditions or circumstances, variety.	Substitution of properties in an object, at a location, or conditions in an environment. Changes can occur to population counts, identities, thematic attributes, spatial, or temporal characteristics.
Movement	A change of place or position; a progress, change, development; passage from place to place, or from one situation to another.	Shift in location of an object over time. The object must maintain the same identity during a movement.

Activities and events may be difficult to differentiate. Nevertheless, activities are goal-oriented and contextualization of spatiotemporal patterns reveals routine activities and some purposes. Changes in routine activity patterns suggest event disturbances or progressing to another stage of life. Several GIS projects exploited human activity patterns with Time Geography. Kwam (2004) examined daily journeys with space-time trajectories to project space use in urban area. Shaw and Yu (2009) developed GIS tools to analyse activities and interactions of individuals in physical and cyber spaces. Both projects examine space-time paths at the individual levels as well as at the group levels for interactions. This line of research advances also push studies of accessibility (especially into healthcare facility) and exposure (especially to pollution) that is, moreover, applied to issues related to social and environmental justice. Similar approaches may be helpful to historical investigation of environmental perception or routing analysis by projecting environmental and social characteristics to the pathways of migration or immigration, troop movements, or transportation routing.

Historians and cultural geographers emphasize the singularity of events. Strictly speaking, an event is not just something that happens; it is a happening of significance. Fraser (2006) stressed the importance of singularity to an event on both "the coming together of prehensions" and "their becoming together in a particular way." Furthermore, she argued that event-thinking seeks to apprehend the relations between actual bodies and incorporeal happenings in states of affairs as well as the transformative power of individual events that create the reality of events in themselves. The singularity of an event is relative in nature and is question dependent. Sunrise, payday, and World War II commonly are all singular events that can serve relative time frames at different temporal scales. Yet, birthdays, college graduation, and weddings are personal. The Geospatial Event Model (GEM) offers formalization of events in GIS to support reasoning about events and their states of affairs as settings (Worboys and Hornsby, 2004). Narrative GIS will need to develop data schemes and analytical functions to discern evidence of actors and environmental conditions coming together in a particular way and to facilitate grasp of the interactions when coming together.

Process is at the center of cultural-historical geography as a "retrospective science" that records, illustrates, and synthesizes geographic patterns to explore geographic processes (Sauer, 1956). Apprehension and appreciation of historical landscape can be acquired through projecting processes that are responsible for the forms and patterns as we observe. Indeed, every empirical concepts of space must be reducible by a chain of definitions to a concept of process (Blaut, 1961). Like event and activity, process is incorporeal and unbounded. Processes proceed continuously, progressively, or eruptively in space and time. An important inquiry about process is to identify distinctive phases and ensuing consequences in process development. A process may promote certain activities or events (such as evolution promoting disruptive events); a process may also be modified by events or activities (such as Licoln's Gettysburg speech energizing emancipation). Narrative GIS will need capabilities to relate processes to forms and patterns as well as relate processes to events and activities to reify incorporeal working of geography into apprehensible contingency to shed light on what may be coming or becoming.

Change and movement are differences resulting from activities, events, or processes. Movement can be considered as change in location, so change is perhaps more fundamental than movement. Persistent identity is critical to change recognition. As mentioned previously, change in administrative boundaries poses the key challenge in historical social science research because the area unit of analysis is no longer constant and so are the spatial representations (i.e. location, size, and geometry) of data assigned to the administrative unit. Many GIS studies on spatial and temporal interpolation methods attempt to handle such issues better. Yet, issues of identity for mobile objects differ from those for area units. From journeys taken by individuals to flows of transactions submitted by multiple agencies, recognizing object identities can be simple or very difficult. Some objects may involve in ontological change, such as a tropical depression becoming a hurricane, and persistent identity may not be possible. With thriving GPS technology, mobile object modelling is a hot research area, but these lines of movement may not fully address the needs of historical research. Narrative GIS will need to ingest useful models and functions from existing developments in change detection, spatio-temporal interpolation, trajectory analysis, and mobile object modelling, and meanwhile develop new conceptual frameworks and functions to relate change and movement to process, event, and activity.

Narrative GIS is intended to respond to historical research needs for temporal GIS. The discussion so far addresses the needs for conceptual frameworks, data schemes, and analytical functions to generate spatial narratives from spatiotemporal data or facts. This is in contrast to geo-narrative research that is being developed under the scope of qualitative and critical GIS (Kwan and Ding, 2008). Oral narration, stories, and diaries are mapped to reveal spatial perception of one's daily activities, spatially referenced ideology or cultural practices, or spatial behaviours and space use on narrative maps with or without multimedia displays

(Caquard, 2013; Mennis, Mason and Cao, 2012; Curtis, 2012; Elwood and Leszczynski, 2012). Both approaches to spatial narratives can bring a spatial dimension to time-centric historical research. Complementarily, geo-narratives spatialize human experiences, whereas spatial narrative generation extracts change and movement from spatio-temporal data to reconstruct process, event, and activities, and through the reconstruction to contextualize these narrative-building elements geographically and historically for rich interpretation on their singularity and contingency. The shift from objects and fields to narrative emphases in both approaches presents a big leap forward in GIS support for historical research.

CHAPTER THREE

INTENTIONALLY-LINKED ENTITIES: A BETTER DATABASE SYSTEM FOR REPRESENTING DYNAMIC SOCIAL NETWORKS, NARRATIVE GEOGRAPHIC INFORMATION AND GENERAL ABSTRACTIONS OF REALITY[1]

VITIT KANTABUTRA[2], J. B. "JACK" OWENS[2] AND ANA CRESPO SOLANA[3]

1. Modelling a Dynamic Reality

We are developing tools that will permit us to move from data to knowledge that will increase our understanding of complex, nonlinear, human systems, such as that of the First Global Age, 1400–1800, and about how such systems are coupled to complex natural systems (Haken, 1983; Puu, 2003; Rosser, 2000). The existing software tools for *social network analysis* (SNA) and *geographic information systems* (GIS) do not provide historians and historical social scientists with an adequate basis for narrative spatio-temporal analysis (Goldenber, Zheng, Fienberg and

---

[1] This chapter is based on work supported by the US National Science Foundation (NSF) for the project "Understanding social networks within complex, nonlinear systems: geographically-integrated history and dynamics GIS" [acronym: SOCNET], which is administered by the NSF Office of Cyberinfrastructure (OCI-0941371; $1,290,704; 2009–2013). This is a collaborative project with May Yuan of the University of Oklahoma (OCI-0941501; $471,193). Any opinions, findings, and conclusions or recommendations expressed in this material are those of the authors and do not necessarily reflect the views of NSF.

[2] Idaho State University.

[3] Instituto de Historia, CSIC.

Airoldi, 2010: 129–233; Owens, 2007a: 4–6; Owens, 2007b: 2014–2040; Yuan, 2008: 1, 4–5; Yuan and Stewart, 2008). Nor have researchers been able to generate cartographic visualizations or other abstractions of reality to help us comprehend dynamic human or natural history as well as we would like (Staley, 2003; Wachowicz and Owens, 2012: 53–79; Wachowicz and Owens, 2013: 127–144).

The characteristics of the *database management systems* (DBMS) we normally use to represent our information, our perception of the real world, form a fundamental barrier to dynamic analysis of systems. In this chapter, we present our database scheme, Intentionally-Linked Entities (ILE), which permits users to model the dynamic reality of the real world in ways superior to what is possible with all other available database management systems. This is a good time to propose a radically new database model (Badia and Lemire, 2011: 61–69). ILE can serve as a general, all-purpose database management scheme, but it embodies characteristics that make it particularly useful for the creation of complicated narratives about social networks within a dynamic geographic environment. ILE also offers a platform for ontological research, which is necessary for the effective integration of information from multiple databases embodying partial abstractions of world history and prepared by unconnected researchers (Doan, Domingos and Halevy, 2003: 279–301; Magnani and Montesi, 2010: 1–32; Shvaiko and Euzenat, 2005: 146–171). An ILE database comprises four components: entities, entity sets, relationships, and relationship sets. An **entity** is a "thing" (usually a noun), a datum that we want to store. Entities of the same kind (e.g., set of "employee" entities, set of "country" entities) are all held together by an **entity set** (e.g., the entity set "employees", the entity set "countries"). ILE implements the "belonging" of an entity to an entity set through data structures involving **pointers** in both directions: that is from the entity object to the entity set object, and from the entity set object to the entity object. Any number of **relationships** can be represented as relationship objects. Each relationship object relates entities that have "roles" to play in that particular relationship. All relationship objects representing the same kind of relationship can be gathered into a single **relationship set**. Each entity has a pointer back to the relevant relationship object, which makes it a simple and efficient task for a user to navigate among the stored entities involved in that relationship. In modelling complicated social relationship situations, ILE permits relationships in which each "role" may be played by more than one entity (Kantabutra, 2007: 258–263)[4].

[4] At that time, ILE entity sets were represented as tables, as in the more commonly used relational database management schemes (RDBMS).

In some respects, ILE can be seen as a direct implementation of the entity/relationship (E/R) model, presented in 1975 and published in 1976 by award-winning computer scientist Peter P. Chen (Chen, 1976: 9–36; Chen, 2002: 100–114). Similar ideas had been used informally before then, and a particularly suggestive paper was published, with an attention-grabbing title, by A. P. G. Brown of the English firm International Computers Limited in 1975 (Brown, 1975: 339–347). Chen became prominent in "conceptual modelling" and Computer-Aided Software Engineering (CASE). Although experts recognize that Chen, with his E/R data concept, created a simple and appropriate model to use for describing the structure of the data in a database, we are unaware of any full implementation of Chen's exact form in a database management system (Chen, 2002). Although they varied considerably from Chen's 1976 model, a group at Bell-Northern Research in Ottawa, Canada, did achieve an implementation during 1980–1981, but we do not know the ultimate fate of the project (Benneworth, Bishop, Turnbull, Holman, Monette, 1981: 478–484).[5]

Briefly:

- ILE enables the fast searching needed for temporal, narrative, and 3D forms of GIS and SNA, where a large quantity of information must be processed rapidly.
- To simplify, *arity* indicates the number of roles (*arguments*) played by entities in a relationship. While some database systems limit arity, thereby reducing the possibilities for dynamic, real-world analysis, ILE supports an unlimited number of roles.
- ILE supports the modelling of complicated relationships and data where descriptive, qualitative, semantic information characteristic of much humanities and social networks research, rather than simply quantity, must be included.
- ILE avoids data redundancy and data fragmentation, which often lead to research errors when other database schemes are employed.

We will clarify this abstract description later in this chapter by explaining the implementation of ILE for the analysis of Ana Crespo Solana's rich database on 18th-century Atlantic trade, which she has created through many years of work in archives in several European countries. We will first examine some well-known relational database

[5] We obtained this citation from Chen, 1985: 174–210, but at that time he appeared to know nothing else about this interesting project.

management systems (RDBMS) as points of contrast and departure for the presentation of ILE.

2. What ILE Replaces: Relational Database Management Systems (RDBMS)

Over the past four decades or so, the relational database management systems (RDBMS) scheme has increasingly dominated all others, until by the turn of the millennium hardly anything else was available in a commercial product. The best known of these is probably Microsoft's Access.[6] The relational scheme constituted a significant improvement over what had been available, and historians and historical social scientists would slowly be drawn to its commercial forms as a way to organize their information. As proposed by Edgar F. Codd (1923–2003) in 1970 (Codd, 1970: 377–387; Codd, 1981),[7] a two-dimensional table forms the basis of organization. Users can easily understand the simple structure. They enter data into cells or fields, which are defined by columns and rows. The resulting tables are linked via relational "keys" expressing a particular data type (e.g. "place" in a GIS database) in two or more tables in the database. The ability to link the tables greatly expanded the computational capabilities over any tabular scheme. The simplicity of the data structure makes the relational scheme popular. Also, the relational model receives support from a widely used "structured query language" (SQL) (Treseler, 2009; Codd, 1990: 371–389)[8]. It dominates work with geographic information systems (GIS) (Peuquet, 2002), and it appears to be the prevalent structure for the data on which social network analysis (SNA) is based.[9]

---

[6] Oracle Database, My SQL, and POSTGRESQL will also be familiar to some users.

[7] http://awards.acm.org/images/awards/140/articles/2485527.pdf (accessed 23 March 2013).

[8] The close interaction between the early developers of the query language, which they then called SEQUEL (Structured English Query Language), is apparent from reading Chamberlin and Boyce, 1974. SQL differs from Codd's Relational model in some important ways. Codd argued against deviations that compromised his relational principles, but to most people, SQL and the pure relational model are similar enough for us to take the two models to be the same, unless we note otherwise.

[9] See the Datasets Wiki, Dynamics Lab (Computational Social Science), University College Dublin http://dl.ucd.ie/wordpress/?page_id=846 (accessed 17 April 2013).

In additional to their seeming simplicity, advanced RDBMS users have also found interesting the algebraic theory upon which Codd founded his creation of the relational database scheme. This "relational algebra" assists in the design of relational databases and in the efficient processing of user requests for information (Codd, 1979: 397–434). Any database textbook will provided a detailed discussion of relational algebra (Silberschatz, Korth, & Sudarshan, 2010, especially chapters 2 and 8). However, the complications introduced by the relational data model itself have made the algebraic theory necessary. For example, good relational design requires that the database be "normalized" so that redundancies in the entries are minimized so far as is possible in the context of relational databases. Codd called the scheme for enforcing these modifications a "normal form", and he published the first three normal forms between 1970 and 1972. Since then, computer scientists have introduced a multitude of normal forms and all their theoretical machinery to combat redundancy problems in relational databases (Fotache, 2006). Most of these normal forms, with the exception of the first few developed by Codd himself (Codd, 1971, republished in Rustin, 1972: 33–64; Codd, 1971; Codd, 1974: 1017–1021),[10] are interesting to database theorists, but they defeat the supposed simplicity of relational databases. This relational database theory does indeed help with forming queries; however, query formation can usually be done just as well with the help of common-sense logic and algorithms.

To illustrate an example of an algorithm for searching in an RDBMS that does not use relational algebra, we offer the following. Start in one relation (table) R1 representing an entity set. Look for certain entries there, say a set S1 of entries. Then, using another relation (table) R2, representing a relationship, figure out another entity set S2 that is related to the S1, using the relationships specified by R2. This approach is likely to use far less computer memory than the SQL one. SQL would cross R1 and R2 first. Then, it would filter the cross product to extract the entries desired by the user. It would finally throw away the rest of that potentially huge temporary cross-product of R1 and R2. Indeed, this throw-away product may be potentially larger than the size of main memory, which would require storage in the disk system. This disk storage can be harmful because access to a single disk may require roughly 100,000 times the access time of main memory. It is hard to compare disk access time to

[10] For a concise explanation of each form, see "Rules of Data Normalization", an archived page of DataModel.org at http://web.archive.org/web/20080805014412/http://www.datamodel.org/NormalizationRules.html (accessed 17 April 2013). Peter Chen provides a humorous account of the scramble to publish the latest normal form in his "Entity-Relationship Modeling" (2002), 2.1.

memory access time because a single disk access can actually result in the retrieval of thousands of records. However, as a rule of thumb, access to disk will slow performance enough to be avoided when possible, *especially for the types of dynamic data analysis and visualization in which we are interested.* Even with the algorithmic approach to an RDBMS, the searching itself can be really slow in large tables. Such computation time, which could be of the order of seconds or more per query on a fast computer, may not seem like much if only one query is needed at a time. However, in some applications, such as temporal/historical GIS, a large number of queries need to be handled in a short amount of time for any comfortably flowing display. For such applications, the relational scheme is inadequate. In ILE, the links, as we shall see, are all ready, practically eliminating the search and greatly facilitating narrative, dynamic GIS and SNA.

Aside from searching, the relational data model offers other significant shortcomings. The two-dimensional table resides at the model's core. This form, with its scheme of columns and rows to establish cells/fields into which data are entered, provides appealing simplicity, but it fails to give users the best way to model the world. Our research project makes this weakness especially apparent because we wish to model complex systems, which are characterized by a large number of links among heterogeneous entities that are reshaped over time whenever there are changes in their sets of links. Inflexibility characterizes the relational model because users encounter difficulties when they wish to alter the number and types of columns.[11] Computer scientists refer to a computer system's ability to handle errors and continue operation as "robustness". Despite the suggestive name "relational", the two-dimensional table, with no pointers, offers a non-robust, inefficient way to navigate relationships. The indexes of relational databases do use pointers to speed up navigation *within each set of entities,* but this solution entails a high cost for insertions and other database management functions: "every index built for one or more attributes of some relation makes insertions, deletions, and updates to that relation more complex and time-consuming" (García-Molina, Ullman and Widom, 2008: 352). Furthermore, indexes cannot help with navigating relationships. Table joins can link records from two or more tables in the database to establish a set, but these joins often inefficiently necessitate

---

[11] See http://www.w3schools.com/sql/sql_alter.asp (accessed 24 April 2013). If the user adds a column, all rows must have some value (if just NULL) for that column. Moreover, adding a column can be quite time-consuming because the whole table will have to be reallocated in memory and on disk.

occupying large amounts of temporary memory and demand significant computation time.

The relational database model frustrates most historians because it forces them to accept assumptions that their own research challenges. The model demands that data come from a semantically homogeneous world, so that representations of the same entity are consistent. Only then can the data fit comfortably into tables, and only then can the tables be linked by unique relational keys. But in the sources of historians' information, the semantics manifest heterogeneity because people created these sources from different backgrounds and different ontologies, expressing varying models of the world. Thus, lots of the information does not fit comfortably into tables. Historians know that they must not lose the variety that is present in their sources or else they will damage their ability to make sense of past reality. Thus, users describe the data in inconsistent ways within the relational database, which creates "errors" within the data model. As Michael Brodie and Jason Liu have asserted, "the chief problems arise from the conflict between the inherent semantic heterogeneity of data to be integrated and the inherent semantic homogeneity of relational modelling tools" (Brodie and Liu, 2010).

Even when the creator of the database schema of a relational database designs it to conform to the assumption of a unique ontology, the same data can occur in many relations. Redundancy is particularly significant when there is an "is-a" hierarchy of entity sets, where a relationship exists between a subtype and some existing abstraction, as with a subclass or another class. Even when a relational database is well designed, no single data structure houses all information about each data entity; not a single row in a table, and not even a single table. We call this situation *data fragmentation*, by which we mean that no single data structure represents an entity. Strictly speaking, a unique table should define each entity set, and referential integrity constraints should control the appearances of the primary key of this table as foreign keys in other tables. However, in practice, database creators often do not follow these principles. Maintaining referential integrity entails a high cost in the time spent on operations. Historians will frequently try to work around the unreasonable assumptions of the relational model, but the resulting data fragmentation makes it difficult to assure that all the information about a single entity is correct. Such assurance is difficult to attain in any event (Helland and Campbell, 2009).

If a database schema does not follow the full relational model, redundancy becomes worse and will result in data anomalies when data are updated or deleted (García-Molina, 2008: 103). Often, one can expect poorly designed relational database schema because the simplicity of tables and the popular language SQL invite design by those who do not embrace or do not understand the assumptions on which the relational model is based.

As Codd pointed out in the early 1970s, data redundancy in relational databases can be reduced by normalization (Codd, 1972). However, many relational databases are increasingly not normalized. Historians often do not normalize because at the beginning of their research, they cannot define precisely the nature of their information, and they will wish to add without difficulty new data they discover, which may be semantically heterogeneous in relation to the data with which they began the database. Great chronological and geographical extents will often mandate such an approach to the development of the database. Therefore, we are not surprised that the magnificent relational database of the China Historical GIS, encompassing the period from 221 BCE to 1911 CE and the history of 20–25% of the world's population, is not normalized. Researchers at Fudan University, Harvard University, and other centres collaborated on the project.[12]

In addition to the anomalies just mentioned, the lack of a substantial, unique place to store each entity can lead to database errors, which of course can cause great practical problems. We have published a description of how the security authorities may have missed an early warning about the so-called "Underwear Bomber", Umar Farouk Abdul Mutallab (Kantabutra, Owens, Ames, Burns and Stephenson, 2010: 39–58). When his father reported him to the US Embassy in Nigeria, an initial search of his name in the database of visa grantees failed to establish a connection because his name was entered incorrectly into another database. Because we know nothing about the databases involved, we do not know if the problem stemmed from a lack of proper normalization or from some difficulty with the integration of information in databases of different federal agencies. Clearly, however, avoiding errors in the updating of information in databases requires work that makes relational databases anything but simple to use. The databases involved could have been designed to force a user to check for similar names and to make the creation of a new person entity a difficult and deliberate act. Such enforcement is possible but difficult and not often done in relational databases because, for the sake of efficiency in introducing increasingly available data, redundant entries are normally permitted.

---

[12] China Historical GIS (2001–2010), http://www.fas.harvard.edu/~chgis/ index.html (accessed 25 March 2013).

All of these problems with redundancy, fragmentation, and the lack of normalization produce databases whose query results frequently contain some erroneous information as "noise". We have created a visualization tool, which will detect such noise when the data have been georeferenced to their longitude and latitude coordinates. Georeferencing is necessary for employing GIS (Hill, 2006; Kantbutra and Owens, 2013).

Finally, its fans have oversold the simplicity of relational databases, which is really only superficial. In her important book on temporal GIS, Donna Peuquet embraces the relational model in part because she finds that the use of pointers complicates too much the user's ability to maintain the integrity of numerous links in *network* database schemes, discussed in section 3 below (Peuquet, 2002: 236–237). However, as a major recent work on database systems makes clear, pointers fill relational databases, and the placement of a relational database on a disc involves precise, low-level physical placement of data and index structures (García-Molina, 2008, chapters 11–13). In other words, relational databases at the implementation level are not really simpler than the databases using the hierarchical and network data models, which relational databases eliminated. Relational databases achieved their popularity largely on the basis of what the typical user can see.

3. What ILE Replaces: Nonrelational Database Systems

We have devoted quite a bit of attention to the relational database model because that is what users of historical databases know. However, other, nonrelational database concepts have begun again to challenge the relational one. We cannot say now if Intentionally-Linked Entities (ILE) will "replace" all nonrelational database systems, in part because ILE could be classified with them. Usually, however, writers apply the term to hierarchical, graph, and object-oriented databases, which have been around in various forms since the late 1960s, eventually as rivals to the relational model. We will not discuss object-oriented databases, which are useful for storing and recovering a variety of data types, including text documents, photographs, and other types of graphics. Their strictly hierarchical structure forms a barrier to combining easily the stored objects and, therefore, hinders their role in historical research (Silberschatz, Korth and Sudarshan, 2010, chapter 22). Many of the current nonrelational models, sometimes called NoSQL (not only SQL), present major problems for almost all users of database systems for historical research, regardless of academic discipline (Helland, 2011).

At a general level, the graph structure of many of these databases is easily understood. It consists of "nodes", "edges", and "properties". Nodes represent entities such as municipalities, government agencies, people, businesses, etc. Properties of a node represent information about it. An edge can be thought of as a line that connects two adjacent nodes or a node to its properties. It expresses the relationship between the two. A query of the database would search for meaningful patterns in the interconnections among nodes, properties, and entities. The general model stemmed from a collective effort known as CODASYL (Conference on Data Systems Languages), whose leading proponent was Charles Bachman (born 1924), the third in our presentation of the major pioneers of the 1970s, along with Chen and Codd (Olle, 1978; Bachman, 1973).

Practitioners involved in database design for certain novel applications exploring social networks and corporate networks have also found relational database systems to be inadequate in terms of their efficiency and their ability to model data well. Some of these practitioners have, in the beginning years of this millennium, fuelled the rebirth of the graph/network database idea, using such newly programmed database packages as Neo4j.[13] Neo4j and other database schemes, some closer to (yet still different from) the relational scheme, are placed in the "NoSQL" (Not only SQL) movement to indicate that SQL (Structured Query Language, to query and update relational databases) is not the sole database type for every application.[14] While these schemes are impressive for some applications, they have not gained popularity among general database users, largely because they sacrifice functionality to gain speed. These nonrelational database schemes do offer an important advantage over the relational one: *relationships are implemented with pointers*. We call such implementation of relationships an *active* implementation, which we contrast with the relational scheme's *passive* relationship implementation, where relationships are indicated by exact *matching* of keys.

Why are some people, and particularly those associated with large corporations such as Amazon, Facebook, and Google, interested in these nonrelational database schemes? For those who want to control lots of unstructured information, there is only one answer to this question: these database designs are faster, a lot faster, because their data models are simpler and easier to extend to the availability of large quantities of information. And they are more flexible because they do not impose the technical restraints of a relational database. However, these nonrelational

---

[13] Neo4j: The World's Leading Graph Database, Web Site, http://www.neo4j.org.

[14] For a list of NoSQL database models, see http://nosql-database.org/ (accessed on 18 March 2013).

database management schemes propose a wild world because they raise serious questions about reliability of results, consistency of data, the difficulty of querying, the compatibility with other applications, steep implementation and operational curves for any entity that wishes to adopt one, and the availability of even minimal customer support.[15] Commodity cluster computing –computing with multiple computing cores linked together – now drives new developments because of the needs of businesses demanding high speed and the ability to handle huge amounts of data. Producing efficient software poses serious, complicated challenges to the adaptability of this solution if it is to reach the desired scale (Thompson, 2013: 29–31).[16]

Some aspects seem good. For example, these database models support open schema, which means that the databases should link easily, as world historians would want. Ultimately, we wish to pull together the available information assembled by researchers in order to enhance understanding of the First Global Age, 1400–1800. However, when the increasing number of databases are independently maintained and then combined in an ad hoc manner, they certainly present serious ontological and semantic barriers to significant information integration. If the linked databases do not align well, how will users write queries, which will be sufficiently "schema agnostic"? The databases may hold the data, but the user will not be able to get most of it in any form that can be used for analysis, within a GIS for example.

Because its promoters failed to provide an effective way for most users to query databases organized on the CODASYL network standard, it lost out to the relational model in the early 1980s. This deficiency became especially clear as machine-based research computation moved away from the mainframe era, when a small number of experts oversaw all computation use to an era of a demand for greater individual research autonomy in controlling databases and querying for analysis and visualization. These researchers wanted to know only what a particular program could accomplish without having to learn how it was done, leaving the implementation to others. Because the developers of the relational model, including Codd himself, were much quicker to provide users with such an environment, including SQL, by the early 1980s,

---

[15] On these points, in the context of a comparison with the relational model, see Leavitt, 2010: 12–14.

[16] Thompson works at *bigdata* (SYSTAP, LLC), an RDF/graph database capable of clustered deployment; RDF: Resource Description Framework. Due to the number of errors in the text and citations, the paper appears to be a draft of 23 January 2013.

researchers flowed in that direction and away from the existing hierarchical and graph/network systems.

The graph/network scheme also lacks some important properties that a good, general-purpose database scheme should have. It does not have an embodiment for entity sets. No physical data structure ties the stored objects together as an entity set. Therefore, graph/network database management schemes are unsuitable for counting queries and serious ontological work, and there is no easy way to perform aggregate functions, such as finding the average, maximum or minimum.

Because current research on these database systems is driven by large enterprises that want lots of data about us, much investigation is directed toward data mining, especially with "natural language" questions, which will pull patterns out of unstructured, ambiguous, contradictory, and generally messy information. This approach could help historical researchers greatly now that more and more sources are available in digital form on the web. For example, one can envision doing some really exciting work by mining the huge, sixteen-volume, mid-19th-century *Diccionario geográfico-estadístico-histórico de España y sus posesiones de Ultramar* (Madrid, 1845–50), which was assembled and published on order of Spain's economic development minister, Pascual Madoz Ibáñez (1806–1870).[17] Yet, the various procedures used now for such data mining would perhaps make the final database unsuitable for many types of analysis that historians would like to use.

Without normalization of the results of database alignment, users will discover a lot of "noise" (apparent errors and information loss), which may make ambiguous the interesting patterns that emerge in the data. However, a normalization scheme may exclude the integration of potentially important information, depending on the query. Researchers in the humanities and social scientists do not want such exclusions because of their uncertainty about the nature of the data they have and will discover during future research.

For efficient searches of high-volume databases, the existing nonrelational database systems appear to rely on pre-defined schema and indices to speed queries. For historians, these limitations would impose two serious problems. First, the researchers would have to know what their data-world looks like, which they often hope to learn through their research. These nonrelational databases must be defined in advance. Second, the frequent recourse to graph-based and hierarchical structures in

[17] Biblioteca Virtual Andalucía (all volumes available for download), http://www.bibliotecavirtualdeandalucia.es/catalogo/consulta/registro.cmd?id=6353 (accessed 10 April 2013).

these database systems will impose arity limitations, which will frustrate historical researchers at every turn once they wish to undertake any significant analysis of a complex human system or the coupling of such a system with a complex natural system. In complex systems, entities typically play many roles and interact with multiple other entities, but nonrelational systems, such as the graph/network databases, limit the number of roles to two. In an attempt to get around slightly the limitation on the number of connections between records, some creators of the newer nonrelational database systems employ complicated schemes, which suggest both problems with implementation and with user acceptance. We do not know when, if ever, these graph/network database schemes will be accessible for historical research by those in the humanities and social sciences, but for now, their existence will continue to create employment for programmers and graduates in computer science and informatics.

4. The Intentionally-Linked Entities (ILE) Database Model

In this section, we present our database scheme, Intentionally-Linked Entities (ILE), which offers all of the linkages for implementing relationships that one finds in graph/networks database programs, and more. In ILE, no contrived structures are necessary to bypass the arity restrictions of the graph/network models because ILE already possesses links from "entities" to "relationship objects" and from "relationship objects" to the "entities". ILE also features the embodiments of entity sets and relationship sets, which one finds in relational databases, but ours permit users to do more complicated work because they can hop directly from one entity to another one that plays roles in the same relationship. In other words, ILE provides a database management system that can be used effectively to study complex systems through the use of appropriate applications, such as geographic information systems (GIS).

To review, we ground ILE on a simple idea: replace all of the by-value linkages of the tables of the relational model with relationship objects, which link entities by means of pointers. Thus, the entities that are related will be *intentionally-linked* by pointers, not by value-matching as in the relational model. An ILE database comprises four components: entities, entity sets, relationships, and relationship sets. An **entity** is a "thing" (usually a noun), a datum that we want to store. Entities of the same kind (e.g., set of "employee" entities, set of "country" entities) are all held together by an **entity set** (e.g., the entity set "employees", the entity set "countries"). ILE implements the "belonging" of an entity to an entity set through data structures involving **pointers** in both directions: that is from the entity object to the entity set object, and from the entity set object to the entity object. Any number of **relationships** can be represented as relationship objects. Each relationship object relates entities that have "roles" to play in that particular relationship. All relationship objects representing the same kind of relationship can be gathered into a single **relationship set** [*relset*]. Each entity has a pointer back to the relevant relationship object, which makes it a simple and efficient task for a user to navigate among the stored entities involved in that relationship.

In modelling complicated social relationship situations, ILE permits relationships in which each "role" may be played by more than one entity. Relationships amongst entities are represented as relationship objects. A relationship object serves to relate entities that have roles to play in the relationship represented by the relationship object. This is done by means of pointers. More specifically, each relationship object has, as its main data structure, an **r-tuple**, where *r* is the arity of that relationship, or the number of roles that entities take in that relationship. *Tuple* is the term for an ordered list of elements. If only one entity plays each role, each member of the r-tuple is a pointer to the entity object representing each entity. The entity also has a pointer[18] back to the relationship object, making it simple and efficient to "travel" amongst the entities involved in the relationship. If any member of the r-tuple permits multiple entities playing the same role, the tuple member can be represented by an aggregate data structure, the simplest of which would be a dynamically-allocated array of pointers to the entities playing that role.

A merchant company would represent one association with multiple entities playing the same role. In this relationship, the first role could be the association itself, a single entity. However, the second role would be played by all the members of the association. An alternative representation of that relationship would be to have one of the roles represent a single person and the other role represent all the associations to which this person belongs, which would greatly facilitate a multilevel social network analysis (SNA) over anything that can be done now. Another example of the use of relationships with multiple entities playing the same role would be the simple parenthood relationship, which is a binary, many-many relationship. In a relational database, we would typically represent this relationship set as a table as in Fig. 1.

[18] More precisely, a data structure containing a pointer.

Figure 1. Parenthood relationships in a relational database

PARENT	CHILD
Dad	Jill
Dad	James
Dad	Joe
Mom	Jill
Mom	James
Mom	Joe
Dad	James
Mom	Jill

The simplicity of the table of the relational model appeals to users, but we note some significant problems. In particular, note the fact that no unique data structure or object represents each entity. For example, "Dad" appears in several rows. In fact, as the default mode of operation, many relational database management scheme (RDBMS) packages permit redundant rows because this procedure eliminates checking for duplicates, which is time consuming due to a lack of complete indices on all key fields.[19] However, even without redundant rows, each entity can still appear in many rows. In this example, because Dad has three children, he would appear in three rows. With no redundant rows, there is still a lot of built-in redundancy. In this case, because there are two parents and three children, we need $2 \times 3 = 6$ rows. In a binary relationship where there are m entities serving one role and n entities serving the other role, and where every pair of entities comprising one entity from one role and one entity from the other role, an RDBMS would record this in m times n rows. In SNA (social networking analysis) application programs, such relationships with high values of m and n occur frequently, causing time-consuming and error-prone data entry operations for many users in history, the historical social sciences, and the humanities.

Fig. 2 shows how the same data relating the parents and children could be represented in ILE. The ILE data structure is much more complicated than the few rows of table in the relational database, but a modern computer supports greater complication, *which the user need not see.*

[19] According to Codd's original definition of the relational data model, the rows in a relation are unique. However, Codd did not consider implementation details. In real implementations, it would be inefficient to require that rows be unique if there is not a good indexing system. Therefore the default convention is to permit multiple rows.

Figure 2. Parenthood in ILE

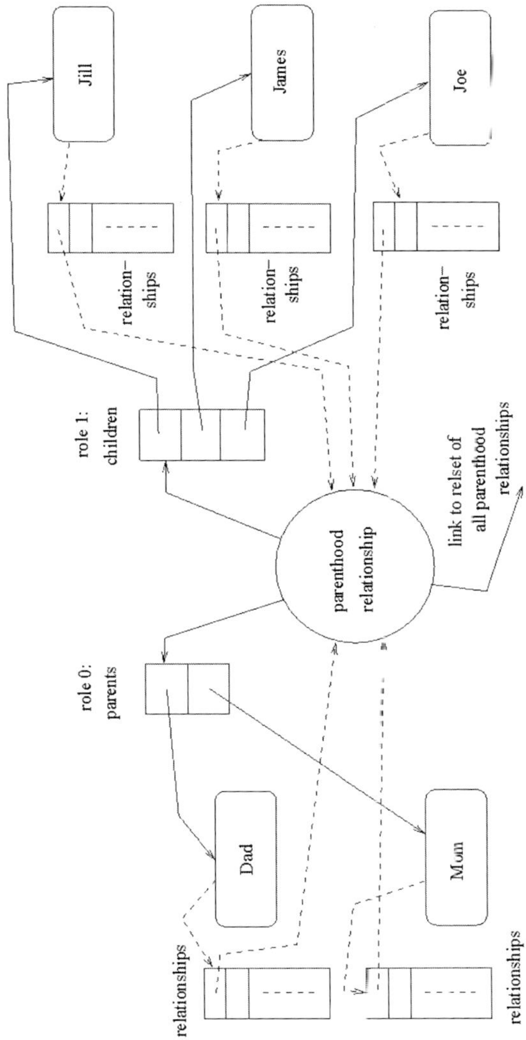

Because a person may have children with more than one other person, modelling a parenthood relationship may become complicated. Consider the case of the actress Elizabeth Taylor.

Taylor had two sons, Michael Howard (born January 6, 1953) and Christopher Edward (born February 27, 1955), with Michael Wilding. She had a daughter, Elizabeth Frances "Liza" (born August 6, 1957), with Michael Todd. During her marriage to Eddie Fisher, Taylor started proceedings to adopt a two-year-old girl from Germany, Maria (born August 1, 1961); the adoption process was finalized in 1964 following their divorce. Richard Burton later adopted Taylor's daughters Liza and Maria.[20]

This parenthood relationship set is more complicated than the one discussed above. Of course, we can always model this relationship just as we would in an RDBMS: each relationship would just have a parent and a child. However, ILE offers another possibility unavailable in an RDBMS: each parenthood relationship has two roles, parents and children. To cover all of Taylor's children, we will employ three relationships, one for each set of children with the same father, as shown in Fig. 3.

The advantage of ILE over RDBMS is more pronounced when we have to represent a *k*-ary relationship (one with arity *k*) with some arbitrary number n_i entities serving role *i*. In ILE, this situation would be represented as a single relationship, whereas in an RDBMS, the number of relationships would be $\mathbf{\Pi_1{}^k=1{}^n{}_i}$ that is, the product of the numbers of entities serving all roles. This relationship value can be a large number of "loose" rows in the relational table.

RDBMS pose another problem. If we misspell a key and if the misspelled version is already an existing key or is valid, we have entered the wrong data. However, if that key is not even supposed to exist, we have unintentionally invented a new entity. For example, if we were to type "Jon" instead of "Joe," we would create a new person. In ILE, new entities are created with a **create entity** statement. They cannot be created simply by the act of creating a relationship. In other words, ILE, or "Intentionally-Linked Entities", could have also been called, "Intentionally-Created Entities".

To summarize, Intentionally-Linked Entities (ILE) possesses all of the advantages enjoyed by both relational databases and graph/network databases, and no obvious drawbacks, except for the complicated implementation, which is no longer insurmountable due to modern computer hardware and programming technology. In any event, implementation need not concern the database user, who will appreciate ILE support for using applications such as GIS and SNA. To clarify what the ILE model involves, we offer the modeling of a real database.

---

[20] http://en.wikipedia.org/wiki/Elizabeth_Taylor#Marriages.2C_romances.2C_and_children (accessed 24 April 2013).

Figure 3. Representing in ILE the complicated parenthood relationships involving the actress Elizabeth Taylor

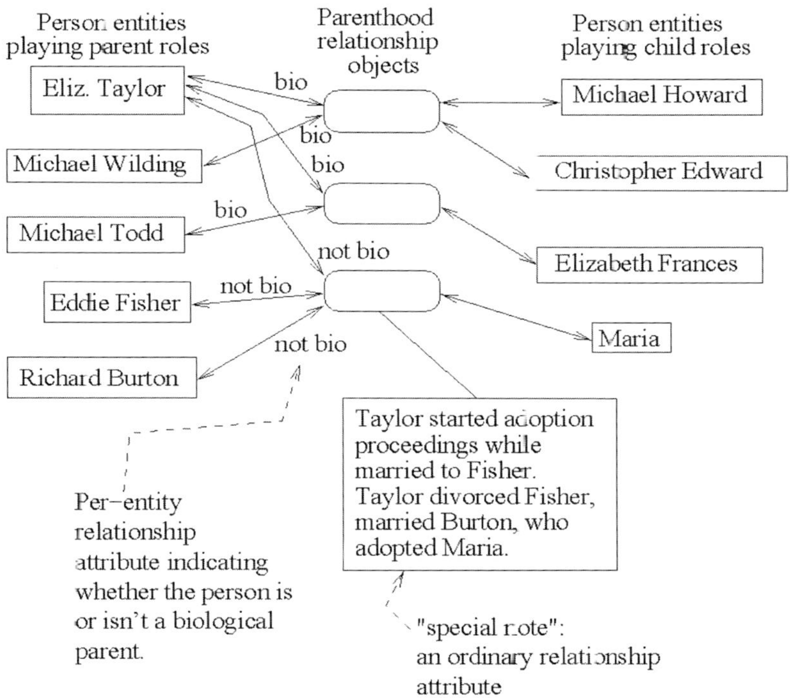

5. Modelling a Spanish Mercantile Database in ILE

In order to show the implementation of the Intentionally-Linked Entities (ILE) model, we have used, with her kind permission and assistance, the database of Ana Crespo Solana. She prepared the "Dynamic Cooperation Networks" database to support a sophisticated analysis of Atlantic commercial networks from the late seventeenth to the late eighteenth century (Crespo Solana, 2010: 181–212). The database represents years of work in the archives of at least five European countries. The database currently consists of multiple spreadsheets, but readers can peruse a lot of the information in several of Crespo Solana's major books.[21] With the assistance of the excellent technical staff in

[21] See list of Literature.

Madrid of the Centro de Ciencias Humanas y Sociales (CCHS) of the Consejo Superior de Investigaciones Científicas (CSIC), Spain[22], she turned the information into a data scheme, which could be used with geographic information systems (GIS).[23] She performed this work with support from the Spanish Ministry of Science and Innovation under the framework of the European Science Foundation's EUROCORES (European Collaborative Research) program "The Evolution of Cooperation and Trading" (TECT). Dr. Crespo Solana served as co-coordinator of one of TECT's five component projects, known as DynCoopNet (Noë, 2011: 3–4; Owens, 2012: 25–51; Crespo Solana, 2011: 11–35, and Crespo Solana, 2012: 191–224).

The database comprises the following spreadsheets:

1. ACCIONES
2. COMPAÑIAS
3. COOPERACION
4. FLOTAS
5. LUGARES
6. RUTAS

Each of these spreadsheets had many fields. Converting the spreadsheets into properly normalized relational database relations (tables) would result in a multitude of tables. These tables would have relationships with each other in complicated ways. As practitioners of digital humanities know, understanding a Microsoft Access™ relationship diagram can be a daunting task. In ILE, a single relationship can capture most, if not all, of this entire database. The relationship is called Actions, modelled after the table Acciones. See Fig. 4.

---

[22] See Chapter 4.

[23] On this work and its results, see "CrespoDATABASEAtlanticTrade [article in English]," http://digital.csic.es/handle/10261/28394 (accessed on 12 April 2013). Interested users of the available data who do not read Spanish should notice that Dr. Crespo Solana has made available a Spanish-English glossary of terms, which was prepared by the Geographically-Integrated History Laboratory of Idaho State University, directed by J. B. Owens. Due to the global financial collapse and Spain's financial crisis, the actual GIS work could not be completed. In addition to the "glossary", the georeferencing of the places in the database was performed in Owens' lab.

Figure 4. The arity-9 relationship Actions

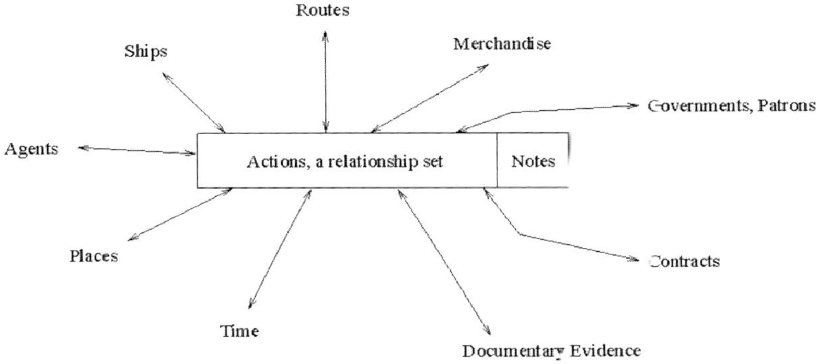

The Actions relationship has the following roles, where some roles can be played by 0, 1, or more entities:

1. Agents
2. Ships
3. Routes
4. Merchandise
5. Governments, patrons
6. Contracts
7. Documentary evidence, information source
8. Time, time interval
9. Places, including port of origin and port(s) of destination

Missing data are permitted. In ILE in general, the user can specify whether or not missing data are permitted, and whether a role can be played by more than one player. So far, the implementation of ILE has been developed in the programming language *Ruby*, and we offer the implementation only in a barebones version. The implementation contains commands for creating database sets, databases, entity sets, entities, relationship sets, and relationships. We have put in place no restrictions in place, such as how many entities can play each role. Users can query the database, but the interface for this purpose must be improved as part of the future work we will do.

One important step taken in the implementation ILE was to design a language for inputting data. We will just look at some of the more interesting commands. To create an entity set, for example, a *ces* command is used. Here is a command for creating the *places* entity set:

ces places {int id || string name | string paren | real longitude | real latitude}

This command creates an entity set of places, where the key attribute is a single integer, id. There are also non-key attributes, that is, its name, an optional **parenthetical attribute** to help identify it (such as the region or country name), and its coordinates. After having issued that command, *cen* commands can be created to create the entities belonging to that entity set, such as,

```
cen places {id => 6 || name => "Acebo" | paren => "Cáceres" \
| longitude => -6.71711| latitude => 40.203002}
cen places {id => 7 || name => "Adal" | paren => "Cantabria" \
| longitude => -3.478889 | latitude => 43.395132}
```

Note that the backslash characters ("\") are currently required as a **line continuation flag**, but an improved command parser would eliminate the need for a continuation flag. We will continue with this example according to a simpler data model, where "actions" is a relationship set with only three roles and *negocio* is an "uninterpreted" attribute of the relationship, instead of an entity playing a role. With this simpler model, we can create the "actions relationship set" with a *crs* (create relationship set) command:

```
crs actions {agents agents_role | places places_role | year time_role ||\
string negocio | string notes}
```

Finally, a relationship in the "relationship set [*relset*] actions" can be created using a *cre* (create relationship) command:

```
cre actions {\
roles::\
agents_role: nombre = "Jacob de Vries e Hijo" \
places_role: id = 41 || id = 910 \
time_role: año = 1772|||\
attributes::\
negocio = "Transporte de azucar de Surinam a Amsterdam" \
}
```

It is permissible for the *notes* attributes to be missing and hence unspecified.

A query involves specifying one of more roles and asking for some or all of the other roles. Because the various entities playing the different roles in a relationship are connected by pointers or data structures involving pointers, *users may execute rapid queries*. They do not need to perform the "table products" or the extensive searches that would be needed in a relational database system.

For example, the command

find all agents using relset actions where place = "Amsterdam"

returns all agents involved in any action that has anything to do with Amsterdam, that is, actions that have Amsterdam as one of the entities playing the "places" role.

While we develop ILE as a general-purpose database, we can think of two classes of applications that could really benefit from this kind of system. The first kind of application is geographic information systems (GIS), especially temporal/historical/narrative GIS and the nascent three-dimensional GIS, or an application that combines both the temporal and 3D properties. In such an application, users need extremely fast searches, especially on portable and small computer systems. Another kind of application that could strongly benefit from ILE is social network analysis (SNA) in the humanities and social sciences, where descriptions and relationships are complicated and data tend to be more descriptive than simply quantitative. For too long, users have not recognized that the database model is the key to doing dynamic GIS and SNA research.

We argue that computer science/informatics neglects database design.[24] This situation hinders many scientific research activities and particularly frustrates our ability to weave social network analysis (SNA) into geographic information systems (GIS) in order to create better abstractions for understanding human history. The neglect of design also creates serious problems for the information integration and researcher collaboration so essential for research in world history. In response, we have designed the Intentionally-Linked Entities (ILE) database management scheme to offer a way forward. In this chapter, we have discussed the three major database models that emerged from the 1970s, without much attention by their creators to the needs of potential users, in order to contrast them to ILE and to clarify the ways in which ILE solves the problems of earlier and existing schemes. Through shaping the ILE design by working closely with a specific group of data users – in this case, by historians of world commerce in the First Global Age as represented by Ana Crespo Solana – we wish to secure the necessary level

[24] Following the position of Badia and Lemire, "A Call to Arms".

of semantic consistency early in the design work, always recognizing the uncertainty about the information such researchers will discover in the future. ILE's flexibility insures that the database design will accommodate any type of information acquired by researchers. In future projects, we will improve querying and other access portal issues for users, and we will enhance the interface of ILE with other applications.

CHAPTER FOUR

AGENTS OF EMPIRE: SPATIAL NETWORK ANALYSIS IN COMPARATIVE STUDIES DURING THE EARLY MODERN AGE[1]

ERNESTO SALAS TOVAR[2],
ESTHER PÉREZ ASENSIO[2],
ISABEL DEL BOSQUE[2], ANA CRESPO SOLANA[3]
AND ROBERTO MAESTRE MARTÍNEZ[4]

1. Introduction

For a number of years Social Network Analysis (SNA) has been applied to historical research. Krebs in his paper "A Brief Introduction to Social Network Analysis"[5] defines it as the mapping and measuring of relationships and flows between people, groups, organizations, computers, and other connected information/knowledge entities. The entities are represented as nodes and their relationships as edges; in the case of social networks the nodes would be people and the edges their relationships. As such, it provides us with a mathematical and visual analysis of the relationships formed between the agents in a network. One of the aims of this paper is to set out the possibilities for analysis offered by visualizing

[1] This investigation has been sponsored by the "*Globa!Net*" project, Spanish Ministry of Science and Innovation, Reference: HAR2011-27694 and, previously: European Science Foundation's EUROCORES (European Collaborative Research) program "The Evolution of Cooperation and Trading" (TECT).
[2] GIS Laboratory, CCHS-CSIC.
[3] Instituto de Historia, CSIC.
[4] Paradigma Tecnológico, PT.
[5] www.orgnet.com/sna.html.

data using a Geographic Information System (GIS). As defined by J. B. Owens, visualization is an abstraction of the real world that allows us to gain insight into it. The greatest contribution to historical analysis offered by the use of a GIS is in organizing information into data models. This is achieved by superimposing different layers containing independent datasets, enabling us to observe visual relationships between the data (Owens, 2007: 2022–2024).

Thanks to these capabilities, like SNA, GIS has become an accepted tool among historians. One of the most recent examples is the DynCoopNet project[6], which analyses the evolution of self-organizing trade networks in the first Global Age (1400–1800). Since the data used by the researchers has a spatio-temporal element, a novel application of GIS was employed to visualize and analyse the information. Most of the data can be georeferenced to a specific date and location, meaning that complex analysis using GIS can be carried out to better understand the organisation, operation and impact of these networks on the formation of what was the first global economy. The development of this historical GIS required the generation of a single, integrated conceptual data model that is scalable and has great potential to meet scientific aims. This model gathered together the information in three databases containing the data assembled by researchers over many years. This information provided the basis for forming the model classes represented by the realities the data expressed, and these classes were used to create a database including all the information relating to individual trade agents, relationships of cooperation and ships. In order to visualize all these elements and their interactions the database contains a number of key tables, namely Agents, Cooperation, Actions and Ships. The first two of these show how the agents interacted with each other over time. The third table records the actions and activities – loans, representation, etc. – brought about as a result of cooperation. Finally, the Ships table, linked to Actions, brings together all the information relating to the voyages made by each of the ships (Pérez Asensio, Del Bosque González, Maestre Martínez, Crespo Solana and Sánchez-Crespo Camacho, 2010: 1–14, here 4–6)[7].

---

[6] ESF: FP: 004DynCoopNet and complementary actions by the Ministry of Science of Innovation: SEJ2007-29226. Dynamic Complexity of Cooperation-Based Self-Organizing Commercial Network in the First Global Age. It forms part of the EUROCORES programme, and it was approved after submission to the TECT (The Evolution of Cooperation and Trading) call for entries made by the ESF (European Science Foundation).

[7] Digital CSIC: http://hdl.handle.net/10261/24908.

In this chapter we will address a number of specific network analysis programmes that do not include a spatial element: Cytoscape,[8] UCINet,[9] and Tulip.[10] The historical data contained in the DynCoopNet database may be implemented in any of these programmes using snGraph (Maestre, 2010) for later analysis. The ESRI GIS tool, ArcGIS, also offers a powerful network analysis package with the extension Network Analyst.[11] Its most notable functions include the ability to georeference nodes and the flows between them or to determine which node has the most connections and locate it geographically. Adding the spatial component to the network analysis carried out using specific programmes is the key aim of this paper. In this regard "it is not only of interest to analyse the characteristics of the elements that comprise them, but also the form in which from a chronological perspective they develop, evolve and enable the progressive modification of all their components in both virtual and geographic space" (Barrientos Martínez, 2007: 8).

Although the Network Analyst extension to ArcGIS is primarily oriented towards the study of urban networks and services (electricity, hydrology, transport, etc.) it offers the possibility of implementing the results of SNA programmes and providing geographical locations to the nodes and the relationships between them. One of the key aspects of the DynCoopNet project where SNA is of great use is in the analysis of relationships of cooperation established among different agents in specific geographic locations, understanding by this term both particular merchants who establish different forms of association and the large monopolies (Crespo Solana, 2000; 2009; 2010a; 2010b: 181–314 and 2010c: 35–50, here: 47; Klooster, 1998). The DynCoopNet project also analysed the life cycle of the vessels that were used in the trade operations of the period. This took into account the voyages they undertook over many years, including ports of departure and arrival and ports of call, and the business transactions or other events that took place. The study of network analysis programmes enables us to compare their functions and how best to implement the results in a GIS.[12] For this reason, throughout this paper we will focus on defining a number of concepts relating to SNA in order to understand the potential of these tools and their application to spatial analysis.

---

[8] http://www.cytoscape.org/.

[9] https://sites.google.com/site/ucinetsoftware/home.

[10] http://tulip.labri.fr/TulipDrupal/.

[11] http://www.esri.com/software/arcgis/extensions/networkanalyst.

[12] The database is available at: "CrespoDynCoopNet DataCollection", Repositorio Digital CSIC: http://hdl.handle.net/10261/28394.

2. Definition of Network Analysis

The development of network analysis is the result of the influence and advancement of a large number of disciplines including sociology, social psychology, anthropology, physics, mathematics, computer science and so on. Before seeking to define a network, we should consider that it is comprised of nodes and the edges that represent the relationships between them. Oliveira and Gama (2012: 99–115) propose a highly simplified definition of a network that serves our purposes: "a network is formed with relational data that can be defined as a set of entities (individuals, groups, organizations or other), with relationships or interactions between them". Thus a network can be several things a) a conceptual model, b) a description of an existing structure or system in the real world, c) a mathematical model or d) a simulation (Luke and Harris, 2007: 69–93). Oliveira and Gama (2012: 100) define the four basic types of networks whose relationships can be studied: information, technological, biological and social networks.

Information networks are based on information exchange between the nodes that comprise them. They exist in order to disseminate knowledge, information or social causes and for business promotion. As an example of information dissemination, the methodology used by systems for citing academic articles presents a directed acyclic graph in which the nodes represent the articles and relationships are established when an article A cites another article B. Another clear example of business promotion are preference networks, which use bipartite graphs that express the individual preferences of consumers. Finally, the most common example of information exchange is the World Wide Web, which makes use of a directed graph in which the nodes represent web pages and the relationships are the hyperlinks between them.

Technological networks are networks that are used for the transport and distribution of services and resources. They are created in specific geographical areas and may incorporate a spatial manifestation such as highways, railways, and maritime routes, etc. Meanwhile there are other, smaller-scale networks such as physical connections between computers. These networks are represented by the Network Analyst tool as geometric networks (hydrological, urban) and transport networks (railways, highways).

Biological networks include certain biological processes, which we will not seek to define in this paper, such as chemical reactions among metabolites, interactions among proteins, regulatory genes, etc. which can be represented and analysed using networks. In fact some computer programmes that perform these analyses, such as Cytoscape, have specific tools adapted to biology.

Social networks are the principal focus of this paper, and have a structure based on the agents represented by the nodes and the relationships between them. These links not only represent relationships but also the flows of information, trade, correspondence, etc. between the agents. By studying the relationships between agents, we can reconstruct the social fabric we are studying. In this sense, one of the most important points to consider in our analysis is the position of the agents in the network. This data provides us with information about the roles each of them play and existing clusters.

3. The Spatial Component of Historical Social Networks

GIS analysis holds great potential for the humanities. In recent years the human sciences have been using these tools to find new solutions to problems previously dealt with by historiography. The use of GIS software which makes it possible to combine geographic information from various sources and bring it together on a map expands the horizons of the humanities. The ability to add thematic information on a real location and analyse the relationships established between these layers of information is the engine of a GIS. The term “spatial” is for some what holds us together and enables us to focus on our interactions with space and with our peers (Getis, 2008: 4). One of the key issues when it comes to applying these analyses to historical networks is time. Since the 1980s work has been carried out on implementing this variable in GIS software in which the concepts of change and movement are key. During the 1990s progress was made in the representation of temporal phenomena and the creation of mathematical models to simulate movement. As such, we believe that it is necessary to work with GIS from a dynamic perspective that takes into account change and movement as part of a system. In this way we can overcome space and time barriers and observe these phenomena from a holistic perspective (Yuan, 2008: 18–24), which enables us not only to learn the position of the elements we are interested in studying, but also their behaviour and movements from a geographic perspective. Recently, GIS analysis has been applied to the representation of spatio-temporal phenomena displayed by interactions between individuals and the flow of human relationships (Yuan and Shaw, 2006:409–430). These two authors introduce the concept of spatio-temporal prisms to analyse these interactions with greater precision.

Within the humanities, traditional history has begun to turn to GIS applications in order to reconsider questions about human relationships in the past. We are deliberately leaving aside archaeology, as the application of GIS software to research in this field is well established (Wheatley and Gilling, 2002; Grau Mira, 2006; Mayoral Herrera and Pérez, 2011). In this regard, Owens has concisely explained the leading role acquired by these technologies in recent years. This is a result of the fact historians work with complex, dynamic and non-linear information and require a technology that enables them to organize and visualize this information simultaneously (Owens, 2008: 35). GIS offer the ability to visualize and analyse historical data. According to Owens (Owens, 2007: 2024), the discipline of history has not made significant use of visualization, which he considers a mistake because it presents an abstraction that can be a very useful means of approaching the real world. Such abstractions enable us to observe the interactions between the different kinds of data and provide us with a method for understanding spatio-temporal data. Despite these qualities, it must be borne in mind that GIS work with static information, which represents a problem for resolving the kind of dynamic changes over time that historians deal with. As such, it is necessary to find a way to represent interactive connections between agents that make it possible to integrate studies at different levels: local, regional and national. This need has led Owens to define the concept of "geographically-integrated history", a term that frees us from the limits imposed by the earlier concept of "History GIS" (Owens, 2007: 2025–2026; Crespo Solana and Alonso García, 2012). This conceptualization is based on the visualization and implementation of different layers of dynamic information that change over time, making it possible to understand the evolution of historical processes rather than small, static portions of information.

Finally, to complete this transition, it is essential that historians break with traditional historical methodology and be prepared to collaborate with other researchers to achieve an interdisciplinary history. This change in mentality will make it possible to observe more accurately the cooperation between agents in the past. To achieve this, it will be necessary to obtain quantitative and qualitative information about the culture and society of the historical moment that concerns us (Owens, 2007: 2030–31). Although much progress has been made in involving historians in the use of GIS, the latter present a number of defects that hinder their use by the social sciences. Owens (2009) gives an overview of these problems. Firstly, the difficulties historians face in handling these programmes. Secondly, a GIS is static and works with time-based data only with difficulty. Thirdly, these programmes can only work with a limited number of variables, which can lead to poor results. Finally, it is difficult to use GIS to represent the disordered, qualitative and imprecise data historians often have access to. The DynCoopNet project has sought to further these efforts by analysing the cooperation of trade agents in the development of complex networks. This cooperation aimed at preserving the flexibility of the trading system in order to control the circuits of capital and information, although it led to the emergence of new forms of cooperation that would culminate in the alteration of the system itself.

4. History of Network Analysis

It is difficult to outline the history of the evolution of network analysis in a few lines. Linton Freeman (2004) has charted this process in several phases that represent the emergence of the first theoretical foundations, and their subsequent development and consolidation. The first period, which Freeman calls the "prehistory of network analysis", runs from the mid-19th century until the 1930s. Earlier still, in the 18th century the mathematician Leonard Euler had proposed some of the theoretical and practical grounds for graph theory (Luke and Harris, 2007). Sociology is the discipline that contributed most to the development of SNA at this early stage. The first sociologists asked themselves how social links are established. In an attempt to answer these questions, thinkers such as Comte and Simmel established a theoretical basis and specific terminology for describing these connections. The key lies in the emphasis that Simmel placed on the position occupied by agents in affiliation networks. Comte, meanwhile, proposed that the interconnections between social agents should be taken into account (Freeman, 2004). In conjunction, the ideas of Comte and Simmel would provide the starting point for network analysis (Grabher, 2006: 164).

The second period in this development began during the 1920s, and forms the prelude to the first network analysis, properly speaking. It was driven by the work done by educational psychologists on social connections, which established influence, interaction and company as key characteristics. At the same time, in the field of social anthropology in Britain Radcliffe-Brown and Nadel were making progress in the development of structural analysis with the understanding that social structures are formed by concrete relationships between individuals (Scott, 1996). The emergence of the figure of the psychiatrist Jacob L. Moreno is considered the starting point for this period of development of network analysis as a methodological tool for social analysis. The novelty of his studies may be seen in his representation of agents using what he termed sociograms, which are a kind of graph in which individuals are represented as points connected by lines. The use of this method enabled the discovery of hidden associations, with the identification of stars, alliances and subgroups (Oliveira y Gama, 2012: 99–115). This period culminated in the publication of the first issue of the journal *Sociometry* in 1937, which created a space for academic discussion in which researchers sharing similar concerns about society participated (e.g. Kurt Lewin, Paul Lazarsfeld and Gordon Allport).

During this stage of development, there occurred a "dark period" between the 1940s and 1960s due to the conflicts between in the scientific community and Moreno's "bombastic" personality. Except for disciplines like geography, biology, anthropology, social psychology and mathematics, which used network analysis, SNA remained fragmented and localized (Freeman, 2004: 42, 120). Finally, in the mid-1960s the publication of the study by Coleman, Katz and Mendel (1966) on the relationship among the connections established between doctors and the diffusion of new medicines put an end to this impasse.

Centred on the Department of Social Anthropology at the University of Manchester, the discipline of anthropology took up the baton to research conflict among groups. Barnes, Bott and Mitchell continued on the path set out by Radcliffe-Brown, focusing on how the structure of social relationships affected not only individuals but also society as a whole. In this sense a lot of attention was paid to the content of relationships while analysing their structure at the same time (Scott, 1996). The theoretical proposals of the "Manchester group" would lead to the development of network theory, providing a common language for making progress in the field (Reynoso, 2008). The third period, during which the discipline became consolidated, is the result of the influence that the British researchers had on a group of sociologists at the Harvard University led by Harrison White. During the 1960s and 1970s they would be at the helm of "the Renaissance at Harvard" through which SNA achieved the status of a general research paradigm (Freeman 2004: 42). White's teaching succeeded in instilling this paradigm in an entire generation that would develop many mathematical aspects of SNA, taking concepts from the social sciences, such as the "social role" of a mathematical form that would allow them to measure and model networks (Scott, 1996). The list of White's students may be considered a "Who's Who" of network analysis (Freeman, 2004: 127).

During his doctoral studies, Mark Granovetter, a disciple of White, asked himself how network theory resolved the question of micro/macro relationships. Specifically, he asked: how does an agent go about operating outside his environment? According to Granovetter, up to that point it is possible to know data about the organization of the community, political structure and social mobility, on the macro scale, as well as what happens within the confines of a small group, on the micro scale. But when it comes to studying how small groups interact to form a larger cluster, we lose track of what occurs. Effectively, this means the relationships between individual nodes influence the formation of networks, or relational analysis (Granovetter, 1973: 1360–1380). Granovetter's studies are very important for a number of reasons. In the first place, because his research into the process of information dissemination marked the beginning of the network focus (Núñez Espinosa, 2008).[13] Secondly, because they enabled the development of more sophisticated and realistic modelling of network structures. As such, they are an attempt to explain social structures and human behaviour (Luke and Harris, 2007: 71). Finally, they are important because they introduce economic activity into the study of networks and a theoretical interest in the economy (Grabher, 2006: 164). This interest is catalysed by the concept of incrustation, which in its economic sense, related to SNA, assumes that all economic interactions are incrusted, or embedded, in social relationships (Reynoso, 2008).

Finally, the 1980s and 1990s represented the consolidation of the period begun at the Harvard University. Two fundamental trends in analysis emerged. On the one hand an eclectic hodgepodge of anthropologists, geographers, social psychologists, political scientists, historians and mathematicians, and on the other a group consisting only of sociologists. Both schools maintained an academic separation that could have led to the disintegration of the discipline. However, a reunification process was spearheaded by Barry Wellman at University of California, Irvine, who thanks to the influence of the writings of Mullins (1973) founded the International Network for Social Network Analysis (INSNA) together with other researchers. This combination of factors paved the way for scientific discussion in specialised journals and conferences on networks, which would reach a critical point with the development of specific software for quickly and cheaply producing and representing analyses.

[13] http://revista-redes.rediris.es/webredes/.

5. Study Proposals on Historical Social Networks

The 1980s represented a major crisis in the paradigms used by historical studies. Perhaps the foremost expression of this process was the disintegration of the final generation of the *Annales School*.[14] The exhaustion of the structure-based theories of Braudel led to the search for new analytical perspectives. In some cases the changes were wrought from within the school itself, with Roger Chartier as one of the lead proponents. These researchers began to take an interest in the small players in the grand structures *Annales* concerned itself with, and began to propose "micro" perspectives. Such is the case of the microhistory of Carlo Ginzburg, the history-from-below of Roger Chartier or the new cultural history which is strongly influenced by American anthropology as personified by Natalie Zenon Davies and Robert Darnton.

Historians began to seek analytical perspectives in new disciplines that Annales had left aside, furthering the interdisciplinarity with which the French school had already revolutionized the way history is done years earlier. In this effort to emphasize individual agency over and above grand structures, history found new horizons in network analysis. As such, historians interested in studying the operation of specific social groups from within were drawn towards sociology and SNA methodology. These first contacts lacked a theoretical background for understanding all the possibilities for applying advanced analyses to history. However, since the 1990s we have witnessed the emergence of a more concrete analysis that has advanced slowly and at different rates depending on the area of knowledge, whether linguistic or historical. Nevertheless, these approaches made from the macro scale remain relatively isolated, almost in the margins of the discipline, particularly in its points of contact with sociology, and regardless of the micro-historical perspective of social relations. In this sense, it is surprising that today it seems that efforts to apply the methodology of network analysis to history remain isolated from each other, with no real scientific dialogue between them (Lemercier, Guzzi-Hebb and Bertrand, 2011).

Traditional political history has focused on analysing power and the state, taking the sources of its power as a reference point. Rafael Guerrero,

---

[14] The *Annales d'histoire économique et sociale* is an academic history journal founded in 1929 by Lucien Febvre and Marc Bloch that placed emphasis on social history and opposed the German historicism of Leopold von Ranke, political history and history based on major figures. One of its strong points was bringing history closer to social sciences in order to resolve some of the problems facing the discipline. Herubel, 2004: 293–312.

who has pursued the line of research initiated several years ago by José María Imizcoz (2009), sets out the change of perspective undergone by this vision of political history: "These analytical perspectives essentially set themselves against the core of traditional political history, the State and official law, displacing the focus towards micro-space and personal factors, in an effort to overcome the confusion between theory of the State, theory of power and political theory" (Guerrero, 2012). With the use of relational analysis – borrowed from sociology – it becomes clear that relationships have a determining role in building social networks. As a result, the static structures and contexts that characterize political history are abandoned in order to focus on the interplay of relationships pursued by agents in order to ascend in the structure of power of the Old Regime. The aim of the different proposals we have analysed in this group is to reconstruct relationships around a central entity, whether this be a person (Imízcoz, 2011), an institution (Artola Renedo, 2011), or a shared identity (Padgett, 2011).

6. What is Social Network Analysis?

6.1 Definition and Basic Questions

Earlier, we dealt with the meaning of a network, but a social network can be defined as the mapping and measuring of relationships and flows between people, groups, organizations, computers, and other connected information/knowledge entities (Krebs). That is, SNA provides a visual and mathematical analysis of human relationships. Beginning with this definition there are certain concepts that must be borne in mind when it comes to understanding SNA. In the first place the entity defined by the SNA is the focus on the structure of relationships, which range from casual connections to close links. SNA presupposes that relationships are important and as such represents and measures formal and informal relationships in order to understand what facilitates or impedes the flow of knowledge, who knows whom, who is connected to whom, and so on. Given that these relationships are not always visible at first glance, SNA may be said to be a kind of organizational x-ray (Serrat, 2009). Secondly, when it comes to carrying out a network analysis the methodological basis of the tool must be borne in mind: relationships and the qualities of the relationships built around individuals and entities. In the case of SNA the principal information taken into account for the research are links and nodes. As such, according to Nuñez Espinoza (2008) it is important to attend to the following points:

1. the population and social group to be studied (characteristics and size of the sample);
2. the relationships and qualities the analysis seeks to bring out; and
3. the information that needs to be obtained to conduct the analyses of the network(s) identified.

This last point makes clear the flexibility of SNA, as it enables us to visualize different types and quantities of relationships, despite the limitations imposed by the methodology with regard to the relationships and attributes we wish to analyse, and the depth in which we wish to analyse them.

6.2. Representation and Visualization

The structure of these networks tends to be represented using graphs based on mathematical graph theory (Tutte, 2001). A graph (Figure 1) is the conjunction of two fundamental elements, nodes and edges. Each edge is defined by a pair of nodes that are also known as "endpoints". The edges may be direct or indirect depending on the nature of the relationship between the nodes, that is, whether it is symmetric or asymmetric, respectively. In connection with the theory of weak connections explained above, the importance of the nodes for calibrating the strength of the relationships with the remainder should be borne in mind. This factor will condition the structure of the data matrices – which form the basis for the graphs – given that if the relationships have no weight and are undirected, the matrix will be binary and symmetrical; while if they are directed and have weight then the values will be found somewhere between 0 and the maximum weight. In both cases these are non-negative matrices (Oliveira and Gama, 2012: 101). The representation of a graph enables us to evaluate the location of the agents in the network, which allows us to observe from within the different roles and groupings that occur in the network. That is, it allows us to answer a number of questions, such as, on the one hand, who are the connectors, the leaders, the bridges, the isolated individuals and the experts, and on the other, where the "clusters" are found and who is in them. In short, who is in the nucleus of the network and who is on the periphery. When it comes to representing graphs we can use several mathematical methods, including lists and matrices. These structures allow us to store graphs in computers in semi-automated form on the basis of adjacency and incidence. Sequential representation is based on the matrices of adjacency that show all the relationships between the nodes in binary mode. Linked representation is based on linked lists of neighbours – adjacency – that introduce the data using its formal definition: as a set of nodes and edges.

The matrices and lists of incidence represent the graph on the basis of its edges, that is, the relationships between the nodes and the edges in a binary matrix. In the case of lists, these are formed using a dynamic vector with linked lists to create an index of edges (Valenzuela Ruz, 2003). Together with representation, SNA allows us to make calculations to examine various questions regarding the network. One of the most interesting aspects for those who analyse networks is the position occupied by the agent in the network and in its structure. Calculations such as degree, intermediation or proximity tell us about the links an agent has with his surroundings, who occupies a central position or who functions as a bridge for the rest.

In the case of Figure 1, we may observe that a calculation of degree has been applied to the visualization of the nodes. This shows us which nodes are the best connected to the network, according to the number of links maintained by each node. Using the Cytoscape software we have chosen to represent this by the size of each node. As such, we can see that node 5 (Cádiz) is the best connected node in this example.

Figure 2 uses the same data as in Figure 1 but implemented using the Tulip programme. In this case we have applied the Fruchterman-Reingold (1991: 1129–1164) algorithm, which functions as a mechanism of repulsion, pushing away the more isolated nodes and bringing closer those that are better connected. While these calculations provide us with valuable information, the key added value of these analyses lies in their spatial component. This provides the geographical space located beneath this network of relationships described with the SNA programmes. As such, our efforts are aimed at implementing the results of the GIS analyses we have briefly described. We fulfil a dual objective: on the one hand we are able to represent the nodes in their geographic and chronological position, and on the other we can visualize what is taking place in the network and observe hotspots, the best located agents both in terms of the network and geographically. In short, the relationship between the territory and those who occupy it.

Given that our paper is focused on the application of GIS analysis to historical social networks, we refer the reader to specialized bibliography regarding the basic measures for SNA[15].

---

[15] Vadis Krebs, "Uncloaking Terrorist Networks" *First Monday*, vol. 7, no. 4 (April 2002); Krebs, "A Brief Introduction to Social Network Analysis" and Oliveira and Gama "An Overview of Social Network Analysis".

Figure 1. Graph of a maritime trade network in the First Global Age (1400–1800)[16]

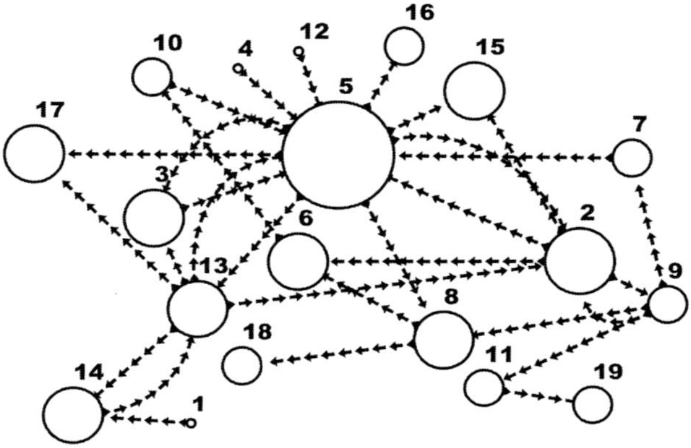

Figure 2. Graph of a trade network in the First Global Age using the same data as in Figure 1[17]

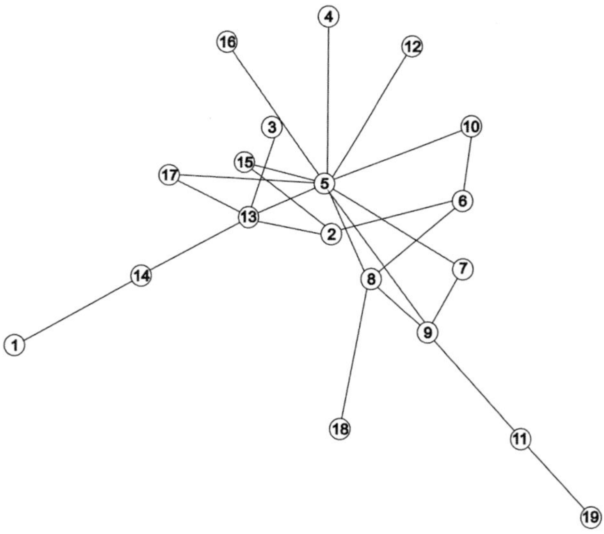

---

[16] Prepared by the authors using Cytoscape. 1. Antwerp, 2. Amsterdam, 3. Brussels, 4. Buenos Aires, 5. Cadiz, 6. Genoa, 7. Gothenburg, 8. Hamburg, 9. London, 10. Marseille, 11. Oporto, 12. Rio de Janeiro, 13. Seville, 14. Valladolid, 15. Malaga, 16. Tenerife, 17. Cartagena, 18. New York and 19. Alexandria.

[17] Prepared by the authors using Tulip.

7. Social Network Analysis Software

The entities that belong to a network can be of many different kinds. From simple individual nodes to groupings of nodes in dyads and triads, by way of links and edges, right up to the whole network. Similarly, the software packages for SNA take the raw data from the lists of edges, and adjacency lists and matrices that may sometimes be combined with their attributes. Most programmes on the market work with data in ASCII text format, although some packages are able to import data from relational databases and store the network entities.

SNA software can be divided into two classes: a) packages offering a graphic user interface (GUI) and b) programmes created to generate programming languages and computer commands. The former are easier for non-expert users and have a rapid learning curve, while the latter have greater potential and extensibility. The most common GUI-based SNA software packages include NetMiner, UCINet, Pajek, GUESS, ORA, Cytoscape and Tulip.

There are other programmes with a proprietary graphic user interface that have been developed for specific uses, such as companies that wish to visualize their internal organization (Orgnet, Keyhubs and KXEN) or the telecommunications or online gaming industries, which use Idiro SNA Plus given its ability to handle large amounts of data.

Finally, networks may be analysed from a spatial perspective, the focus of this paper. The ESRI program, ArcGIS, offers the Network Analyst extension that enables representation and analysis of geometric and transport networks. We shall proceed to analyse the properties of this tool as a complement to those used in specific SNA programmes. Our software comparison will focus on four specific packages, UCINet, Network Analyst, Cytoscape and Tulip (Table 1). Of these, the first two are proprietary software while the latter two are open source. Despite the fact that proprietary software programmes are easier to learn and still have a greater number of users than the free software, open source programmes are making much more rapid progress in terms of functionality and features. Furthermore, these programmes more often include user manuals and tutorials that continually improve in terms of intelligibility and applications. As we saw above, visualization in spatial analysis programmes is a powerful tool for representing the real world (Owens, 2007). As such, visualization in SNA programmes allows us to better understand data and efficiently communicate results (Freeman, 2004). SNA software allows us to modify several aspects of visualization such as colours, size and other properties of the network in order to better display the conclusions shown by the analysis. Most SNA tools already incorporate these features so that users can produce the best possible visualization. In this paper we will focus first on comparing SNA programmes, of which Cytoscape has the best level of visualization. Meanwhile, data visualization is interactive, meaning that users can generate new data on the basis of the original data matrix. For example, it may be that we want to make visible only a section of the information in a very dense network. Most programmes enable us to do this by filtering the information according to chosen categories. According to the information contained in Table 1 we can draw some conclusions about the analysis tools that we have considered in this section. Perhaps the most important aspect of the three programmes we have presented lies in the possibility two of them offer of working with open source and the fact they work on any operating system. Meanwhile, open source allows software updates to be developed more effectively, allowing users to improve their analyses as the project progresses. In the case of Cytoscape, version 2.8 has recently been released in its eighth year of development, which while respecting the functionalities of a network analysis tool (interactions between nodes via edges), has added two new features. Firstly, it enables users who are not experts in programming to map images using nodes, which significantly increases its capacity and flexibility for visualization. Secondly, it is now possible to write equations in the spreadsheet in which the data is found in a similar fashion to programmes such as Excel (Smoot, Ono, Ruscheninski, Wang and Ideker, 2011:431–432). Tulip, at the time of this study, had just released version 4.1 of its software for all operating systems. As announced on their website http://tulip.labri.fr they have solved some issues present in previous versions.

The two most important new features are corrections to the spreadsheet when managing properties, and access to new plugins, not only for users of Windows and MacOS but also Linux, which were not available in previous versions. The user interface, as with other GUI-based programs, is very intuitive and allows visualization of multiple windows with partial and complete information about the network elements. UCINet (Stephen, Everett and Freeman, 2002) is the only one of the three tools that is designed specifically for social network analysis. Its structure includes the original programme together with Spreadsheet and NetDraw, which are extensions for working with data matrices and visualization graphics respectively. It is the only one of the three programmes that has a charge, with a trial version that after 60 days requires payment for the license. Network Analyst is a very powerful tool primarily aimed at solving problems of mobility in urban and interurban networks. For the present case the features offered by this tool for the representation and visualization of trade networks in the 18th century are: modelling maritime routes by georeferencing the points passed through by the ships based on information taken from nautical charts and the number of ships using each route, assigning different thicknesses to each route (Figure 2); locating in space the nodes representing the ports and agents that control maritime traffic and ports of departure, arrival and call; calculating the amount of goods being transported each year by these vessels on each route, and specifying whether these were slaves, commodities, gold, silver or coinage. Furthermore, we can establish the direction of travel on routes in order to invert them and perform these calculations in the opposite direction.

Figure 3. Georeferenced West Indies route showing the direction of travel of ships.[18]

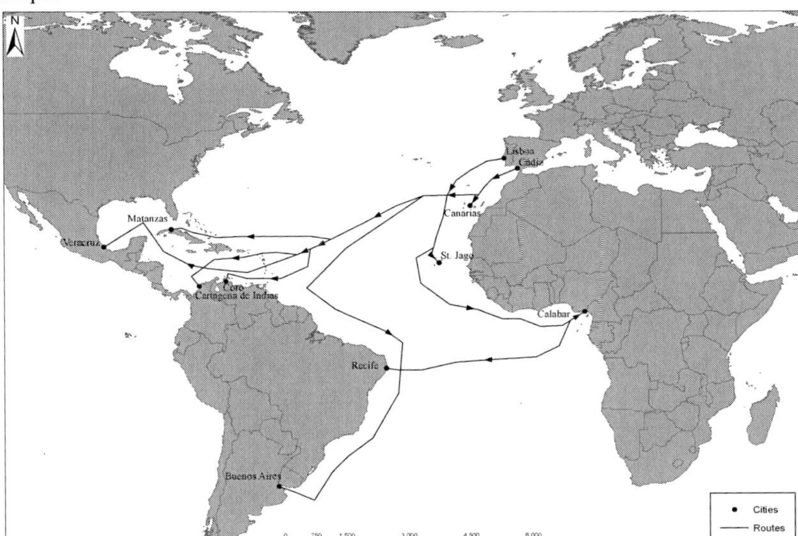

One of the possibilities offered by the DynCoopNet project data is the determination of transatlantic trade hotspots using data based on the number of connections implemented in the ArcGIS Spatial Analyst. Moreover, we can use SNA programmes to determine which agents are best placed within the interplay of trade relationships. These two pieces of data allow us to state whether agents who are better placed in the structure

[18] Prepared by the authors, *Puertos*: Ports; *Paradas intermedias*: Ports of call.

of the network are also located geographically in these hotspots. Moreover, we can implement the relationships the SNA programme created for us in Figures 1 and 2 in the GIS in order to obtain a spatial visualization. Figure 4 shows the spatial dimension of the data used in Figures 1 and 2 together with the connections between them.

Figure 4. Map of trade relationships between cities in the First Global Age[19]

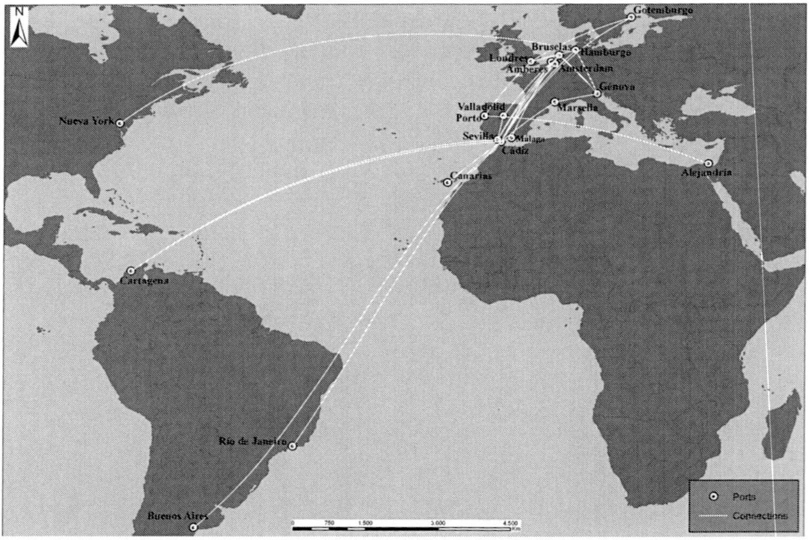

[19] Prepared by the authors; *Ciudades*: Cities; *Conexiones*: Connections.

Table 1. SNA programmes. Functions, formats and capabilities[20]

Software	*Application*	*Data input formats*	*Visualization and representation formats (output)*	*System specifications*	*License*	*Description*
Cytoscape	Tool for integrating complex and general data for visualization and analysis	SIF (Simple Interaction Format), GraphML, XGMML, GML, KGML, SBML, BioPAX, Excel and text tables (.csv, .tab and tab delimited tables) Other programmes such as Pajek, igraph, Graphviz.	SIF, XGMML, GML, GraphML, Cytoscape Session (.cvs), vector/bitmap images (.jpg, .png, .pdf, .ps)	Cualquier OS que soporte Java	Open Source (LGPL)	Cytoscape was originally developed for bioinformatics research, but is now presented as an independent analysis platform. In addition, a large number of plugins for different users come with the programme, which greatly extends its functionality. Also, it enables connection to external web services to obtain information to work with from public databases, other networks, etc.
Tulip	Analysis and visualization tool for relational data	Tulip format (.tlp), GraphViz (.dot), GML, txt, adjacency matrix	Tulip format (.tlp), GML	Any OS (Windows, Linux, Mac OS)	LGPL	The tool provides the developer with a complete library of plugins supporting the design of interactive applications for relational data visualization that can be adjusted to the problems the user wishes to solve.

[20] Prepared by the authors based on data from INSNA and ESRI.

UCINet	SNA tool	Excel, DL, .txt, Pajek, .net, Krackplot, Negopy, proprietary format (##.d & ##.h)	Excel, DL, .txt, Pajek, .net, Krackplot, Mage, Metis, proprietary format (##.d & ##.h)	Windows, MacOS and Linux (using BootCamp, VMFusion Ware, Parallels or Wine)	Shareware	A comprehensive SNA tool. It can work with a maximum of 32,767 nodes (with some exceptions) although in practice analyses of 5,000-10,000 nodes can lead to delays in calculations. It includes the analyses mentioned above, basic graphics and statistics. It also includes powerful matrix analysis routines such as matrix algebra and multivariate statistics.
Network Analyst	Extension to ArcGIS (Spatial analysis)	.txt, .xls, .shp, .csv	.shp, .pdf, .jpg, .gif, .png, .txt, .csv	Windows	Paid licence	A tool whose main function is to solve problems for routes connected in a network. Its main features are: to calculate the shortest path to solve a problem; learn if the zones of influence of a particular service cover all potential users; establish optimum routes for loading and unloading goods.

8. Implementation of Data

The design of a GIS necessarily involves a stage prior to the implementation of the data: that of modelling. This is a fundamental and indispensable phase in the creation of any quality system that involves the implementation of a database, regardless of its nature, and the model is in itself its materialization, the way the data is organized (Del Bosque González, Fernández Freire, Martín Forero and Pérez Asensio, 2012). The first step, therefore, for approaching a GIS project requires undertaking a conceptualization, abstraction or simplification of reality or the object of study. To facilitate this task both graphical languages such as standard UML (Unified Modeling Language) and CASE (computer-aided software engineering) tools are frequently used. These are applications that reduce the time and economic costs of developing a project by facilitating the generation of all its lifecycle documentation. The resulting conceptual model is independent of both the hardware used in the system to be developed and the software, including the database manager that will be used to store the information itself. A semantic distinction can usually be made in a GIS between two types of entities or datasets: those that are properly geographic (with characteristics related to the space they occupy such as position, geometry, or topology) and those that are thematic, which in the case of historical research projects such as the present case usually occupy the greatest amount of information and may involve high levels of complexity due to the multitude of relationships that can be established between them. Once the information has been modelled its physical implementation is carried out by entering the data (thematic and geospatial) into a database that can respond to the principle of a relational, object-oriented or object-relational database. Whichever of these is used, the classes or entities previously defined in the conceptual model together with their relationships are translated into tables with attributes and relationships between them. In the example of the historical research presented in this paper related to social network analysis, the resulting complex conceptual model has been transferred to an open source database, PostgreSQL, which basically handles the alphanumeric or thematic aspect of the data, and uses an extension called PostGIS for managing the properly geographic information, thereby maintaining a certain separation between the two contexts. Once the data has been entered into the database, which in the case of the DynCoopNet project was carried out using a data entry interface developed specifically for this purpose, the subsequent utilization and analysis of the data is performed with different platforms, both GIS and any of the network analysis tools discussed in this paper that do not handle the geospatial component. Moreover, a simple and rapid way to upload data from the DynCoopNet database to the network analysis programmes is the tool implemented as a package in Java snGraph (Maestre Martínez, 2010). This software was developed precisely in view of the special topology of the networks that we want to represent, and it enables simple visualization and sharing of the data and the creation of models to exploit databases quickly and efficiently or export them to programmes that allow more in-depth analysis. The most important aspect of this tool lies in its multiplatform capability, since it is not a network analysis tool, but rather a bridge between systems. As such the data stored in PostgreSQL can be exported to any of the programmes we have discussed above for performing the analyses required. Depending on the type of SNA programme to be used, the import format of the data may vary, although text format (.txt, .csv, etc.) is very common for this purpose. The way of handling the information may also vary (e.g.: Tulip requires nodes to be entered on the one hand and the relationships between them on the other, while in Cytoscape this differentiation is not required, facilitating the task of data processing). The main problem faced by the DynCoopNet project is that not all the information contained in the documents can be used. In fact, the documents that have been used to transfer the information into the database share the same trade agents that will be part of our network analysis. As the authors explain, the agents are not just the people who run the West Indies fleet and negotiate the goods to be transported. They are all those involved in transatlantic trade, including representatives, privateers, pirates, capitalists, brokers, etc (Pérez Asensio, Del Bosque González, Maestre Martínez, Crespo Solana and Sánchez-Crespo Camacho, 2012: 158–160)).

9. Conclusions

Social network analysis opens the doors to the relationships established among a group of people. Knowledge of these relationships allows us to isolate agents and understand their behaviour and the reasons for their social rise or fall. This new analytical perspective offers us a tool for evaluating the degree to which the interplay of relationships among agents in a network has shaped historical changes. The application of these techniques, which originate in sociology, is currently coming into its own among historians. As we have seen, they offer us a new analytical perspective on aspects that traditional historical studies have ignored. This is also the case with Geographic Information Systems, which have gradually been introduced among historians in recent years as a tool for visualizing spatio-temporal phenomena. Moreover, the visualization capacities of GIS allow us to observe changes in a territory over time, which may be added to the evolution of the relationships among agents in the historical networks under analysis. In this way we are able to observe not only changes in a territory but also the movements made by these agents to achieve a better position both in the network and in relation to trade hotspots. This is the most important point we have dealt with so far, which is none other than the vast possibilities offered by implementation of the results of historical network analysis in a GIS. Territory becomes a key element for describing the interplay of relationships, in our case of trade agents in the First Global Age. This inclusion of the spatial component is the next step that should be taken by SNA in its integration with historical analysis. Finally, the dialogue between historical sciences and these technologies still has a long way to go. It is a journey that demands collaboration across disciplines in order to propose new analytical perspectives and applications of these techniques geared towards resolving issues surrounding historical relationships.

CHAPTER FIVE

NORTHERN PORTUGUESE COMMERCIAL NETWORKS AND THE GEOGRAPHIES OF TRADE IN THE EARLY MODERN PERIOD

AMÂNDIO JORGE MORAIS BARROS[1]

1. Introduction: Trading Networks and Global History

This chapter discusses the nature, the performance and the achievements of the commercial networks of Oporto in the sixteenth century and outlines a recent topic of research: the social and economic mechanisms of international trade from an Atlantic port city through the action of worldwide mercantile networks, studied from an interdisciplinary outlook. The highly topical character of trade networks studies is attested by the recently research fields proposed by the Anglo-Saxon historiography, spread over three main lines: first, the interest on social networks and on the subject of social trade; second, the so-called Global History (Meyran, 2008:10–15) developed from the notion of the *Great Divergence*; third, the discussion of intercultural esxchange as a result of merchants intervention overtaking geography, "narrowing" the space, and introducing modes.

In the first case, the interest for the social dimension of trade dominates. The model, primarily proposed by Douglass North (1990) brought to the attention of economic historians the importance and influence of institutions for economic development and growth, and is resumed by Ogilvie (2011) which analysis is based on the supportive role of institutions and solidarity mechanisms between merchants and their ability to influence the prices and market relationships. But again, the analysis is somehow restricted because is mainly focused on the study of

[1] CITCEM, Polytechnic of Porto-School of Education.

medieval and Early Modern merchant guilds, which did not exist as such in all regions of Europe, and pay no attention to the organizational contributions anticipated by Iberian traders namely their capability of contract enforcement through the (informal and customary) norms inside merchant consulates and communities.

In the second case, the research program focuses on the study of relationships between kingdoms, territories and continents from the Early Modern Era to the present time (Global History) searching the origins of the development gap between the continents; in the words of its great theoretician, Pomeranz, this is the *Great Divergence* (Pomeranz, 2000), and indeed started the historiography interest in China. Since then, the focus of focus of analysis has shifted from Atlantic history to Asian history. Like every historiography trend it has virtues and shortcomings. It had the merit of placing trade as a key issue for the study of global connections throughout history. In this regard, and as proposed by B. Bailyn (2005), trade and merchants ignited the first globalization, streamlining every economy and market from local to national and international, and turned them into agents of change before industrialization. The main criticism which may be made of this theory is the general absence of the Castilian and the Portuguese merchants and settlers in the activation of global trade relationships.

In the latter case, cultural exchanges introduce current issues, such as consumption, the relationship between consumption and market and between consumption and marketing (of course, within the narrow limits of Early Modern societies), including the economic, cultural and symbolic value of goods and merchandise exchanged from 1400 to 1800. A special place is given in this model to the symbolic notion of commodity, as defined by Bourdieu, and applied to goods that functioned as complimentary gifts, proof of friendship, and distinguishing the attribute of power. But also to the market cultural function in the context of ways and fashion that were introduced by the demand and supply dynamics. Research undertaken by Hancock, Barros and Pereira (2009) makes clear that the Madeira and Douro wines were changing as fashion, taste, and demand changed.

As has been said, these historiography trends ignore or underestimate the role of Iberian agents and networks in the origins and development of globalization movements; such underestimation may be explained by the lack of published primary sources for Portuguese history, general ignorance of recent research, and the choice of a chronology that usually starts with the late 16th century, i.e., coinciding with the rise of northern European states and kingdoms. This chapter is methodologically open to proposals from all of the three schools but fundamentally aims to change their narrow view by introducing new actors and new data in the process. I will attempt to analyze the networks organization model and present their organization charts, to identify merchants and to measure their role within the structure through short biographies describing significant commercial episodes, social connections and political aspirations. Methodologically, I will deal primarily with the city's notary documentation, which recent studies acknowledge to be the best tool to monitor social and economic relationships inside historic communities, town council minutes (*vereações*) and municipal ordinances, and commercial correspondence exchanged with other European businessmen. The information collected from these manuscripts lays the foundation for a comparative study of the integration of Portuguese networks in the evolutionary process of the European economy in the Early Modern era.

This evolutionary process has in the perception of space and in the construction of trading geographies essential elements of these networks' consolidation. In this sense, the key-assertions of this chapter will have strong cartographic support; thus, geography – the historical maps included and the readings they suggest – will serve as a tool the better to construe network dynamics and its economic, political, social, and cultural settlement and evolution in time and space, hence the city and its port achievements (Denègre and Salgé, 2004).[2] Besides what's been said so far about it, I chose not to go further in the debate about networks theories although I don't disregard its significance; instead I prefer to present practical examples duly supported by additional scientific literature.

The working methodologies proposed and the innovative themes introduced in this chapter seek to demonstrate the following thesis:

One: with regard to Early Modern long-distance trade and the tangible results it achieved in some areas, ground-breaking economy operating mechanisms were active before industrialization, and, in the words of De Vries (De Vries, 2008), they were driven by the "industrious revolution" phenomenon. Consequently, as the examples in this chapter will prove, it is impossible to comprehend fully the construction of the Early Modern economy, and the supposed First Global Age, without recognizing the contribution of the Iberian merchants to the process; nor can we

---

[2] The maps were drawn by at *Oficina do Mapa* by Miguel Nogueira to whom I owe, once again, special thanks for his willingness and competence, producing high quality historical maps that enrich my own work; Oficina do Mapa/FLUP is a mapping centre based at the University of Porto http://web.letras.up.pt/oficinadomapa. Further acknowledgements: Ana Crespo Solana, Susana Barros, Carlos Cardoso, Diana Cardoso.

understand this structural change without acknowledging the role of the Atlantic world to which the Iberian peninsula networks were attracted. However, we should be careful not to overstate their ability to interfere in markets; on the contrary, we must recognize their limitations especially in the origins of overseas expansion. In the Cape route and the overall Indian trade, the primary investments were guaranteed by European trading firms against extensive privileges; roughly between 1550 and 1630 their action was too important to be ignored. It is certain that long-distance commercial routes were in their infancy; trade was still very speculative and only available to a few investors who could bear large investments and have them immobilized for several years; markets were tight. We have to be cautious about using expressions such as "modernity" or "mass consumption" or "globalization" in those centuries. Yet in spite of this, commercial networks were effective and capable of taking advantage of the information channels they controlled and the human resources at their disposal to get involved in the production, transportation, and distribution processes which showed the high level reached by Early Modern trade.

Two: these later remarks introduce another point: network organization was the most effective way to succeed in business and participate in the dynamic for change of 16th-century trade.

Three: trade and commercial networks were the preferred organizational model for Oporto merchants who combined it with the strong position they had in the Atlantic to succeed in Europe and to be considered as partners by European trading companies.

Four: commercial networks forged new informal economic relationships avoiding the legal framework defined by the Portuguese and Spanish crowns and in most cases operating outside that official boundary in a "free" approach.

Five: these networks had a social structure based on the social actor's theory and embodied in concepts of trust and reputation; they destroyed the conventional space, overcoming political, religious, and cultural barriers; the result was a profound expansion of trade and economic connections. However, as stated before, the notion of Early Modern global trade is debated by scholars; besides, we must not lose sight of the historical individuals who tend to be eclipsed by the idea of people in relation.

Six: their business and methods contributed to the rise of new market centres such as the United Provinces, thanks to the co-operative practices they promoted against the political limitations of their day.

Seven: finally, because of these networks among European merchants, there emerged big economic hubs as well as small and medium sized ports. Studies show that these networks and small and medium sized ports played an active part in the economic integration of states and regions.[3]

2. Portuguese Trading Networks: The Elites

Although familiar to a new generation of Portuguese historians, these ideas are still far from being assimilated by most Portuguese and international historiography which repeats and insists upon old clichés such as the linear evolution of trade, the bipolar relationships between centre and periphery, the notion of primary and secondary centres, and remains generally unaware of the history and dynamics of ports other than the imperial ports of Lisbon and Seville. Hilario Casado Alonso in his study of Early Modern Castilian networks (2012) has begun to replace these notions by a new research agenda which proposes multi-faceted approaches; economy in the Early Modern period is no longer seen as linear, resulting from central state political programmes, but as the result of multiple transactions between multiple partners in a complex process in which the social dimension of trade had a very substantial role. This study of Early Modern Oporto networks follows Casado's proposals and is pertinent because it details the development of the city from a secondary port to a hub of commercial interest to merchants, ships, and ports of northern and southern Europe; it offered goods that would soon become demanded by the masses, such as sugar, and the port offered northern European agents direct access to overseas markets.

Portugal, Castile, the Iberian peninsula and its overseas territories, and the European routes mentioned formed a huge space where the Iberian networks ran through in the context of Portuguese and Spanish monarchies (before and during the Iberian Union, 1580–1640, with all the consequences of political project) and their economic links which need to be thoroughly studied. On the other hand, to talk about trade networks means to recognize affirmation strategies and to address the phenomenon of the elites in the Early Modern period: is to verify the existence of successful social groups who gained reputation and politically influence through trade and business. Historians commonly refer this model of

---

[3] I must say that we must not takethese concepts for granted: "free" commerce may be an anachronism – and in most cases it is – in the Early Modern period; on the other hand, merchants and networks sought to be always close to the power and, as discussed in the event of such networks it were; finally, that the disengagement to which I refer had more to do with specific aspects of impossibility (and complex not only caused by the legal framework imposed) participation in trade of the Cape route and the consequent search for alternatives in the Atlantic.

upward social mobility to nobility and crown service; this study will show that there was a similar and strong phenomenon involving merchants and traders. We will get to know better a group of businessmen recognized for their commercial achievements and affirmed as elite (social, cultural, and economic) – first in their own home city and, shortly thereafter, by the crown, which would eventually guarantee them substantial advantages.

The process was even more complex. From medieval times, the city has always had merchants, recognized by the central state as having a special status granting them prerogatives such as the urban government, and the prestigious privilege of "citizens". Being part of a group has consequences at any time in history, e.g. in the context of social representations (Blockmans and Janse, 1999),[4] the exercise of power, and its base of support. In a merchant environment such as Oporto in the 14th and 15th centuries, a restricted group of traders – the local historiography calls them *bourgeois*, which is not incorrect – emulated nobility and acted accordingly inside the city walls, developing a code of dress and manners and a way of life very similar to the one we expect to find in the royal court or within noble houses. Anchored in local and regional interest (as would remain the case for centuries to come), the elite followed the same path as that of other mercantile elites in the Iberian Peninsula and in parts of Europe (Poujade, 2008: 147; Tognetti, 1999: 97; Puglia, 2005: 14–18; Gurevic, 1989: 184): trade incomes were used to buy land, like the old-fashioned medieval landlords (Morais Barros, 2013: 304).

Medieval merchant families, considerably enriched through shipping and European trade, gained municipal power, but in the 16th century they lost their leading economic role to the newly converted Christian traders who soon became the more dynamic economic agents of the city. Although they never reached the local government (the merchant aristocracy represented society's traditional values and were always supported by the crown) (Dias, 1937: 39), their international contacts ensured for them an effective management of influences, because of which they were able to be directly or indirectly linked to the Atlantic projection of the Spanish monarchy. The urban Oporto of the 16th century was transformed into a political arena where rivalry between the two elites was undisguised.

4 For general social representations, see Cornelius Castoriadis, *Instituição Imaginária da Sociedade*; note his statement that every social group has a representation of itself and the thought that autonomous societies are those in which their members are aware of this fact, which at a certain point fits the attitude of the merchants studied in the present text.

Early modern businessmen had their origins in the complex late-15th-century process of settlement of new converts in the city when about thirty families of Castilian "*conversos*" obtained royal permission to live in Oporto (Moreno, 1978: 7–38), and in the half-century that followed they were able to extend their web of interests to the most attractive economic sectors: dynasties of new-Christian merchants such as the Paz and Bentalhado families leased the northern Portuguese custom-houses; the Vila Real, Nunes, Gomes / Vaz, Henriques, and Vitória, among others, rented churches, tax benefits and economic rights and taxes, and all of them appear to have taken over the maritime movement of this port. From this standpoint arose both an "entente cordiale" in rivalry with the old bourgeois elite of the city.

As I stated, the city administration was out of their reach, unlike what happened in some Castilian regions in the 15th century (Martínez Millán, Carlos Morales, 1990: 149–161), but they undoubtedly influenced and were identified as such: by trade and finance many of its members achieved the esteemed and privileged status of Oporto citizens, created a devotional confraternity[5] through which they exhibited a public image of conformity with the religious canons in place and strengthened solidarities between them; they lived in houses, rich in architecture and interior furnishing along the streets of S. Miguel, Belmonte, Ponte de S. Domingos and Flores; they were culture consumers importing books, painted boards and paintings from Flanders and Italy, silk cloth, tapestries, counters, and bezoars from India and the Far East, and some of them even wrote erudite books such as Bento Fernandes on algebra and commercial mathematics, and Gabriel (Uriel) da Costa on theology and philosophy.

Their social influence, and even their inclusion in political networks when associated with the traditional aristocracy, and even their survival as a group were justified as soon as they became indispensable both to local and central powers by providing them with financial resources to govern when budgets dwindle: funding the municipal administration projects (for instance the 16th-century harbour project), paying the king's armadas and arming his ships. Therefore, the question of the elite rivalries must be nuanced. Often, an atmosphere of peaceful coexistence and of opportunistic alliances interesting for both parties was felt. To new-Christian merchants and networks it was highly advisable to maintain good relationships with those who ruled the city. It is a stance known in the European world of business, according to which vulnerable rival groups in a city or port (either abroad or at home but lacking security

[5] The *confraria*, confraternity or brotherhood of *Nossa Senhora das Neves* (Our Lady of the Snow).

against potential enemies) seek this type of alliance and relationship (Soldani, 2011: 36–37). Some of these alliances were more than passing; the royal prosecutor in the city, Francisco Dias, who witnessed them, wrote about the (polemical and social commented) mixed marriages where old-Christian families benefited from new-Christian dowries and business advantages: Diogo de Paz's daughter married António Leite; António Ribeiro's daughter married the son of the king's custom-house factor Diogo Fernandes das Póvoas, and finally Henrique Homem Teles married one of Henrique Gomes's daughters (Dias, 1937: 20).

Map 1. Merchants and elite distribution in Oporto in the Late Middle Ages and 16th century

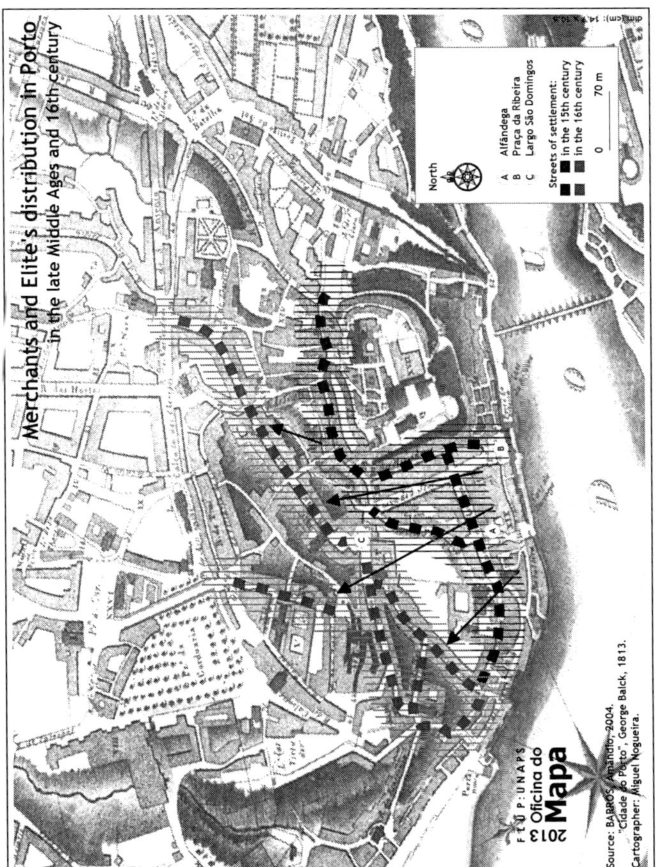

3. Networks and Ports

When studies of pre-industrial trade based on networks model began to be developed historians realized that there was a lot to re-examine; besides the aforementioned notion of bilateralism they noticed that Early Modern economy did not emerge from trade controlled by states, nor was it defined by ideas of superiority of some monarchies and states (the northern European states and kingdoms) over others (the south European kingdoms); instead, and depending on contexts and spaces, they realized that it would be possible to define trade as a complex, multi-connected process in which the roles (dominant or secondary) changed according to the interests, the means of access to markets, the sources of wealth and the quality of the relationships established between various different agents.

Thus, historians have begun to shift their attention from the relations between states to topics such as inter-connections between ports (the major and the so-called secondary ones), or to merchants not connected to the state economic framework; by doing so, they have illuminated multiple aspects of trade that were hitherto unknown. While medieval individual merchants could provide simple goods and services, the more complex Early Modern economy required a coordinated group effort. Port history, for long just a simple register of ships' entry and departure, has become the history of port societies (e.g. HISPORTOS and the current research project on the Casa de Vélázquez); on the other hand, the history of international trade is no longer about diplomatic treaties and political decisions between monarchies and states (although these cannot be ignored) but has become one of merchants and merchant societies whatever the context (Crespo Solana, 2009).

Historians have begun to abandon the study of the individual merchant who decided everything, in the line of the traditional political historiography (a theme that has been losing popularity in favour of the study of the firm) (Lanero Fernández and Ortega Montes, 2007: 146–148) and now deal with merchant communities. Thus this text is devoted to what Elisa Soldani defines as the "dialectic between family and company" (Soldani, 2011: 28–31); in the case under consideration here, we are dealing with a set of merchants and businessmen united through marriage alliances, family, and confessional ties and solidarities, acquiring skills to integrate extensive and demanding circles; men who did not choose to live in the seclusion of their city and port but spread their interests, members, and representatives to an extended (world) market through the financial fairs of Castile, the multifunctional northern and central European cities, the traditional (and still powerful) Mediterranean trading posts which were about to become comfort zones for Portuguese new-Christian refugees who later escaped from the Inquisition, and finally the Portuguese and Castilian overseas ports.

Commercial networks proved to be most effective for the development of trade and markets in the Early Modern age. In a period marked by strong competition, rivalry and conflict both in the European continent and overseas – "there were about one dozen European nations involved in the colonization process. While some of these were too small to have any real impact, there were still five or six nations who established some major colonial relationships" (Engerman, 2009: 13). The multinational commercial networks were systematically unrelated to political troubles and conflicts, and, through cooperation mechanisms (which included actions that crossed the outer fringes of the law), managed to keep the business running and became inescapable drivers of commercial development. For the Portuguese they were essential to the stability and efficacy of trade in Europe; in fact they relied on these commercial networks since they didn't have extensive merchant colonies abroad that could collectively decide trading strategies (as, for instance, in the case of the Spanish) (Casado Alonso, 2013: 355–374).

Departing from the most important ports and national economic centres (such as Lisbon, Oporto, and Viana), establishing affiliates across the continent in commercial, financial, and exchange markets, such as Antwerp, Florence, and Medina del Campo, Lyon, Paris, Rome, Ancon, Venice, Salonika, among others, and overseas trading in dyes, slaves, and sugar on the Atlantic islands, the West African coast, Brazil, Spanish Indies and Portuguese India, these institutions devised mechanisms designed to facilitate the movement and support of agents, goods, money, and information over long distances.

In the Early Modern age, Oporto was able to manage and tie together information in order to develop an Atlantic commercial project. Known as a wine export port – which is a wrong idea at least until the end of the 17th century – the city was the home of commercial companies which successfully worked to turn the city into a base for Atlantic ventures. Being an Atlantic city and port with an interesting merchant fleet in the 14th and 15th centuries, and merchants dedicated to international trade in Flanders, the British Isles, and the Mediterranean, in the 16th century and after a period of certain erasure caused by the forced removal from the Cape route business by reason of the royal monopoly, the merchant networks ignited an expansionist project that made this port acquainted with dynamic and demanding business such as the slave trade from African posts to the Spanish Indies where they were involved for a long time, trade in the Atlantic Islands and, further on in the century, the sugar business in São Tomé, and of course, in Brazil (Morais Barros, 2013).

From a political point of view there was also a context that explained the rise of these networks. It is impossible not to make a comparison between the trade flows in Castile before and after 1580. The relationship between the Oporto networks and the commercial house of Simón Ruiz, a privileged observer of this movement, is easily detectable in the extant mercantile correspondence. This example can help us assess the importance of the new trade and business order. As happened with the Ruiz firm after Cateau-Cambrésis (3 April 1559) and the strengthening and deepening of her economic relationships with France (Brunelle, 1991), after 1580 the Portuguese networks followed the same path and extended (to an unprecedented level) not only the ties with this firm, but also with all areas included in the Castilian geography of trade in Europe, America, and Asia, thus providing an essential contribution to the development of economic ties between Iberian spaces. Finally, the way networks intervened and the co-operative and collusion mechanisms they developed brought them information and circulation channels that in the later 16th century and the early decades of the 17th century would give them the means to continue active when the decision-making centres were moved to the northern European ports. This is one of the more innovative aspects of this research. When we look at the Iberian Inquisition processes involving new-Christian merchants, and also some commercial lawsuits in the Castilian Court of Valladolid, this ability to integrate and co-operate becomes very clear. This also applies to France, in the ports of Nantes and Rouen (in addition to medieval ports such as La Rochelle, Honfleur and Harfleur) and the financial centres of Lyon and Paris. Oporto's transformation into an international and open commercial hub is clearly visible on the following maps showing the evolution of its commercial geography and hence its status in the centuries between the late Middle Ages and the Early Modern period.

These maps portray the dynamic shifts and growth of a maritime-based economy. In the first phase, traditional commerce based on the transportation and sale of foodstuffs, salt, and leather from the city hinterland, and textile imports developed through intense coastal shipping especially in the course of the 15th century. In the 14th and 15th centuries this port offered maritime transportation; the shipping industry was solid and operated mainly in Europe at the service of ports and international trading communities. This geography depicts yet another reality that came out of the late medieval crisis: the enlargement and acceleration of market integration, a process in which the Iberian peninsula and its ports actively participated.

Maps 2 and 3. Oporto commercial evolution, 15th and 16th centuries

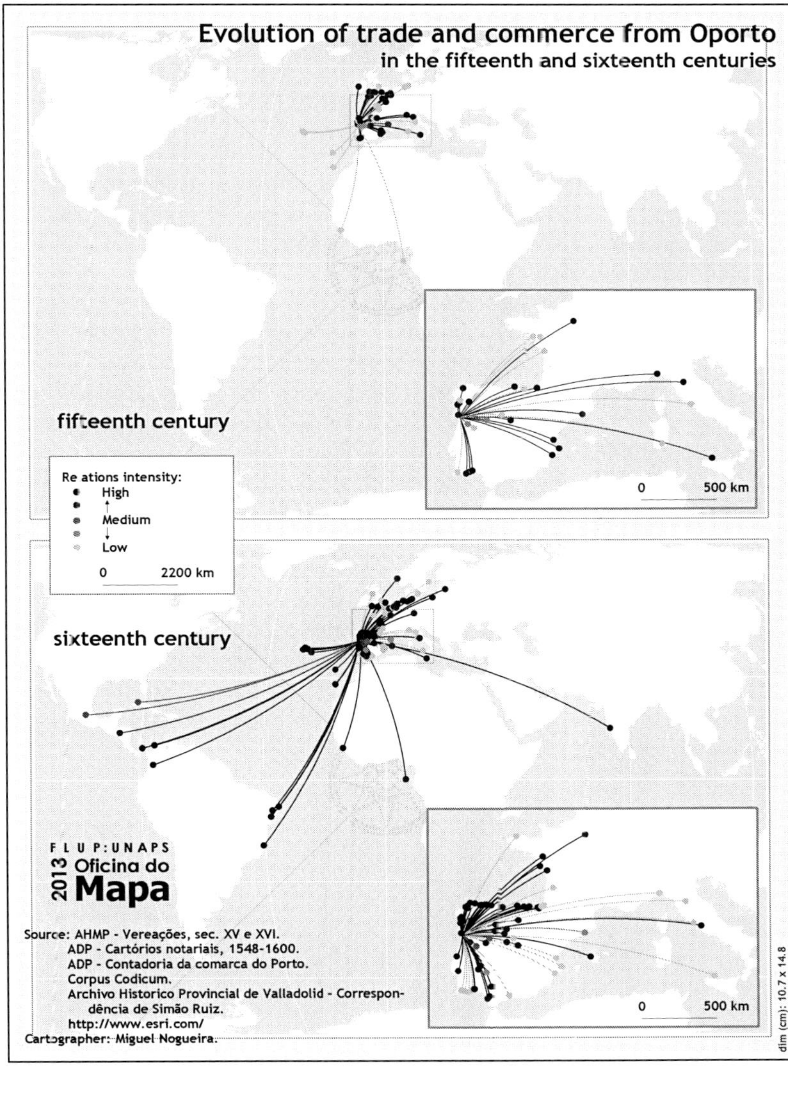

In the 16th century, although it continued to be chartered abroad, the fleet's main operations were directed elsewhere and their nature changed. The second map shows how an economy integrated around the trading affairs based in the Atlantic and deep relationships with the strong distribution ports in Northern Europe consolidated its competitiveness. Trade expansion required – and simultaneously allowed – a commitment to a branch of investment banking. In 16th-century Oporto we have the first reports of banks and exchange agencies linked to the mercantile networks and trade, responsible for carrying out financial transactions and accelerating cash circulation on a regular basis, a subject to which I will return later. This demanding activity was not possible in Oporto without those merchants and networks, which were cosmopolitan and accredited in European finance centres, and were primarily responsible for the commercial vocation of this Atlantic port city.

Thanks to Iberian overseas expansion and the new inflows of precious metals, a phenomenon incisively felt in the Iberian Peninsula, and accompanied by similar ones on a lesser scale in Lombardy, Liguria, Tuscany, Veneto, part of the Low Countries, and South Germany (Casado Alonso, 2012: 18–19) – which ran through the Iberian networking channels despite all the legal restrictions – there was scope for larger scale business, financed by the new inflows of overseas gold and silver. Porto was included by Phillip III in a group of ports well-known for silver contraband.

The following map, which is closely related to the previous one, expresses very clearly how through the action of well organized merchants an area that could have been considered peripheral until the mid-16th century, progressed: the north-western Portuguese sea harbours hitherto gravitating around two major (and central) areas, Lisbon and the Tagus Valley and the valle del Duero, in Spain, were transformed into an area of centrality.

Maps 4 and 5. Evolution of geo-economic status in Oporto, 15th–16th centuries

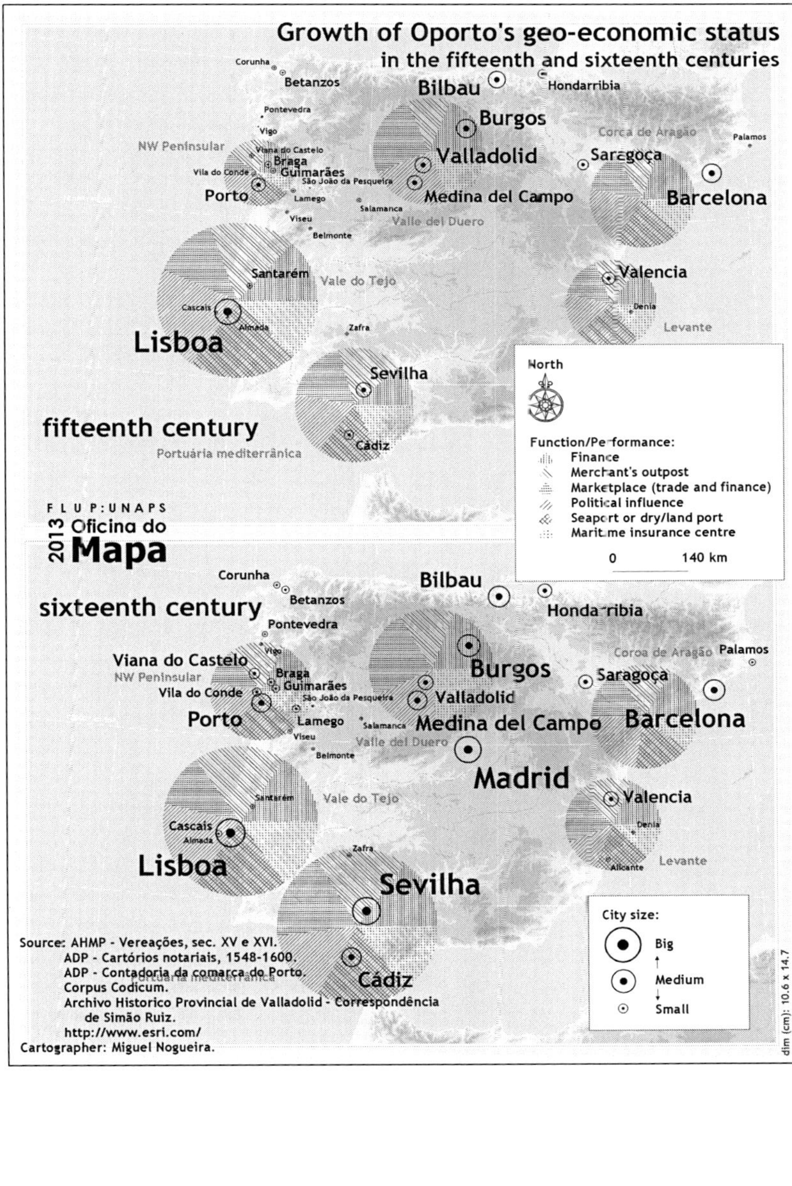

4. Nature and Development of the Portuguese Networks

In general, networks were self-organized and followed membership criteria based on family, creed, origin, and professional ties; their role was the result of strategies planned within the structures that they had created themselves. The origins of this form of merchant organization date back to the medieval Mediterranean, even if not exclusively, and they still applied in the Portuguese ports in the 16th century. They were part of a culture of trade – in particular as regards the recognition and protection of the merchant and his business – which early on became part of an interesting heritage of cultural and social relationships.

Oporto commercial networks integrated new-Christian merchants; they gained notoriety in the construction of the Atlantic economy, throughout the 16th century. However, religious creed did not prevent them from associating with agents of other creeds either because they needed or because they wanted to do so. In the first case we have to note that, by law, the itinerancy of new-Christians was limited, and merchants had to trust the performance of old-Christians who, on their behalf, or at their service travelled across Europe and overseas. When the new-Christian community of Antwerp moved to Ferrara, in the 1530s and 1540, the merchants transferred their goods and financial capital to the names of their associates in the so-called Pepper Consortium (such as the Italian Guicciardini family) in order to avoid confiscation by the commissioners of Charles V in Lombardy; the same strategy was followed by merchants in Oporto who used to rely on old-Christian businessmen to import and export textiles, gold, and silver (Andrade, 2006: 65–108; Andrade, 2011: 5–17; Barros, 2013: 297–322). In the second case there are business relationships (in Portugal, Africa, West-Indies and India and Southeast Asia) between new-Christians, old-Christians, Jews, and Gentiles. Such strategies, with the alliances with old-Christians mentioned above, mean that we have to study long-established concepts like those of crypto-Judaism or sincere conversion. Not all of the new-Christian merchants remained in Judaism, not all of them refused a sincere conversion to Catholicism, and some of them suffered a crisis of religious conscience (as happened with Gabriel da Costa). This flexibility of religious belief might allow some of them to continue their international projects and economic integration.

Internationally, how did these networks work? First of all, we should be cautious and note the lesser dimension of the Early Modern Portuguese networks phenomenon when compared with similar cases from Flanders, England, Genoa, and Castile, for instance; structurally limited by the kingdom's size, sparse population, and relatively poor economy, especially before the expansion of the 15th and 16th centuries, Portuguese networks could hardly aspire to a leading role in international trade, although in some cases they made themselves noted. It is even difficult to talk about Portuguese merchant colonies abroad, as one can for Castilian colonies in Nantes, London, or Antwerp. Nevertheless, the strategies of organization, the overall functioning and development of the Portuguese mercantile networks resembled that of their Castilian counterparts; moreover, economic and social historians have recently identified mechanisms of co-operation, of rivalry and of competition. "Formed by individuals grouped by their status as subjects of the same king or political authority" (Casado Alonso, 2012), the existence of networks is explained by the pursuit of efficiency in business, especially when that business acquired a global dimension. As for their members abroad, they needed associates, business partners and proxies in the foreign ports and economic markets where the Company had interests, and those agents were recruited from family and religious groups, among friends, neighbors, and people of trust and reputation (of competence). Amplified and internationalized in the 16th century, trade developed this network phenomenon, which must be understood on at least two (complementary) levels. First, at a local level where the networks were incorporated by individual merchants united by family ties and minor partners; they articulated interests and interdependencies, and developed business in what we could call the primary sector, exploring, for example, plantation cultures such as sugar and being involved in small transactions with the hinterland. The most they expected was social recognition at home and sometimes engagement in commercial ventures promoted by the crown; that involvement could project them to the second level. This other level was international, even multinational; in it, the same networks joined (and broadened) their interests to the interests of the biggest European companies, becoming members or correspondents and benefiting from the financial, political, and economic coverage they could offer them.

One case shows this kind of evolution clearly. In 1552 there came onto the Oporto mercantile scene the businessman Manuel Henriques. We can assess his importance by the significant number of times he is mentioned in the commercial letters of Simón Ruiz with the Portuguese and European merchants. Henriques was one of the Diogo Bentalhado's sons and was part of a generation of new-Christian merchants responsible for the decisive step that pushed the city into the path of economic internationalization through the Atlantic sugar trade. Not coincidentally, the first document where he is noticed reports a sugar transaction involving the northern Portuguese port of Viana which is systematically mentioned as a strong partner in this trade and in the 16th- and 17th-century colonization of Brazil; the Bentalhado/Henriques firm had also (along with others in Oporto) interests in the neighbouring ports of Galicia, and in Cantabria and Biscay. As was the practice in those times, he entered the service of the family firm very young (at fourteen to sixteen years old), and became acquainted with distinguished merchants such as Vila Real, Vitória, Gomes, and so on. In the next eight years or so he chartered ships to the Atlantic outposts of São Tomé, Cape Verde, and Bahia, negotiating for sugar, slaves, and Azores dyes, and dealing with Antwerp. In the late 1560s he left the city and settled in Antwerp – where we notice him in 1570.[6] From the Brabant he benefited from the family and the existing new-Christian community contacts to establish an influential business firm – dedicated to finance and business representation – which was a valuable asset for Portuguese and Iberian firms. Henriques had powers of representation and managed the business of many Portuguese and Castilian merchants with whom he corresponded; he negotiated on behalf of these merchants and their networks and managed their international banking accounts, using financial instruments and techniques resources to transfer funds, received and accepted bills of exchange and moving money through the most important finance fairs in Europe. We are still far from knowing the significance of this middleman role he played, like others did, making himself (themselves) indispensable to the point of keeping prospective buyers off each other and maybe influencing prices, and raising transaction costs, plus receiving a fee for his (their) work. He still had relatives in his home-city and he also placed a son (Francisco Henriques[7]) in Paris, in charge of the transactions with Lyon, and another one (João Henriques[8]) in Rome, responsible for banking the savings of Oporto merchants like Simão Vaz. Until his death, in 1596,[9] Henriques provided valuable services to Iberian merchants and networks, moving funds belonging to Simão Vaz, Manuel Gomes da Costa,[10] and Diogo Pereira of Oporto, Francisco Rodrigues de Elvas and Pêro Godines

---

[6] ADP – *Po1º*, 3ª série, liv. 34, fl. 117.

[7] His murder in Paris in 1589 had a great impact on the accounts of Pêro Godines and Manuel Gomes da Costa (already established by them in Lisbon), for he kept a deposit of 7000 Portuguese *cruzados*, i.e., 2,800,000 Portuguese *reais*. AHPV, *Cartas de Simón Ruiz*, c-137-177_01.

[8] AHPV, *Cartas de Simón Ruiz*, c-137-188_01.

[9] ADP – *Po1º*, 3ª, série, liv. 110, fl. 126.

[10] Costa was the son of Simão Vaz and he actually went to Lisbon as mentioned above.

of Lisbon, and Simón Ruiz of Medina del Campo, to name just some of the most important. As with some famous European merchants, his house in Antwerp was open to guests and associates; Pêro Godines lived with him for a time (as he says in a letter to Simón Ruiz). He systematically used the postal services available and was always very well informed about the evolution of the economic and political situation in Europe; he sent comprehensive reports about the state of the markets, commodity prices, the repeated blockades of Antwerp, and his own suggestions for alternatives to this port to his associates in Oporto, Lisbon, Seville, Bilbao, Medina del Campo, and so on.

Regarding the organization of the Oporto networks, at the top of the pyramid, that is, at the companies' decision-making level, we find limited administrators; within the networks, the power levels vary, as we can see in the organizational chart of Simão Vaz's network presented below. Note that we are often tempted to interpret lightly, and sometimes to excessively generalize the historical documentation relating to these associations. Historians may thus perhaps run the risk of making judgements that are not very accurate. As in other areas in which history borrows tools from other sciences, we should be very careful and clear when it comes to dealing with concepts such as network nodes and meshes, particularly when we examine the links among those aforementioned in the commercial letters, in the issue of influences and how they were carried out (for example, for matters related to donations and gifts, offerings, deference and proof of friendship), and in the notion of trust, among many other aspects. Sometimes, we tend to confuse an effective member of a company with an overseer or a correspondent or simple agent of that company, or assign him a responsibility that he did not in fact have. Similarly, and because of the repetition of data concerning geographical references, there is a risk of confusing the territorial basis of a network with the financial centres it uses. Therefore, care should be taken when reviewing the nature of the links between the people we deal with: the fact that certain names are repeated, or relationships between them are recurrent, does not mean that there are no rivalries between them;[11] that the permanencies, in the names, practices, and place names, do not imply that the network was a static entity with activity defined at its foundation and remaining unchanged through the whole of its existence. The network grew; sometimes solving problems required organization chart adjustments by hiring new people or reassigning (promoting or demoting) the functions of the existing one; the methods of work also evolved; and

[11] As in Simão Ruiz's well known company, whose biggest rivals were the Genovese, but who nevertheless continued to do business with them.

its geography of interests too. Finally, and allow me to make this comment, we often tend to forget that these networks comprised people who are, as we all know, the core raison d'être of history; we tend to forget the historical person, with needs, qualities and flaws, competent or incompetent, honest or cheater. In this kind of narrative we tend to point to successful ventures and forget that some business failed, maybe due to competition, to tough economic conditions, or to the incompetence, differences, or dishonesty of the partners involved. In 1557 the brothers Tovar, Henrique and Duarte – new-Christian merchants settled in Porto and in Antwerp – reached an agreement on money deposits made by Henrique de Tovar "of Flanders" on his brother's behalf in the latter city and, at the same time, he forgave his nephew Jerónimo de Tovar for defaming him concerning their business.[12]

Having correspondents, business attachés, overseers, and simple representatives in Antwerp, Lisbon, Seville, Nantes, Trancoso, Funchal, Angra do Heroísmo, Garachico, Palma, S. Tomé, Mozambique, and Salvador da Bahia, and repeatedly maintaining relationships with them, does not make them influential members, decision makers of the network, or even members of the network. Looking at the activity that Simão Vaz and some other Portuguese networks maintained with Brazil, the documents show us, first, the migration of people from within the network (mostly brothers, sons, and nephews) to that territory to arrange sugar mills or sending goods to Portugal. At the end of the century, this tendency to migration seems to suggest that the management of commercial functions was handed to commissioners (the abovementioned *feitores*, foremen/taskmasters or overseers), whilst the younger members of the firm no longer travelled overseas. They remained in Portugal from where they accomplished missions in Lisbon, Madrid, Seville or in other Portuguese or Castilian economic centres. For example, the case of Bento Dias de Santiago, brother of Miguel Dias – the head of the network and both merchants in Porto – who spent several years in Pernambuco setting up the Camaragibe mill (Barros, 2012: 311–336).

As mentioned earlier, at the decision making core, the networks are restricted and subject to many disruptions. This does not mean that they did not relate and collaborate with each other, and did not have an enormous influence, at various levels. Yet, we cannot look at Medina del Campo, Lyon, Seville, Antwerp, or Milan, banking and exchange centres at the time, as the main nodes of the networks that have their headquarters elsewhere (for example, in the Iberian, Breton, Normand, and Flemish port-cities). Nor can we consider that just because business was conducted

[12] ADP, – *Po 1º*, 3ª série, liv. 15, fls. 28-28v.

with a merchant and banker such as Simón Ruiz, governor of Medina del Campo, he was an effective member of the network. Just because the same agent was used repeatedly in a specific port does not mean he was working exclusively for the network. Dealing with the Genovese did not mean the absence of a profound relationship of rivalry and even enmity, as is the case, as mentioned before, in the relationship with the Castilian crown and the subordination of competing companies, such as the company of Simón Ruiz (Elliot, 2002: 259–260). And, to cut a long list short, even within the same network it was common to find competition and confrontation, driven by underlying hierarchies and power games (Priotti).[13]

Table 1. The trade network of Simão Vaz, Porto, 1570–1600

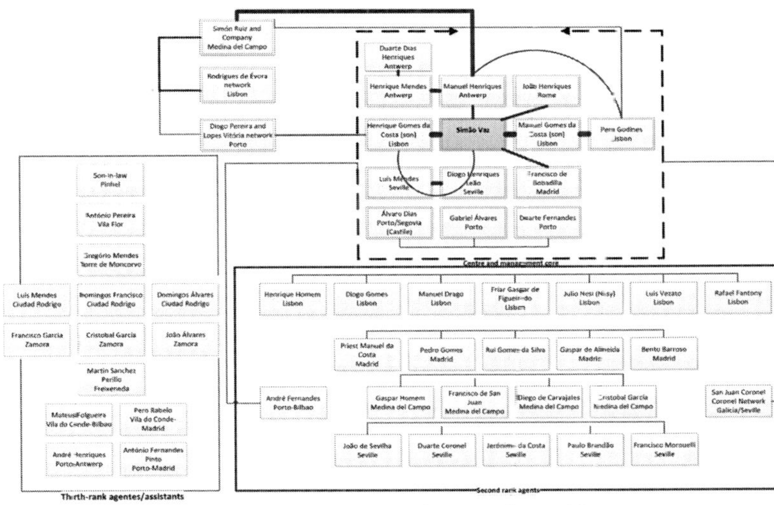

Source: ADP – Notary Records; AHPV – Simón Ruiz Archive

In the Portuguese ports in the 16th century, the merchants were organized into companies, and, at the same time, maintained short-term contractual relationships with traders in order to do business in various areas, including textiles and wines. Moreover, it is curious to see how these men, who excelled mostly in international sea trade, and who called themselves "coarse-mannered merchants" (*mercadores de grosso trato*), entered into numerous associations (over one or three years) with more

---

[13] *Réseaux marchands en conflit et hiérarchisation des centres économiques. L'Atlantique franco-espagnol, 1560-1570*, available in http://www.usc.es/estaticos/congresos/histec05/b24_priotti.pdf.

modest merchants in businesses that explicitly avoided the traffic of goods by sea ("that do not risk sending goods by sea, only by land") (Barros, 2006: 49–70).

However, these merchants excelled in international trade and they managed a universe of contacts on a worldwide scale, helping to elevate the status of their home-ports, the international reach of their businesses, and the complementarity among them.

The operability of this company is clearly visible in the following diagram and map of its activity.

Diagram 1. The geography of Simão Vaz' trade network, 1570–1600

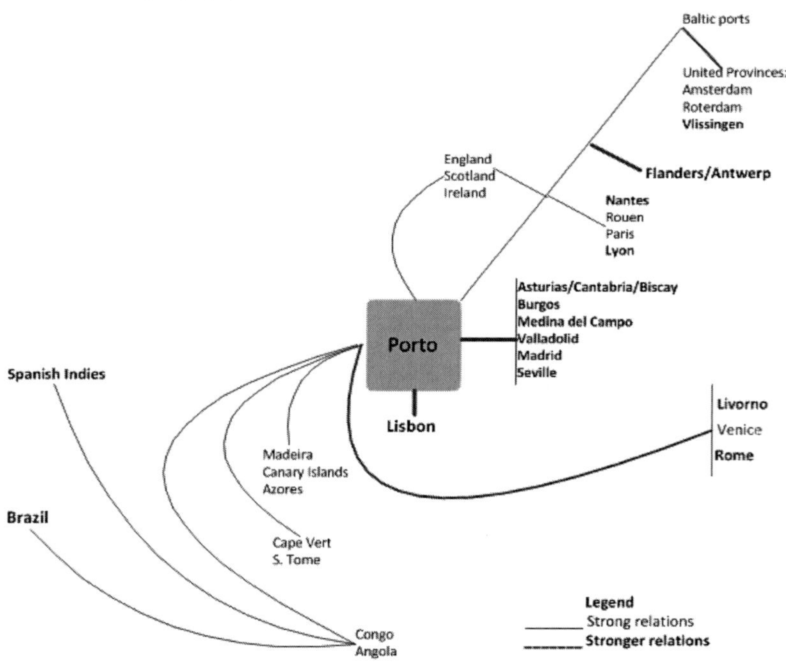

Source: ADP – Notary Records; AHPV – Simón Ruiz Archive

The impact of this movement is reflected in the progress of port activities because of the incentives given to the transport routes of dyes from the Azores, African slaves to the Spanish Indies, and to the development of the sugar mill economy in Brazil. Note that we are referring to a reality mainly centred on the Atlantic. However, the Indian and Asian horizons were not entirely excluded from these ports, due to the widespread migratory flows in Portugal in the 16th century; nor were they far from the minds of the merchants. It was precisely due to the networks in which they were involved and which they managed that business from the East became linked to the city (albeit indirectly and derived from the role of financing the crown that these merchants assumed from the early 17th century), when, in 1615 and 1616, Manuel and Henrique Gomes da Costa, sons and successors of Simão Vaz, armed four vessels (the *Nossa Senhora do Vencimento*, the *S. Julião*, the *Nossa Senhora da Guia,* and the *Nossa Senhora da Penha de França*) to sail to India, in an operation that cost them 110 million *reais*![14]

This widespread activity, which is recurrent in almost every major commercial network of Oporto, leads us to the question of finance and to the international management of political and social influences in order to get better results in trade and business; both can be seen in the actions of Bento Fernandes and António da Fonseca's firms in Oporto and in Rome.

Bento Fernandes is mostly known for the book he wrote about algebra and commercial mathematics (Barros, 2013)[15] and such a cultural skill was an attribute of the mercantile elites; active since 1552 – he belonged to the generation of Manuel Henriques – he saw the increasing importance of the sugar economy in Oporto with the beginning of the exploitation of Brazil which would triumph shortly after his death in 1570.

Following a tradition not exclusive to Portugal, Bento started as a merchant dedicated to the import and distribution of English,[16] Spanish,[17] and Flemish[18] textiles and clothing; he kept his shop in Ponte de São

---

[14] The *naos Vencimento* and *S. Julião* cost 25,500,000 *reais* each; the *naos Senhora da Guia* and Penha de França, 20,000,000 *reais* each, 20,000,000; see *Livro em que se contem toda a Fazenda e Real Património dos Reinos de Portugal, India e Ilhas Adjacentes*, ordenado por Luiz de Figueiredo Falcão, secretário de El-Rei Filippe II, Lisboa: Imprensa Nacional, 1859, p. 205.

[15] *Tratado de Arismetica*, printed in 1555, is one of the most important scientific books of the Portuguese renaissance; it is a masterpiece of the merchants' cultural values and crucial to our knowledge about the form of mercantile associations, commercial techniques, and accountability.

[16] July 22 de Julho, 1552: he received 72,800 Portuguese *reais* of English cloth from António de Freitas (he was partner of a venture with Duarte de Tovar, Jácome Rodrigues, Bento Rodrigues, Diogo Rodrigues and Francisco Nunes, who ordered 300,000 Portuguese *reais* of English and Flemish cloth; ADP – *Po 1º*, 3ª série, liv. 6, fl. 248.

[17] On 10 December 1557 he bought from João Navarros 25,000 Portuguese *reais* of Castilian cloth; ADP – *Po 1º*, 3ª série, liv. 15, fl. 33v.

[18] On 28 April 1565 he bought from Melchior Rodrigues, bishop's collector, 65,500 Portuguese *reais* of Flanders cloth; ADP – *Po 1º*, 3ª série, liv. 20, fl. 50v.

Domingos street (famous for being near new-Christian firms) where he was visited by buyers and traders from regional Portuguese towns like Amarante, Aveiro, Braga, and Lamego, among others. This was his major activity until the late 1550s. For the years 1558–1565 there are no documents available[19] so we cannot trace the evolution of his social and economic status as he gave more attention to other businesses. In any case we know that everything changed when the family and business ties with António da Fonseca's network strengthened.[20]

Born in Lamego (in the heart of the Douro Valley) to a new-Christian family, Fonseca was the brother of Jacome and Jerónimo da Fonseca, Isabel Nunes da Fonseca and Genebra, and Rui Fernandes, author of the well-know *Descrição do Terreno ao redor de Lamego duas Léguas*.[21] Organized in a powerful business network, the Fonseca, Nunes, and Fernandes families underpinned their interests in the main ports of the kingdom: Jácome, Jerónimo, António da Fonseca, and Rui Fernandes in Lisbon, and Genebra da Fonseca in Oporto, where she married Bento Fernandes in the 1530s. From these ports their influence reached the Portuguese communities in the Mediterranean, especially the one based in Rome. When the atmosphere for new-Christians became insecure with the Inquisition and the first prosecutions, Jácome took refuge in Rome near the Papal court (by then sympathetic to new-Christian interests under Pope Julius III), where he set up an influential banking network dedicated to papal bulls and licences and expedition loans and financing, active from at least 1543 (Novoa, in press). Threatened by the papacy of Paul IV (Giovanni Pietro Caraffa, 1555–1559), who was openly hostile to the new-Christians, Jácome left to Thessaloniki (Salonika) after he appointed his brother António da Fonseca as his successor in the company management; from then on António became the family protector, receiving and keeping at his place the family as was the case of his sister-in-law, widow of Rui Fernandes (Corpo Diplomatico, X: vol. 8: 311–312). Settled in Rome by 1556, he discreetly consolidated his position, which was already considered the most powerful among the Portuguese merchants when Pius IV (1559–1565) was elected pope; it was said that he influenced this election by buying votes, and before that he was accused of being involved in the death of Caraffa. In December 1562, crowning all his power,

[19] The notary books of those years were lost.

[20] António da Fonseca was his brother-in-law.

[21] *Description of the territory two leagues around Lamego*, an important document dating from around 1531–1532. Some additional data on these merchants can be found in Barro's introductory study to this book.

António da Fonseca was appointed administrator of the Santo António dos Portugueses Church.

It was probably then that the connection with his brother-in-law in Oporto grew deeper. This was no coincidence – these men hardly did anything by chance – but instead coincided with the city's commercial and economic growth and the general need for funding. In the five years between 1565 and his death, all Bento Fernandes' work came to be dominated by the order of bulls and papal documents, and the corresponding banking and exchange operations to get them, always in association with and under the direction of Fonseca, and putting into practice all his banking abilities. Before 1566, Bento received (by António da Fonseca's intermediation) a notarized power of attorney from Cardinal Alexander Farnese to be administrator of his property in Portugal, among which was the Benedictine monastery of Santo Tirso with all its jurisdictional rights and churches (in Oporto and Entre Douro e Minho province), which he did until the eve of his death in 1570.[22] As said, years when the city's economic life took off; and the attention paid by men like Antonio da Fonseca, the expansion of banking and trading transactions, the growing number of people devoted to commerce and wealthy enough to auction tax incomes and ecclesial benefits, the development of international and overseas trade, the establishment, funding, and functioning of shipping and freight companies, and the circulation of merchants, factors, and employees between Oporto and Brazil, are the best proof of that.

The circulation of merchant letters, bills of exchange, credit, loan, and funding requests, and business ventures were combined at Bento Fernandes' shop in Ponte de São Domingos Street, as happened at an even higher level at the office of Simão Vaz in Belomonte Street, or at the house of Domingos Lopes Vitória in São Miguel Street. David Igual Luis reminds us of this concentration of different functions (banking, finance, trade) in the same person (Eklund, Herbert & Tollison, 2004: 703).[23] To

---

[22] ADP – *Po 1º*, liv. 30, fl. 172v and following.

[23] Among many examples: ADP – *Po 1º*, 3ª série, liv. 21, fl. 18v; idem, liv. 34, fl. 175v. There were two Alexander Farnese in those days and both were related to the north of Portugal by their friendship with the bishop of Viseu, D. Miguel da Silva, who was ambassador of kings Manuel and João III. The first one became Pope Paul III (1534–1549), and his protégé (*nepote*) Alexander Farnese, 1520–1589, (cardinal, grandson of the Pope and his favourite, son of Pietro Luigi Farnese, duke of Parma) was appointed bishop of Viseu, by then vacant, when Miguel da Silva fell from grace; with the bishopric rents he supported the Portuguese bishop until his death in Rome in 1556. From then on he benefited from those rents. The fifth Lateran Council recommended the protection of family

point an example related to Bento Fernandes, in June 5, 1565, the knight[24] André Golias requested a loan of of 200 ducats of *fiorini di camera* ("câmara velhos", cameral florins) to be sent to António da Fonseca in Rome which he would deliver to Sebastião Veloso in order to get the benefice of the São Tomé church, Braga, on behalf of Gonçalo Mendes d'Antas. Bento passed him three copies of the insurance policies that covered the money and Golias promised to pay the rates current in Roman banks as soon as Veloso sent the letters, plus 2.5 per cent for his work.[25]

Diagram 2. Bento Fernandes in the network of bankier António da Fonseca and associates, 1552–1570

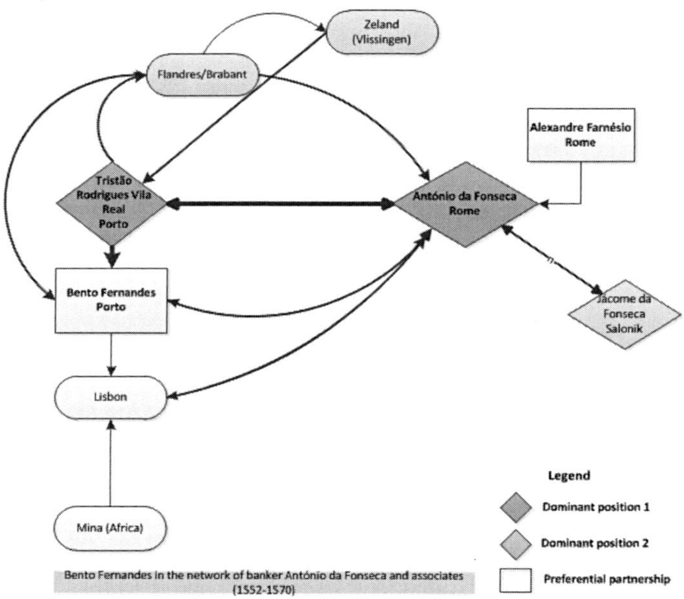

I did not find Oporto bankers who were not new-Christian: the family network of Simão Lopes, son of Álvaro Eanes, active in the city maritime trade, developed in parallel with a banking business funding trade ventures, transacted bills of exchange and marine insurance through its

members to prevent them falling into poverty and the nomination of cardinals' *nepoti* was sometimes justified by that. The benefices could be considerable: Farnese accumulated 64 benefices plus the Roman vice-chancellorship.

[24] *Cavaleiro-fidalgo*, in Portuguese.

[25] ADP – *Po 1º*, 3ª série, liv. 20, fl. 178v.

integration in the international commerce of Azores dyes and the contacts with Manuel Henriques, through whom Dutch maritime transportation was mobilized.[26] The banking branch of the firm was probably launched by Álvaro Eanes and was founded in the bond he developed with António da Fonseca, who was his business agent in Rome for the dispatch of bulls and papal letters;[27] Eanes was active at least between 1572 and 1585.[28]

To conclude the bankers topic, I mention Simon Vaz who himself had a very long history in capital and investment movements on an international level. Like the preceding ones he was involved in the business of bulls and acquiring papal licences, which it seems were a good cover for new-Christian capital transfers, with his cousin Diogo Henriques (brother of Manuel Henriques), involved in international money transfers from Oporto[29] and correspondent of João Pascoal, a banker based in Madrid. Around 1590, if not earlier, Vaz became a financial agent for the Habsburg crown and commended to pay the wages of the army of Philip I (II of Spain) in the northern provinces of Portugal and Galicia;[30] he also provided funds for the archbishop of Braga, exerting influence and social protection through the networks of Bento Fernandes, António da Fonseca, and Tristão Rodrigues Vila Real.

---

[26] Ver o remate de contas de pastel desta companhia em ADP – *Po1º*, 3ª série, liv. 82, fl. 82v e ss.

[27] ADP – *Po1º*, 3ª série, liv. 82, fl. 99: a credit of 300 *cruzados de câmara velhos* (coin from Rome), to be negotiated with António da Fonseca in Rome; credit for the nobleman (knight) Rodrigo Vilela, of Amarante (in the northern Portuguese province of Trás-os-Montes; the priest António de Queirós would be responsible for the money transfer. Banking and finance attracted several Oporto merchants; besides those already mentioned, Luís Mendes also got involved in the late 16th century (ADP – *Po2º*, 1ª série, liv. 12, fl. 10) in association with the powerful Ximenes family from Lisbon and Antwerp.

[28] ADP – *Po1º*, liv. 79, fl. 3, mentioning new money transfers to Rome by intervention of António da Fonseca commercial firm.

[29] Partner of Manuel Rodrigues de Elvas, from Lisbon, paid 100,000 *reais* to the widow of a pilot of the *Indian Route*, who lived in Vila do Conde, in his name. ADP – *Po1ª*, 3ª série, liv. 112, fl. 88v.

[30] ADP – *Po1º*, 3ª série, liv. 107, fl. 60v and following. Acquittance of João Mantilha "payer of this city's soldiers" (2 pagador da gente de guerra desta cidade"), to Simão Vaz, Diogo Henriques e João Pascoal, 27 September 1595, for the loan of 564,800 *maravedis*. In a short period of time other loans, equally large, followed (liv. 107, fl. 72v, fl. 105v, fl. 136, etc.).

Map 6. Commercial network of Simão Vaz, 16th century

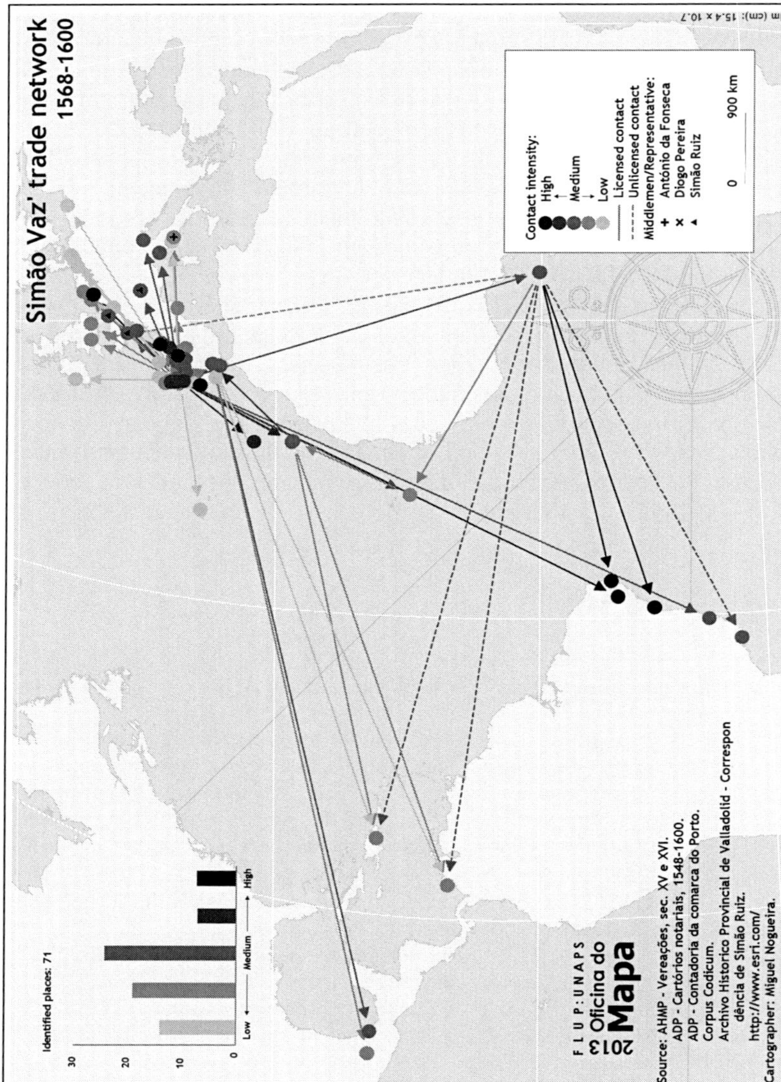

Map 7. New Christians' Commercial Network in relation with Bento Fernandes (1552–1570)

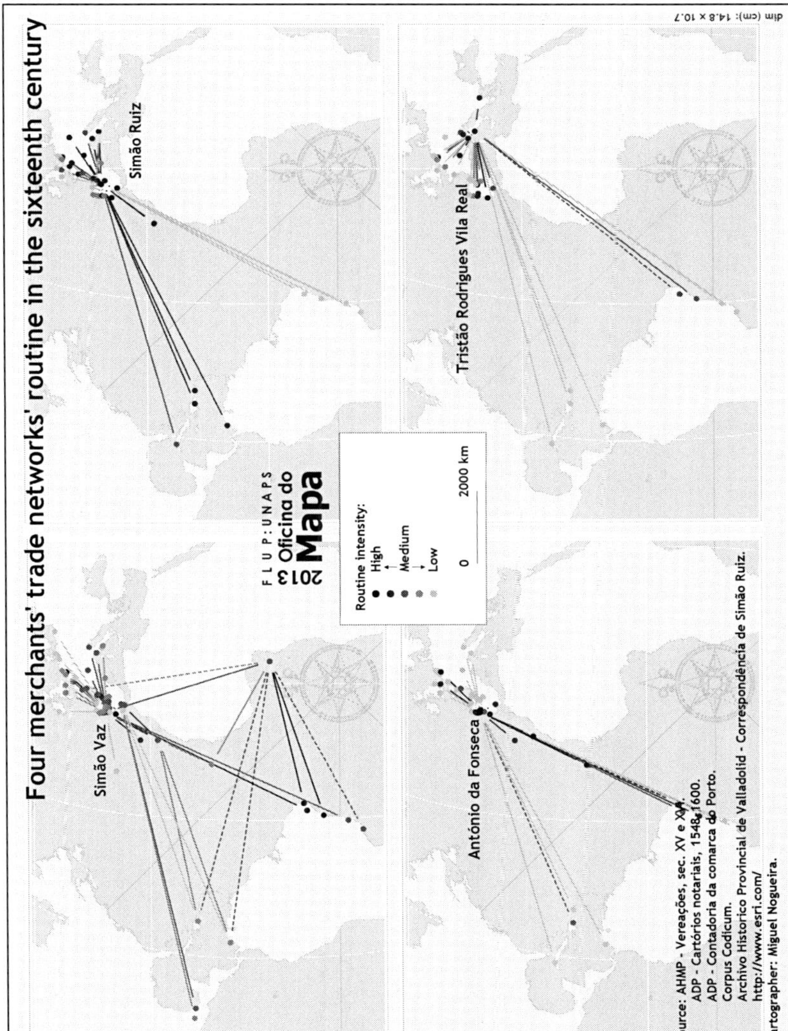

Map 8. Commercial networks of four merchants in the 16th century

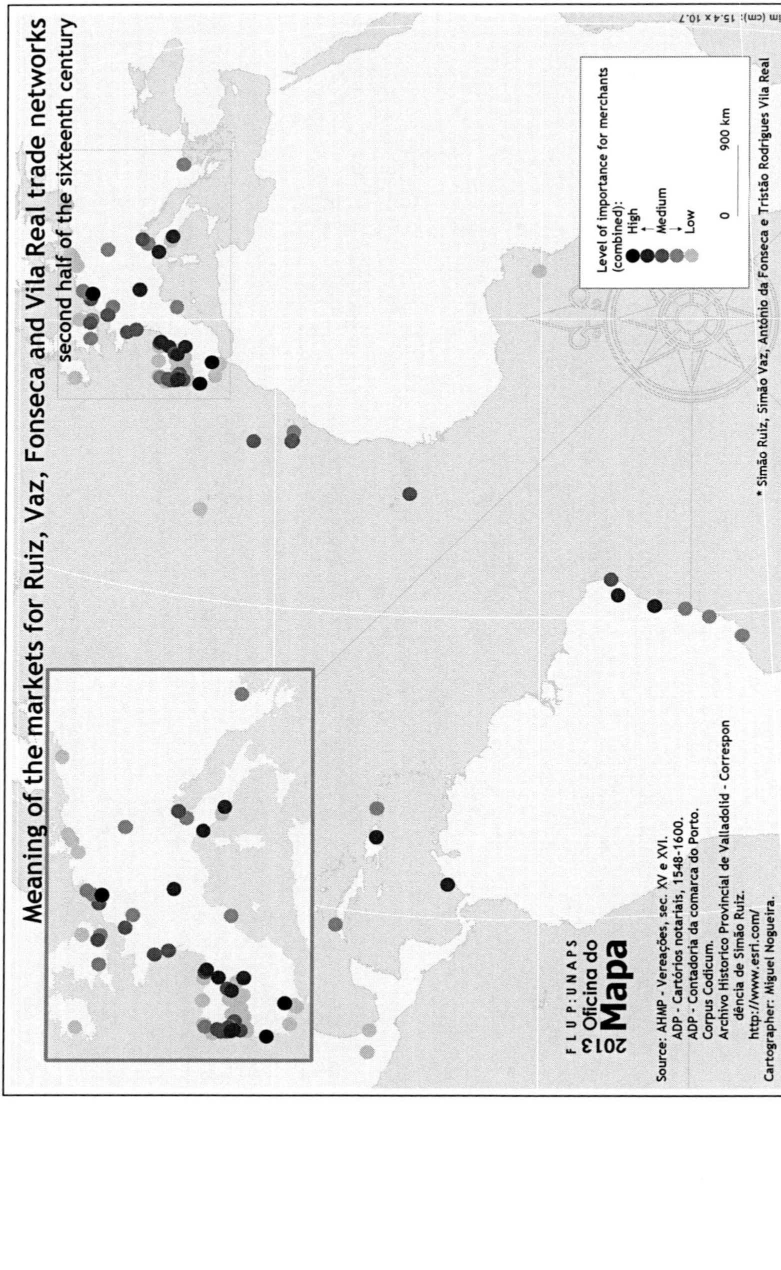

Diagram 3. B. Fernandes and A. Fonseca's network and their protectors

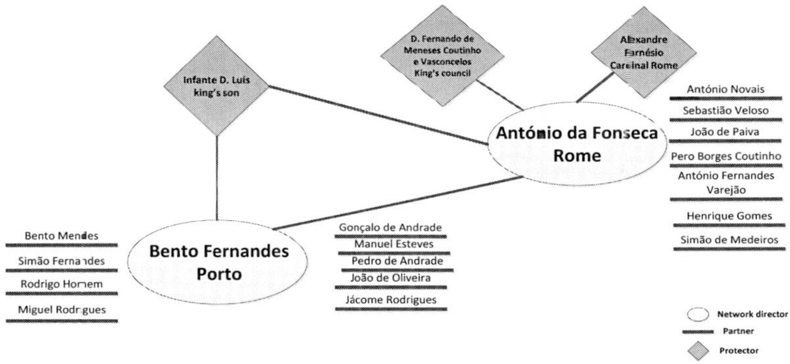

For several years I collated data about Simão Vaz; he seems to have been the most complete businessman of Oporto in the second half of the 16th century. Taking advantage of his bank he financed massive sugar sales through the ports of Viana, Oporto, and Lisbon and distributed in Burgos, Nantes, Rouen, Livorno, and Venice; on 5 June 1595 he accepted a bill of exchange from João Gomes, a sugar retailer, up to the amount of 2500 Portuguese *cruzados*[31] guaranteed by Simón and Cosme Ruiz of Medina del Campo and payable in Lisbon; Vaz transferred to the merchant the amount of 2000 *cruzados*[32] which he used to pay for the sugar he received from Martim Jácome, of Viana.[33]

I cannot assess the significance of these banking operations in 16th-century Oporto. Moreover, no historians talk about them or even mention the existence of banks in the city. It is safe to say that the Oporto banking and finance sector could never be compared with the major private and public banks presented in the money and financial fairs of Antwerp, Medina del Campo, Lyon, Geneva, Piacenza, Besançon, Frankfurt, Leipzig, etc., but considering the high number of references to it, the quality of the merchants engaged in it, their nationalities (Portuguese, Castilian, Flemish, Italian), the extent of its geographical reach, and the business it supported, it is my opinion that 16th-century Oporto banking proves the existence, in the city, of an international money market, a private financial sector, and a source of capital investment that supported and justified her growing internationalization. Thus, no matter its weight and meaning, the subject requires further research.

31 One million Portuguese *reais*.
32 800,000 *reais*.
33 ADP – *Po1º*, 3ª série, liv.

5. The "Destruction of Space"

Managed and financed as we have seen, the networks redesigned geography, or better, created their own geography through an actual "destruction of the space" clearly visible in the previous diagrams. Starting from the Iberian peninsula with constant movement between ports and inland centres more convenient to their interests, such as those that resulted from the rise of Sephardic merchants and their settlement strategies; note that this map deliberately does not mark the political boundaries between Portugal and Castile; the dynamic vectors of circulation to and through both sides are conspicuous.

This "destruction of space", crossing political frontiers and disregarding cultural and religious limits, commended itself to the Iberian merchants and networks if they wanted to succeed in the competitive Early Modern world of trade.

A good example of this approach and behaviour is provided by Diogo Pereira. Born and raised in Oporto, the first notice of him dates from 1580 and the last one reports his death in Madrid in 1617.[34] Between the two dates we have a life full of events, commercial projects, desires (and realizations) of greatness, trading monopolies, and a fair number of lawsuits against him.[35] He started as a business agent for Simão Vaz in Medina del Campo and Bilbao, and later he was the representative of the company he had in partnership with his cousin Domingos Lopes Vitória in the same places. Later, he became the head of his own extended network. He constantly moved between Oporto, Bilbao, and other Basque and Cantabrian ports selling sugar, monopolizing the cod-fish supply to Bilbao and some Brittany and Normandy ports. Associated with the Henriques family and partner of Manuel Gomes da Costa and Pêro Godines in Lisbon, he reached the network of Manuel Caldeira and entered the universe of slavery and the slave trade. Through that he jumped into finance and banking and dealt in silver using his connections in the major Iberian ports to smuggle it from the Spanish Indies where he had representatives. He was challenged by the Spanish crown authorities in Bilbao, Canary Islands, and Madrid but we have no proof that he was ever convicted.

---

[34] AGI, Contratación, 950, nº 29, 1621 (bienes de difuntos: Diego de Pereira y Felipa Nuñez).

[35] At least seven lawsuits against him are kept in the ARCV, between 1601 and 1617.

Map 9. New-Christian and Jewish communities in commercial network with Oporto, 15th–16th centuries

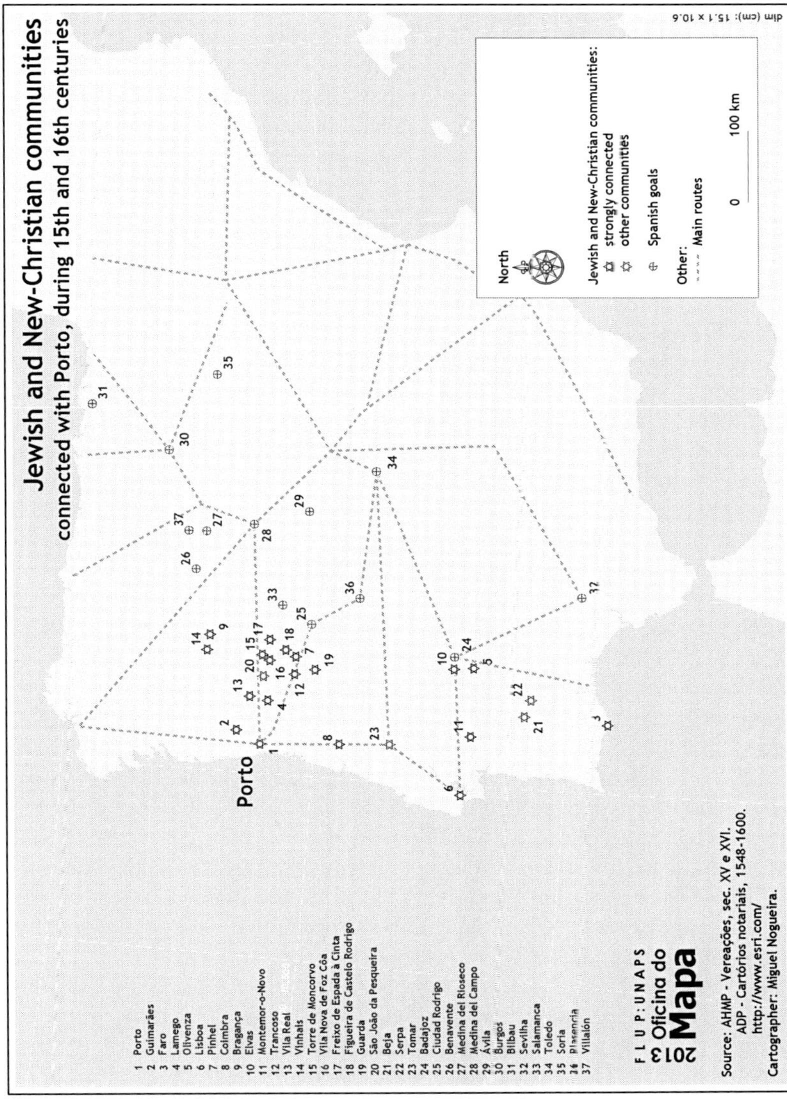

Working in his firm and in close relation with him we find (as we would expect) family: sons and nephews who took care of his business wherever needed, but mainly in Oporto, Bilbao, and Madrid. To give a little more precision to his life path and the way he crossed frontiers, I mention the following facts, starting by the notice of his decease: "he died from the present life in the city of Madrid, and in Madrid and other parts of Castile, Biscay, Galicia, Asturias, Flanders and the in the kingdom of Portugal, in many cities, towns and places elsewhere the said Diogo Pereira left … many unfinished business and accounts, and many demands … and today there are many and growing debts of money, merchandise and commodities, many bills to be paid, and infinite debts of slave licenses, for which proxies and attorneys were empowered among whom were Simão Pereira and António Pereira, nephew and brother of the deceased, present at the court of Madrid".[36] This expression, which had also been applied to Diogo Pereira during his lifetime, while indicating place name and dwelling, also reflects a strong relationship with government officials. In fact, Pereira extended his business to the administration and logistics of the Habsburg and Portuguese empires, becoming naval contractor responsible for the construction of galleons in Oporto shipyards, buying and selling slaves in Guinea, Congo, Angola, Nueva España, Cartajena de las Indias for which he used ships from Seville, Lisbon, and Oporto, and being paid in silver bars.

If with slave-trade Pereira skirted restrictive decrees (*pragmáticas*), with sugar-trade he openly defied bans and royal prohibitions. In 1601 he was criminally charged by Diego Perez de Cernuegla, Bilbao customs-house lieutenant, who accused him of having sent a ship from Portugal to Calais and from there to the "rebellious islands", i.e. the Netherlands, Zeeland and England, and he was actually arrested for some time; and, although it became evident that he refined sugar in the Netherlands, France, and Italy, and sold it from there, the fact is that his defence-attorneys presented the argument that he was "a man of rectitude and punctuality, a person of many businesses which he treated with plenty of justification and good faith, and everyone in all the major markets of this kingdom and elsewhere where he traded were satisfied"; they won the case and got him released from jail without any penalty.[37]

Diogo Pereira's case was not isolated; one simple look at the diagram of Simão Vaz's company, or to Bento Fernandes' relation with his relative's networks, is enough to realize this. But the origin of Oporto internationalization is, as was said, in the *conversos* settlement process, or

---

[36] Idem.

[37] ARCV, – *Registro de Ejecutorias*, Caja 1928, 70.

even before. A 15th-century lawsuit opposing the municipality and the city's Jewish community shows the importance of international solidarity networks: "despite all the wars, many carracks and ships could get to Pisa and other Italian cities, as happened with the *nau* Giralda, and this because the Jewish communes everywhere helped themselves" (Dias, 1983: 356–357). These models were still operative in the 16th century and were an indispensable business asset. Tristão Rodrigues Vila Real's generation[38] was the first to set directions and commercial practices, later reproduced by Manuel Henriques, Luis Mendes, and many others; they were the first to co-operate – in the way that most interested them – with regions which would later become rivals of the Portuguese (crown) interests, such as the United Provinces of Holland and Zeeland. But what were those interests? Were they the interests of the individual merchants and their networks or the interests of the state and the elites linked to it? It seems to me that the right answer is in the spirit of the first interrogation; because the strategies of merchants were driven by the existence of an "international order of money and business" not sympathetic to political and military rivalry that used to result in exasperating disruptions of traffic. And the best proof of this behaviour is in the continuity of the relationship networks maintained with shipping agents from Zeeland, Holland, and England before and after 1580.

Moreover, the archive data presented in this chapter – and remember the information about sugar refining in the Netherlands which is quite significant – are crucial to correct some statements about the presence of United Provinces ships in business held as exclusive, as in the same case of sugar trade. Evaldo Cabral de Mello (2002: 106–109) claims that Dutch participation occurred only in later stages believing the flag of most hulks (*urcas*) departing from Recife between 1595 and 1605 was actually from Hamburg and their final destination was actually Antwerp. The notary documentation of Oporto allows us to go deeper into the issue by showing that the flag they used (not only the Hamburg flag but also Swedish and Danish flags) was a flag of convenience (in fact many of the vessels registered in Hamburg were Dutch) and the final destination was Amsterdam, Rotterdam, or Vlissingen. When the company of Simão Lopes closed its 1586 accounts, it referred to the partnership with the international networks of Manuel Henriques, the Coronel of Baiona (Galicia), or the Marquina and "Carcoeira" from Burgos and they all were interested in dyes from the Azores and Brazilian sugar freight in Dutch

[38] I mentioned this merchant when I talked of Bento Fernandes. Vila Real was the mastermind of Sephardic business in Porto; he committed suicide when he was arrested by the Coimbra Inquisition.

ships named *Salomão* (Solomon) and *Lucrécia* (Lucretia).[39] In 1595, some Oporto merchants, including Pedro Aires Victoria, chartered the "Swedish" hulks (actually they were Zeeland hulks) named *Palomba Azul* (Blue Dove) and *Caçador* (Hunter) to load sugar in Brazil and bring it to Vlissingen on their behalf.[40] Two years later (1597), Gaspar Jorge sold to Diogo Henriques the sugar loaded in Brazil (Bahia) in the Dutch hulks *Fortuna* and *Caçador*, which the buyer would be responsible to receive it in "any of the ports of Holland, Zeeland and Germany" through proxies he had in Antwerp.

This documentation, however, confirms Mello's ideas (based on the studies of Jonathan Israel and Eddy Stols) about the leading players in the sugar trade, stating that they were not Dutch but Portuguese Sephardic, given the numerous and easily identifiable names registered in ship chartering contracts. And the general absence of foreigners in those contracts and the role of some Portuguese merchants as front men of the Dutch from the second decade of the 17th century raise the question of the origin of this dependence which could be older; however, we must look into Dutch archives to check if there are not proxy contracts between Dutch and Portuguese.

What concise interpretation can we draw from this data? In my view, not one but several; first, the use of north European fleets by the Portuguese networks means they acknowledged the potential of their shipping agents, and that was the result of a systematic practice and acquaintance through the frequency of contacts they had in various routes; secondly, because that option was assumed as a business decision took to get cost savings that explains why merchants used it so often even in the face of royal prohibitions, in a process that was clear in Malaga, Seville, and Cadiz in the 17th century when the Iberian merchants were the first to fight against embargoes on the United Provinces navigation decreed by the crown (Crespo Solana, 2009: 4–5); thirdly, the action of men like Tristão Rodrigues Vila Real, Henrique Gomes and his son Simão Vaz, António da Fonseca, Manuel Henriques, and their successors, maintaining sustainable commercial relations with Amsterdam, Rotterdam, Vlissingen, Hamburg, Bremen, Emden, and other "forbidden ports", confirms the claims of Clé Lesger that the rise of Amsterdam (and hence of the United Provinces) was a process of long duration.

---

[39] ADP – *Po1º*, 3ª série, liv. 82, 82v e seguintes.

[40] ADP – *Po2º*, 1ª série, liv. 4, fl. 11-13.

6. Introducing Global History

Map 10. Dimension of Oporto trading ventures, 16th century

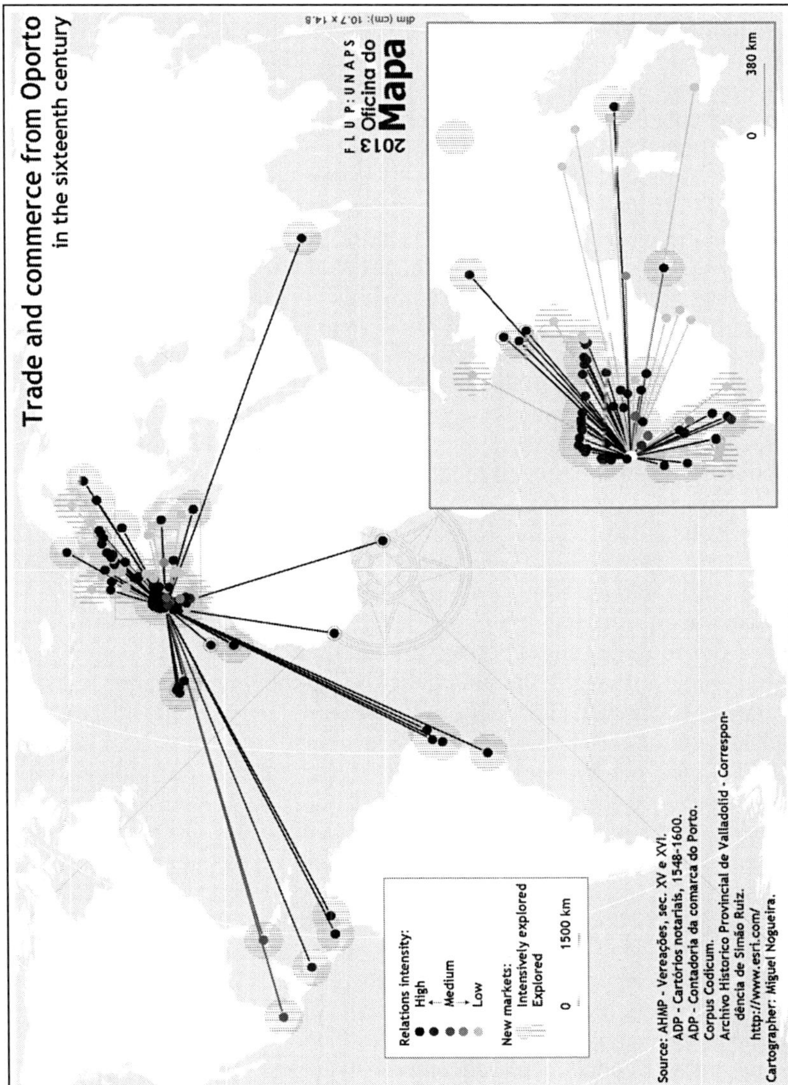

This map, depicting the dimension of Oporto trading ventures in the second half of the 16th century, could function as a corollary of what was stated in the preceding pages, and as a fair conclusion of this study. Here we have a commercial economy related to all the major Early Modern referential hubs, and extended at a global scale; here we have an Atlantic space adjusted to the interests of its users, but above all, articulated by the regularity and intensity of the relationships they established. But this social and economic phenomenon, integrating history and geography, with complex and modified trade relations, and not ot mention spatial and political constraints, still leaves many questions unanswered.

First of all, one already mentioned: we don't know yet the specific weight of these networks and their ventures in the overall Early Modern economy. From a single port, like the one studied here, it is impossible to assess the importance of these commercial organizations' activities in Europe's economic history, and it is time Portuguese historiography invested in comparative research to fill that gap, paying attention to the figures and estimates registered in notaries' and merchants' account books, which has not been the practice up to now.

Secondly, and urgently, the concepts of globalization and free initiative, which are frequently claimed for these Early Modern networks and businessmen, must be reassessed. Unquestionably, the operations of these private agents, or "free agents", worked at a higher level interfering in different geographies, undaunted by restrictions and embargoes imposed by religious, ethnic, cultural, and political differences, acting, if we wish, at an informal level, and self-organized (a characteristic often pointed out to define these networks that I find obvious and redundant). The outcome influenced the emergence of multicultural, multiethnic (in environments beyond Europe), multinational, and (highly debatable) stateless relations. But this is a dilemma pointed out some time ago by Jorge Pedreira: the dilemma between what was the result of political and imperialistic projects and what came from purely economic forces (Pedreira, 1998: 434). This raises the question of the effectiveness and impact of networks – I keep discussing concepts of globalization – which leads us to the comparison between the Portuguese Atlantic and Asian networks, concluding that Oriental business such as the spice trade was much more valuable when performed by crown agents (although I do not discuss here the economic viability of this enterprise nor its weaknesses, fraud, smuggling, ineffectiveness of regulatory mechanisms, nature of contacts in the supply, distribution and destination markets, etc.) and were virtually nil when done by private individual agents.

The impact of private networks in a hypothetical global economy was, on the contrary, higher in the Atlantic imposing models (e.g., economies of plantation, slave labour-based with a view to mass production and consumption), interfering in the reception, processing and distribution markets, accelerating economic interdependences, and making a part of Europe's population relationship to commercial products as sugar entered a new phase. The aforementioned Pedreira points out this phenomenon when he introduces the concept of Atlantic empire, although it seems to me that it was an empire *sui generis* built on complicities and games of interests defined in ports and from ports by the will of multinational merchants rather than by royal decrees.

Speaking of the "Eurasian and African world-system" between the first and the 16th century, Philippe Beaujard (2009) considers that it was created around "five multicentre hearts": China, western India, Persia, Egypt and Europe. Combining what has been written on this chapter and looking at Map 12, we see how the Iberian merchants, their networks, and their business partners achieved the expansions of the "European heart" towards the Atlantic, and in the 16th century began to win the centrality that history would grant.

Driven by necessity by virtue of a restrictive royal monopoly that cut them away from the Indies' *armadas* business in favour of the big foreign merchants' companies who financed it, the Atlantic merchants struggled to build a new business commercial and economic space at a time when the Atlantic was anything but attractive when compared with the appeal of the goods brought by the huge carracks of the Cape route. Nevertheless, the strength of their organization, the quality of their relations, and the on-going and reinforced business they managed proved to be the most resilient when the crown and the state sought to interfere; at that time, the economic and social international solidarity chains worked. Organized in multinational networks, the merchants kept their prosperity without much effort by transferring their interests and fortunes to the rising and organized mercantile states of the United Provinces and England; for them, the shock was not great; moreover, they had been preparing this path for at least a century.

CHAPTER SIX

TRADE NETWORKS IN THE FIRST GLOBAL AGE: THE CASE STUDY OF SIMÓN RUIZ COMPANY: VISUALIZATION METHODS AND SPATIAL PROJECTIONS

AMÉLIA POLONIA[1],
SARA PINTO[1] AND ANA RIBEIRO[2]

1. Theory and Methodology

The strategy and work plan of DynCoopNet were centred on the study of co-operative networks linking various locations and agents during the Early Modern age. The main goal was to study the historical mechanisms of co-operation among self-organized trade networks during the period between 1400 and 1800. The project set out to prove that, in the period from the 15th to the 18th centuries, the world economy was increasingly characterized by widespread collaboration that went beyond the boundaries of countries and continents. It was made possible by new means of global communication and the building not only of formal but also of informal networks, frequently multinational. The main assumptions of this project were that co-operation tied together several self-organizing networks, and that the world economy became, at this period, a dynamic, open, complex, non-linear system. Thirdly, it was assumed that the history of any place could not be understood without examining how it was connected to other locations and to the system as a whole.[3]

---

[1] Universidade Porto-CITCEM.

[2] CIDEHUS-Universidade Evora.

[3] Extracted from *DynCoopNet Project Proposal* submitted to EUROCORES - TECT program (European Science Foundation). See also Owens, 2008.

According to DynCoopNet, co-operative self-organizing networks were characterized by a diffusion of authority; they served as a source for the creativity and innovation necessary to respond flexibly to the disruptions in commodity, information, and capital flows. DynCoopNet's strategy and work plan eventually would allow a revaluation of worldwide dynamics from a new perspective, centred not on the structures, not on the systems, but on the agents.

Co-operation refers, precisely, to interpersonal exchange. Co-operation exists when an individual assumes a costly behaviour which brings him and others a bigger benefit (Hamilton, 1964a, 1964b). It is "a behaviour that provides a benefit to another individual (recipient), and the evolution of which has been dependent on its beneficial effect for the recipient" (West, 2011: 232). Accepting these definitions means to understand co-operation as an oriented interaction to achieve a certain goal, through a certain strategy at a specific moment. It is an arithmetical puzzle of costs and benefits, beliefs, expectations and motivations, between collective and individual interests.

Economists' definitions of co-operation focus on two fundamental characteristics. The first defines co-operation as a collective action of individuals who aim to share a certain task, lucrative for all partners (Jesus, 2009: 80). The second understands co-operation as a social process where individuals, groups and institutions act in a concerted way to reach common goals. Economic approaches, in this sense, focus not only on economic characteristics of co-operative relations (cost *vs* benefit), but also on the social characteristics of partners and their relations. Co-operative behaviour is thus driven by goals, expectations, and motivations which imply a collective or dyadic interaction between individuals and between these and the systems within which they interact over time (Axelrod, 1984: 6). In this sense, co-operation implies a social structure conditioned by historical time and space (Ribeiro, 2011: 10–14). History provides thus an excellent field to debate evolutionary models of co-operation.

Besides the theories of co-operation, self-organization theories became also a useful tool for a more accurate understanding of the dynamics under scrutiny. Self-organization is seen as a process where some form, order, or co-ordination arises out of the interactions between the components of an initially disordered system. This process is assumed to be spontaneous: it is not directed or controlled by any agent or subsystem inside or outside the system. Self-organization occurs and it can be studied in a variety of physical, chemical, biological, social, and cognitive systems. One of the most detached authors to connect the principles of self-organization with world systems functioning is the political scientist George Modelski. He stresses the connection of the dynamics of self-organization with the concept and the dynamism of world system evolution (Modelski, 1990; Modelski, 2008). According to him "Dynamic physical, biological, and social systems evolve in such ways that order increases so that several parts are mutually adapted in what are evolutionary processes" (Modelski, 2004). The large-scale processes of the world system, as well as the social organization of humanity – which are evolutionary – may also be seen as self-organizing. Powered by a set of influential mechanisms, self-organization is thus seen as a critical feature of the world system (Jantschs, 1980). According to this understanding, one needs to highlight not only the features of disruption and unsettlement that occur in stressful periods, but also the creative and positive aspects of reorganization and the greater inclusiveness of larger harmonies. They are seen by Modelski as a necessary component of the world system, and a factor to take into consideration for its functioning. One could apply the same theoretical principles to the world system of the First Global Age.

Both theoretical approaches require the use of concepts and methodologies frequently imported from other sciences rather than history. We could mention the concept of networks; the very concepts of co-operation and self-organization or, in the methodological field, the social network analysis and the agent-based analytical approach. Abundant secondary literature has been produced on these subjects by a wide range of disciplines – sociology, economics, mathematics, and anthropology.[4] As for network analysis, it is consensually accepted that social network theory views social relationships in terms of nodes and ties. Nodes are the individual actors within the networks, and ties are the relationships between the actors. There can be many kinds of ties between the nodes. In its most simple form, a social network is a map of all of the relevant ties between the nodes that are being studied.

Taking this as a theoretical premise, the understanding of a system depends, to a great extent, on the configuration of these complex networks. Starting from the position of the individual agents in a network and looking at their connections, one can characterize and understand the system, e.g., understand the functioning of a given social phenomenon.

---

[4] For network theories, see for instance: Watts, 2006; Latham, 2002; Carrington, 2005; Knoke, 2008; Barabasi, 2003. For the economic approach on this subject see Grabher, 2006; Hill, 2002; Jackson, 2003. For the concept of co-operation and the models of analysis, cf. the accurate panorama provided by the international conference *Evolution of Cooperation – Models and Theories* (Luxemburg, 15–18 September 2009).

Following Wasserman and Faust, it is possible to accept that "network models conceptualize structures (social, economic and so forth) as lasting patterns of relations among authors" (Wasserman, 1999: 5). It means that the unit of analysis is not, in fact, the individual, but the entity consisting of an assortment of individuals and the existing links between them. According to these assumptions, the positioning of individual agents in networks (whether political, economic, or social) is an essential procedure to understand the functioning of the system as a whole. If one is to apply these assumptions to our analysis, which is centred on self-organized networks, the actor network theory principles[5] or, more accurately, the agent-based modeling theories also reveal themselves as a useful tool (Wasserman, 1999; Haythornthwaite, 1996). In fact, agent-based modelling is required when individual behaviour is non-linear; when agent interactions are heterogeneous and when middling will not work, because the system is generally unstable (Bonabeau, 2001) – which are all conditions recognized in the historical reality under study.

On the scope of the research program of DynCoopNet, social network theories and methods became thus crucial, when trying to understand the functioning of the global system of the First Global Age as based on self-organized networks. The perception that a spatial representation was as well essential in order to highlight the dynamic complexity of the commercial and financial networks active on the ground arose concomitantly. One of the project's underlying assumptions was that "networks happen not only in socially, economically and historically defined spheres, but also in geographically determined areas" (Schulte-Beerbühl, 2004: 17). Moreover, the first element of the acronym of DynCoopNet, *Dynamic*, asks for a space/time approach. As stated by Yuan, "some disciplines, such as geography and landscape ecology, emphasize the spatial dimension of world knowledge, and other disciplines, such as history and climatology, take timecentric approaches to organize evidences of reality. However, it is the space-time integration that provides the explanatory power to understand and predict reality. Dynamics is by definition an integration of space and time" (Yuan, 2008).

Geography provides the spatial framing for network building and dissemination (Owens, 2009; Polonia and Owens, 2010). In the last years, social sciences have launched new research questions based on a new way of understanding space, at the same time that geography is playing an important role in historical research. However, this "spatial turn" cannot be seen as a true novelty, but instead as a rediscovery of space and its

[5] About actor-network theory, see: Serker, 2006; Law, 1999; Latour, 2005; Murdoch, 1998.

relation with people. Many of the classic histories of the last half-century and more – since Fernand Braudel's *Mediterranean World* – were spatial in the sense that the spatial relations are those that best explain the patterns of change over time (White, 2010; Braudel, 1983).

Additionally, new perspectives have emerged in historical sciences thanks to the massive databases that have become available. Nowadays social sciences are familiar with statistics and graphics. Databases are being built by historians to gather a large number of data sources and therefore allow this information to be analysed with new computational tools. In fact, the trend within the humanities and social sciences towards digital visualization tools reflects the fact that nearly all historical data have some spatial component associated with them (Torget and Wilson, 2009). Visualization methods and spatial projections became, thus, essential tools to pursue DynCoopNet's main scientific goals.

In order to experiment with a real application of this theoretical and methodological framework, the Portuguese team of DynCoopNet (from now on DynCoopNet-Pt) initiated research based on the trade and financial network of Simón Ruiz. Simón Ruiz was a new-Christian merchant, centre of a commercial and financial company operating from 1553 to 1607 which extended its influence all over Europe and overseas. Simón Ruiz and his company seem to be an excellent case study. He was born in the Castilian village of Belorado, in 1525 or 1526. He began his commercial activity around the age of 25, i.e. in the mid-16th century, as an agent of Yvon Roncaz from Nantes, in the trade of textiles in the trade fairs of Medina del Campo. These fairs were considered focal points in the circulation of people, goods and credit (Braudel, 1985: 502; Casado Alonso, 2011), encouraging Ruiz to participate in successive trade partnerships. The first of these involved him with Juan de Orbea, treasurer of Aragon, his brother André Ruiz in Nantes, and Ochoa Lanier in Bilbao (Lapeyre, 1955: 60–62). Trying to improve his profits in olive oil, spices, indigo, salt, and wheat trade, Simón Ruiz started to place trustworthy agents in places of vital geographical importance to his business. He also fostered an informants' network in places with international relevance in the world of trade and finance – Lisbon, Genoa, Rouen, Rome, Venice, or Lyon (González Torga, 2005; Casado Alonso, 2008: 54). Simón Ruiz's company was not only important for its trade. Medina del Campo was also an important exchange place, and the merchant soon became a banker. Between 1576 and 1588 he was one of the main creditors and *asientista* of the Spanish king, Philip II. In many ways, this man matched the portrait of the archetypal 16th-century merchant banker. Simón Ruiz's key role in central places in the European economy transformed him into a model, not only within the Iberian peninsula, but also on a European level. In this sense, the potential scope of Simón Ruiz's business network and the variety and heterogeneity of his partners justified the DynCoopNet-Pt choice of his trade and financial network as a representative case study of the functioning of the mechanisms of co-operation in the second half of the 16th century (Ribeiro, 2011).

The documentary inheritance left by his company makes it possible to describe the co-operative relationships the team was looking for. On the other hand, they displayed how this agent network was built, showing how his partners, representatives, collaborators, or informants took position in an evolving network. The archive of the company, belonging to a private foundation, is now available at the Valladolid Provincial Archive, allowing DynCoopNet-Pt to use serial data sources. The team choice was oriented towards bills of exchange (about 9000 out of about a total of 21,000 were analysed), and commercial letters (as a sample, about 300 commercial letters were examined, addressed to Portuguese merchants, out of a total of about 15,000, Portuguese and non-Portuguese).

Bills of exchange appeared as a mercantile credit paper in order to solve the risks and difficulties of cash transportation. In the 15th and 16th centuries, bills of exchange became a method of payment, as well as the most frequent way of obtaining loans. Medina del Campo became a pivotal place in Europe as the place of well known financial fairs, as a centre of acceptance for exchanges, payments, and liquidations other Castilian fairs made with foreign commercial centres. Medina became the most important financial centre of the Castilian crown, attracting the most important European bankers and financial brokers of that time. To understand how DynCoopNet-Pt shaped its empirical work, it is important to understand the nature and the functioning of these data sources. Bills of exchange may engage four different players: the deliverer (who previously advances the money to the taker); the taker (the one who writes the letter of exchange and orders the payment); the beneficiary (who receives the final payment); and the guarantor (the one who pays the money).

Figure 1. Example of a bill of exchange[6]

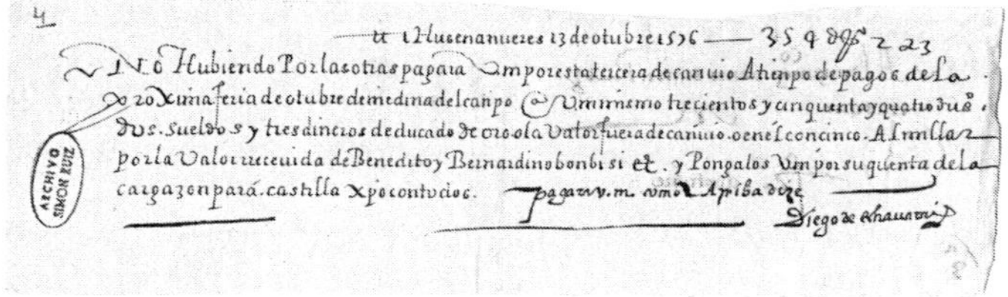

With this basic structure, bills of exchange permit the identification of the name and the attributes of the agents; their spatial location; their relationship and connections; the circuit of money and, sometimes, the trade that motivates the respective payments. As some of these agents used legal representatives, some letters of exchange may involve more than eight agents. They may also refer to their location, professional and social status, titles of honour, or even family relationships. This information allowed the team to identify the main agents of single or repeated transactions and to reconstruct the networks that sustained such commercial transactions. It also allowed us to identify the evolving status of each agent in a network and to point out some criteria that sustained the constitution of the networks.

Figure 2. Example of a merchant letter[7]

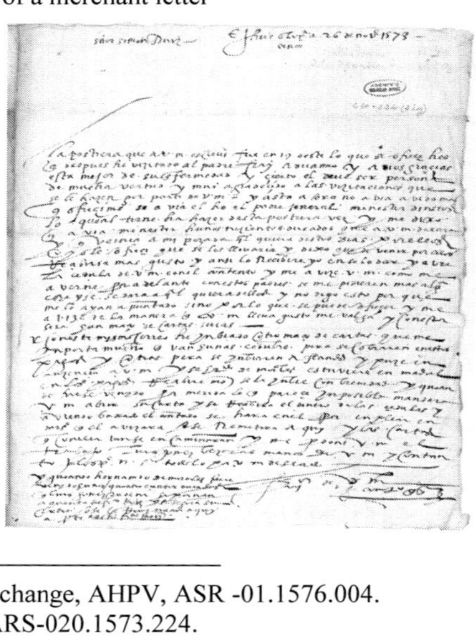

[6] Source: Bill of exchange, AHPV, ASR -01.1576.004.

[7] Source: AHPV, ARS-020.1573.224.

On the other hand, merchant correspondence was one of the most important features of 16th-century trade activity. It became vital for the spatial enlargement and agent dissemination across the most important points of business. Formally, the letters do not have a standard structure. They present themselves as a more complex and richer universe of information. Extracting data from them to feed a standard database becomes a major problem. Besides the individual characteristics of the agents, this source allows the researcher to establish relationships among the commercial partners, as it gives information about the business world: prices of goods; trade circuits; features relating to insurance, freight, and freight rates; exchange rates; qualitative and quantitative information about goods; international, regional, and local markets; weights and measurement units. Merchant correspondence also enables us to identify certain events and conjunctures which conditioned the set-up and the evolution of the network: news on political crises, rise of new dynasties, wars, international conflicts, peace treaties, climate changes, calamities, market flows and investment opportunities as well as business risks could be mentioned.

This data source brought to light details about relations between agents who sustained the ties of co-operation. They indicate why some were able to influence both the functioning of markets and the behaviour of other agents, particularly those connected to co-operation mechanisms: friendships, help, trusts, and reputation are some of the elements which can be extracted from these sources. They also provide data which indicate the intensity, direction, and durability of co-operation relations, as well as the typology of the existing relationships. The data related to the geographic location of the agents were used to allow geo-analysis and the use of geo-visualization methods.

A database of the information gathered, based on Timelink software[8] designed to support multilevel research, is now available online[9] and a lexicon, together with a thesaurus of the levels and the mechanisms of co-operation, is available to be applied to other themes under the scope of analysis of cooperation (Polonia, 2010a, 2012). From this research resulted two PhD dissertations: *Mechanisms and Criteria of Cooperation in Trading Networks of the First Global Age. The case study of Simón Ruiz network, 1557–1606* (Ribeiro, 2011) was the result of a study focused on the mechanisms of co-operation, based on network analysis. Simultaneously, the location of the agents and the relations between places in Simón Ruiz's spatial network enabled Sara Pinto to understand the

[8] For more information on the software please access http://timelink.fl.uc.pt.

[9] The database is available in http://timelink.dyncoopnet-pt.org/mhk.

spatial configuration of this merchant company network, and its evolution over time during the company's existence (1557–1606) (Pinto, 2012). *The Simón Ruiz Company: Spatial analysis of a mercantile company in the 16th century* was the result of this data exploitation and visualization.

The analytical results of the project's empirical work thus ranged from the constitution and reconstitution of networks and their evolution, projected in space/time representations, to the appraisal of network models: network topology, connectivity, and density, and the appraisal of the relative positioning of each individual in the network. Besides that, the research allowed us to identify categories of partners involved in networks; the identification of variables which interfere with the model; the identification of contextual conditions which interfere with the network constitution and evolution; the identification of variables which influenced the constitution of the networks (risk patterns, risk sharing, rivalry, expectations, motivations, personal memories); the identification of the value of trust and reputation for co-operation; the categorization of co-operative behaviours; and the identification of the mechanisms of co-operation, including punishment or reward.

The following points, directly connected with the two PhD theses mentioned above (Ribeiro, 2011; Pinto, 2012), intend to provide evidence of the paths and the outcomes of the DynCoopNet-Pt research project.

2. Network Functioning

To increase the efficiency of long-distance trade, co-operation among merchants and businessmen like Simón Ruiz emerged as a means to satisfy the needs of different agents who, alone, could not invest in such a costly and complex venture. The structured network of Simón Ruiz persisted and evolved dynamically from 1553 to the beginning of the 17th century. Inside that structure, and applying the theory of network analysis, rules of functioning evolved to sustain the health of trading and financing among partners from different nationalities, religious beliefs and socio-cultural backgrounds. Which network characteristics allowed the most profitable and efficient functioning of different trading and financial realities?

This network had its life cycle and evolved through time. Active for more than 40 years, different types of relations were fostered within this network, important for the success of business operations. Most of the links established between individuals in this case study were based on financial issues (credit, creditor/debtor, exchange partners, etc.) and commercial operations (selling/buying). However, kinship, friendship, sociability (acquaintance, living with), trust (formal or informal representation or intermediation) and distrust (suspicion statements), co-operation (brief or lasting business partnerships), defection (refusal to co-operate) and cheating (deliberate deceiving) relations were also revealed in this research, showing that Early Modern business comprised multiple overlapping types of links to build up a set of fruitful connections.

Figure 3. Simón Ruiz network visualization

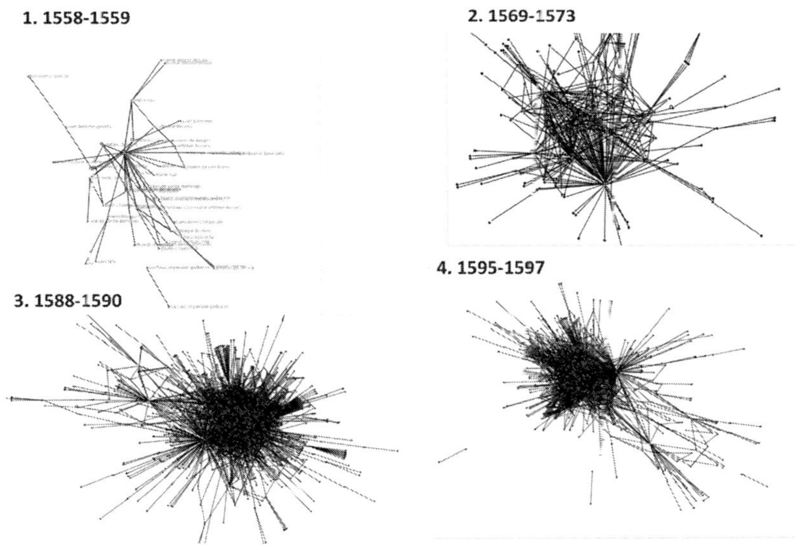

These representations indicate the variability of the number of agents and links over time in Ruiz's business network. Still, its configuration is similar throughout the second half of the 16th century. It is a star-shaped representation, due to the excessive centrality of Simón Ruiz. Since the research is based on his company's archive, he is mentioned in almost every documented act. These visualization outcomes are sketches from a partial reality. This limits the following results, as an "ego-network" consists in the existence of a focal node (such as Simón Ruiz), who is directly linked to most other agents present in the network (Knox, 2006). The (mis)representation of reality could give the impression Ruiz was determinant for contact between relevant individuals. However, the probability of direct links between those agents should be considered, as each of the individuals involved in the network had their own network neighbours who do not figure in this sample.

If the first years of Ruiz's independent business activity show a very centralized structure, that image is diluted as time goes by. The emergence of star-shaped sub-graphs of smaller dimensions reflects a strong position of certain individuals in the network. Those individuals are not only those who were linked to a great number of other agents but they also perform a central role in the network structure. They provide intermediate contacts and business between important or less important individuals in the company activity and if they were not present a significant part of the network would be disconnected. These smaller centres vary according to different time periods, which also show the degree of variability in the network's hierarchy.

Between 1565 and 1568, Simón Ruiz and the Bonvisi of Lyon are two central hubs which are not only directly connected, but use other agents placed in Lisbon, Medina del Campo, or Antwerp to facilitate their relationship. These were António Gomes de Elvas, Fernando de Morales, Diego de la Torre, Luis Gomes de Elvas, Andrés Ruiz, Francisco de Mariaca, and Francisco de la Presa. The latter dominated contacts with Spanish merchants, who after the bankruptcy of 1557 were invited to participate in the financial life of the kingdom. Through their collaboration, these merchants took advantage of the Ruiz-Bonvisi circuit to reach the large European trading houses in places like Lyon and Antwerp. Giovan de Ugalde was an important piece in this puzzle since he was the extension of Simón Ruiz in the northern Biscay region, from which the trade to Antwerp departed. The Bonvisi sent him money to forward to Antwerp where it was needed. In this case, the circulation of money was something like Simón Ruiz – Bonvisi – Giovan de Ugalde – Fernando de Frias Ceballos.

The other network axis, towards Lyon, by Agustin Spínola and other Genoese bankers in Medina del Campo and Seville, was dependent upon Genoese like Stefano Lercaro (Medina), Juan Battista Spínola (Lyon), Nicolo Grimaldo (Medina), or Lazaro Doria (Medina). They acted as commissioners in the transfers from Simón Ruiz and other Castilians to Lyon, but the final destination was to the heirs of Buenaventura Micheli and Geronimo de Arnolfini. Two central Castilian agents played such a role: Diego de Castro and Geronimo de Salamanca. The former controlled the connections with Lyon and from Lyon to Antwerp.[10] Gerónimo de Salamanca appears during 1566 connecting Simón Ruiz to other Genoese houses in Madrid and in Lyon as their debtor. And houses like Lercaro and

[10] Bills of exchange AHPV, ASR-01.1564.040 and ASR-01.1566.046.

Sauli are intermediating and conducting Simón Ruiz money to the Bonvisi beneficiaries.[11]

In the period of greater expansion, 1574–1580, the relations between agents became more focused on Simón Ruiz himself, as the density of the network decreased, even if it had never been high at any period of the firm's activity. Only 1.68% of the agents were connected to more than 20 neighbours (Ribeiro, 2011: 100). Ruiz's privileged trading partners were the Portuguese Gomes de Elvas, not only allowing Ruiz to be part of the distribution of oriental spice in Castile, but also as valuable partners to the redistribution of spices in Antwerp. Luis and António Gomes, Manuel Gomes de Elvas, and António Fernandes de Elvas were also related, as well as Simón Ruiz, with other agents in Antwerp who were in touch with the most connected agents such as Filipe Jorge, Rui Nunes, or Nicolau Rodrigues de Evora. Other central figures were the Bonvisi of Lyon, linked directly to Simón Ruiz, and some representatives of Simón in Spain, such as Lope de Arciniega in Madrid, Hernando Correa de Velasco (also in Madrid, but dealing directly in and with the court), and Antonio Vasquez, a banker. In conclusion, the Simón Ruiz network was, at this point, centred on the axes Lisbon-Madrid-Antwerp and Madrid-Lyon.

Bankers were key individuals, when comparing the number of connections agents had with their occupational attributes. The Bonvisi, Antonio Vasquez, or Nicolau Rodrigues de Évora were essential to avoid the collapse of part of the network. In this period, banking activity was indispensable for the participation of the Ruiz firm in the *asientos* business, not only financing but also being able to place the money in turbulent Antwerp. They were also vital to the banking activity of Simón Ruiz himself, when he acted as payer or taker, mainly in conjunction with other Castilian agents. In these cases, important banking figures such as Gaspar Suarez, Juan Ortega de la Torre, or Francisco de Bobadilla act as intermediaries in the payment of bills of exchange to the beneficiary or the deposit of a certain amount from the donor to the taker. For example, on 5 February 1579 Simón Ruiz was sending 413 écus to the Portuguese Fernão Mendes de Sá and Diego Aleman, who were in Lyon. For that he made a deposit in the bank of Juan Ortega de la Torre from where the taker, in this case, Ventura de Marriega, picked up the amount in question.[12] In 1574, Simón Ruiz acted as a payer of a bill of exchange in Medina del Campo, but he did not pay it directly to the beneficiary, João Fernandes de Castro. Instead, he did it through a deposit in the bank of Gaspar Sanchez. The

---

[11] *Vide* e.g. AHPV, ASR-01.1566.009.

[12] Bill of exchange AHPV, ASR-02.1579.198.

beneficiary went to the bank and raised the amount from Simón Ruiz's account.[13]

The centrality of bankers in this network activity enhances the relevance of credit as an essential dimension of long distance trade in the Early Modern period. Credit was the sustenance of these men's activities when the investments in a journey delayed liquid profits or when there was a need of active capital for the supply of goods, freight, or insurance. The last period reflected in the sources was a time for handover and renewal of the company. From 1592 onwards, Simón, already old and ill, decided to definitely associate his nephew to his businesses, even the financial activities, creating the company of Simón y Cosme Ruiz (Lapeyre, 1955: 73). They are the hubs of the main star in the graph above. If Simón Ruiz had 94 different partners in the network, Cosme Ruiz Embito amounts to 90 partners. However, they are not the only ones. Considering that the total number of nodes between 1595 and 1597 was 643, it means that Simón and Cosme were only directly linked to a relatively small number of partners, accounting for about 14 per cent.

They are linked with two smaller stars, centred on Teutónio Nunes and Hieronimo da Fonseca, both placed in Rome. In the centre of the graph there is a dense nebula. A network topology like this implies a stronger interaction between partners, which strengthened the network activity and the collaboration between agents. Regarding the importance for the network resilience, older established partners are the ones with greater numbers of contacts.

Out of all the agents apart from Simón and Cosme Ruiz, those with more than 20 connections per node constituted a group of only 12 individuals (1.9%), and only two of them were new partners in the network. One of them was Paolo Bonvisi, the first family representative of the Bonvisi of Lyon in Spain. He was acting as a beneficiary and payer in Madrid and Medina, associated with Simón and Cosme Ruiz. Benedeto Bonvisi had died and Paolo was now the new partner of Estevan, Antonio, and Bernardino Bonvisi in Iberian affairs. The revenues from the Castilian finances were of major importance, since the decline of the Lyon fairs had forced the firm to leave that city, and they had begun to exchange in Piacenza, where Bernardino, Antonio, and Estevan were based at this time. The other new agent with significant linking was the Castilian Christoval de Aldana, with 47 connections with different agents. He had a partnership in banking with the Genoese Carlos Velutti (with a total of 43 links). They ensured the capital transfers between Simón and Cosme Ruiz and Portuguese partners in the *asientos*, such as the Veiga and Rodrigues de

[13] Bill of exchange AHPV, ASR-01.1574.023.

Évora, or the Jorges in Lisbon.[14] In the years 1595 and 1596, Aldana participated in 458 acts.

Antonio de San Roman, also in Lisbon and ensuring the same kind of operations with almost the same protagonists, was part of this group.[15] Juan Battista Gallo and Juan Luis de Vitoria were the agents best positioned to ensure the opposite circuit from Medina or Madrid to Lisbon, also linking the Portuguese capital to Simón Ruiz's company.[16] These top agents performed the roles of takers and payers in numerous exchange acts. The circulation of money seems to have been privileged between Castile and Lisbon. On the one hand, the main partners of Simón Ruiz's *asientos* seem to have been the Portuguese. It was important to transfer money from all partners, and Lisbon did not have a direct means to send capital to Antwerp. In such ways, merchants and bankers transferred money first to Medina or Madrid and from there to Antwerp. The profits of such operations were also transferred from Medina to Portugal. On the other hand, Lisbon was the main door to reach Portuguese colonial markets, either by buying colonial products and reselling them in other European markets, or for ensuring European goods to meet the demand of the European population overseas. In this sense, it was important to have trustworthy and successful agents like the ones identified to ensure such circulation.

Comparing the evolution of the number of acts described in the records and the number of nodes (individuals identified as operating in the network), two evolving rhythms were identified – before and after 1574 – despite the tendency for a regular increase in the number of agents in the network.

Simón Ruiz's network, in the first years of his autonomous activity as businessman, began with few partners. This scene of gradual and slow growth of the network was maintained until 1568–1569. In the first half of the 1570s the tendency was for instability, reflected in lower levels of partners and activity. After 1574, the network clearly expanded and apart from some short-term disturbances, it never went back to the previous number of nodes. After 1580 it reached a number of nodes between 300 and 400, with the exception of 1589, when the number of agents in the network exceeded 450 members.

---

[14] Bill of exchange AHPV, ASR-05.1595.403.

[15] Bill of exchange AHPV, ASR-05.1596.118.

[16] For example, Bill of exchange AHPV, ASR-05.1595.195 or bill of exchange AHPV, ASR-05.1596.125.

Figure 4. Comparative evolution of the number of acts and the number of network nodes per year, 1558–1597

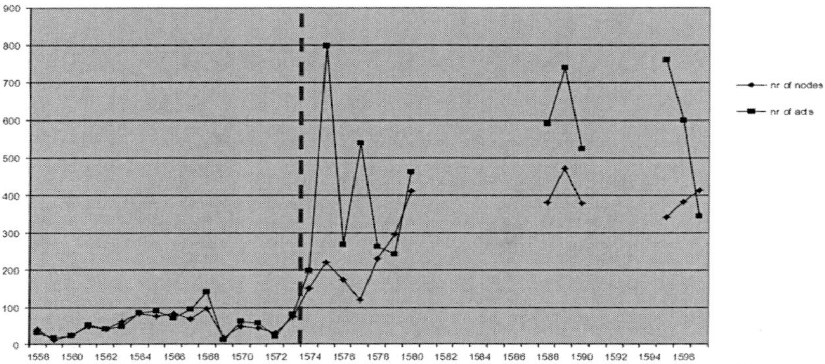

1574 was a key year in Simón Ruiz's business activity. For the first time, the Castilian merchant was formally part of the Castilian public debt deal in exchange for silver and rent revenues. Until then, trade was the primary activity of Ruiz. After 1574, the network witnessed a growth boom and began to be more and more dependent on financial businesses. In 1575, the number of transactions reached 800, while the number of partners did not grow on the same scale. This tendency was maintained until the end of Simón Ruiz's life. It seems that the change to banking and public loans had enforced the expansion of the network, but not in the same proportion as the number of trade or financial acts registered. Among the types of business in which Simón Ruiz was involved, the financial ones and the *asientos* were riskier. The amounts invested were equally larger and implied the danger of bigger losses. Diversifying the partners' network was designed to minimize the risks. This strategy gains visibility in two ways. First, the amount in debt or to credit was split in several bills of exchange addressed to the same beneficiary(ies), using different takers and different payers. For instance, between 26 and 31 July 1578, Simón Ruiz sent ten bills of exchange from Madrid to Lyon, to pay to Benedetto and Bernardino Bonvisi. However, the amounts in each bill are different, as are the takers and the payers, to increase the probability of success in the payment of those bills of exchange.[17] The second strategy highlighting the growing number of agents is particularly prominent in the firm's activities after 1580. When Ruiz had become a well-known banker in the Spanish court he began to finance not just other merchants or the monarch

[17] AHPV, ASR-01.1578.233 or AHPV, ASR-01.1578.236.

but any individual needing credit. However, the values were split among bills of exchange of really small value, usually less than 50 *reales* (Ribeiro, 2012).

According to some management specialists, a firm using this strategy acquires a set of unique information which helps it take decisions on a firm basis and reduces the level of trust placed with a small number of agents. The firm becomes less dependent upon individual agents (Beckman, 2004). At the same time, the predominant type of activity within the network seems to determine the frequency of contacts with strategic partners.

The evolution and predominant business strategy of the network is profoundly dependent on exogenous political, military, economic, and financial events. Margrit Schulte-Beerbühl also invokes this explanation in order to explain the expansion and failure of German commercial houses in London (Schulte-Beerbühl, 2007: 307–347). Unstable periods, such as the bankruptcies of 1575–1576 and 1596–1597, generated a trend of instability in network activity. In those periods, the number of partners was also affected. In the period between 1575 and 1577, the number of nodes shows a continuously decrease, whereas in 1596 and 1597, regarding the same contraction in the network activity, the number of partners grew. This differential behaviour is obviously related to the network's resilience to the crisis itself. In the first period, the firm's past activities were not profoundly embedded in financial affairs. The network structure, based on trading partners, had proved to be resistant to trade fluctuations and did not suffer extremely with the financial crisis. The answer of reducing the number of partners is marked not by the firm's health, but by an external market uncertainty. It was the reason for reinforcing existing relationships whose quality was already known. One can understand that, since "organizations seek stability and trust in inter-partner relationship" (Beckman, 2004: 262). Even so, in the overall period the network saw a great expansion in the number of partners.

Instead, in 1595–1597, the causes of the crisis derived directly from business activity in which the network was implicated, like the Spanish public debt. The reduction of the number of partners was not immediate, as in 1575, but it was a tendency from 1590 onwards. Levels of activity decreased significantly, but the number of partners rose, dividing risk among a multiple set of agents, exchanging smaller values. In 1597, the higher values circulating in the network in each act reached around 3000 *scudi*, but only appear three times in 345 acts. In 1580, at a time of network expansion and opportunities (Ribeiro, 2011: 97–110), 24 out of

432 bills of exchange had surpassed the value of 3000 *scudi*, reaching a maximum of more than 8000 *scudi*.

Table 1. Normalized exchanged values distribution in Simón Ruiz's network – 1580 and 1597 (in *maravedies*)[18]

Values	1580 %	1597 %
Less than 50,000	5.7	19
50,001–100,000	8.2	13.6
100,001–500,000	55.4	54.8
500,001–1 million	19.5	10.6
1 million–2 million	8.3	1.5
2 million–5 million	2.2	0.5
More than 7 million	0.7	
Total	100	100

On the other hand, comparison of the value distribution indicates an overall decrease in the values circulating in the network in 1597. The percentage of bills of exchange of less than 100,000 *maravedies* increased while letters above one million *maravedies* almost disappeared. Times of no financial disruption in the firm's activity, like 1580, favoured bills of exchange of greater value since there was less risk of investment loss or non-acceptance of bills. Levels of risk, within or outside this population, influenced the evolution of this social structure and configuration. However, it shows a quick and strong adaptability.

Figure 5. Variability of partners in different periods of Simón Ruiz's network

Periods

#	1	2	3	4	5	6	7
1	35	9	8	2	4	2	1
2	9	124	41	16	20	10	3
3	8	41	209	37	46	25	17
4	2	16	37	120	45	15	9
5	4	20	46	45	606	143	81
6	2	10	25	15	143	658	163
7	1	3	17	9	81	163	787

Periods:
1- 1557-1559
2- 1560-1564
3- 1565-1568
4- 1569-1573
5- 1574-1580
6- 1588-1590
7- 1595-1597

[18] Source: Bills of exchange of the Fondo Simón Ruiz, AHPV.

The darker squares in the figure above represent the total number of nodes in the network during a certain period of time. The other fields indicate how many of the former agents remained in the network in previous or later periods. Ruiz changed his partners frequently and the percentage of newcomers is never less than 70 per cent in any of the periods analysed. This is a costly strategy, but, according to rational choice theory in microeconomics, strongly rational, since contact with newcomers was perceived as the best way for new gains and business opportunities (Granovetter, 2005; Allingham, 2002). On the other hand, by varying partners as much as possible Ruiz could reduce levels of uncertainty, since new and possibly contradictory sources of information could be precious in a world where information circulation determined success in long-distance trade affairs. Comparing historiographical approaches based on institutional economy, these results are surprising, as transaction costs could be reduced by long-term partnerships where trust should be easily enforced by mutual understanding and well-known predictable business practices (North, 1985; Schulte-Beerbühl, 2008). If this strategy was prevalent in the whole chronology, the behaviour of these agents in stressful periods requires further explanation. The number of older partners was bigger in those times (periods of 1569–1573 and 1595–1597), helping to reduce risks by strengthening ties with former partners, whose past collaboration had been well recorded. When the network was living prosperous moments the number of older partners significantly decreased relative to the newcomers.

The evolution of the network depended strongly on endogenous strategies, as well as on the exogenous dynamics which conditioned partners' choice. The end of the 1560s and beginning of the 1570s was a tumultuous time in historical conjuncture. War spread all over Europe, conditioning financial and commercial operations. The Antwerp market stagnated, mostly due to the rebel's blockage of the Scheldt and British corsair attacks in the North Sea (Van der Wee, 1962, II: 251–252). The conflict not only disturbed and delayed valuable cargos such as spices, but also interfered with textile production and devalued money by 70 per cent due to silver abduction by Spaniards and Rebels (Vasquez de Prada, n.d.: 125). Castilian fairs suffered from successive delays and cancellations, and money was short in Medina del Campo and Seville for trade reinvestment and financial revenues (Ruiz Martin, 1986: 278). Between 1568 and 1577 was the period of the great number of private companies' defaults in Medina del Campo and Burgos (Abed Al-Hussein, 1986: 237). Lyon had lost its regularity, followed by money disorder and speculation (Gascon, 1971: 669). In this adverse scenario for the company's business, Ruiz tried to assure business, even at minimum levels, since his preferential markets were almost stationary and liquidity jeopardized. He chose to rely on Francisco de la Presa, the Bonvisi of Lyon, António Gomes de Elvas, Hernando de Morales, Luis Gomes de Elvas, Francisco de Maluenda, Agustin Spínola, Nicolau Grimaldo, and Fernando de Frias Ceballos to maintain a presence in Lisbon, Antwerp and Lyon.

In the 1590s the problem came from within the structure, but the strategy was also to enlarge the number of former partners. On 29 November 1596 a new bankruptcy was declared by Philip II. The king cancelled the debts he had been repaying, suspended the guarantees made by the Hacienda and dictated the end of interest for every contracted debt. Simón Ruiz was profoundly embedded in these financial operations. The answer to this demanding situation in terms of liquidity seems to have been the focus on trustworthy partners, with whom the company had enjoyed repeated trading and financial contacts. The period between 1595 and 1597 registered the greatest percentage (26%) of former partners still being active in the network – the Bonvisi, the Veiga, the Rodrigues de Évora, or the Jorges from Lisbon, for instance, relevant partners in the *asientos*.

Still, being compulsorily and permanently tied only to these agents would not only represent a cut-off in future business revenue growth, but also total dependence on the same agents, any one of which in case of default would seriously compromise Ruiz's activity. A stable yet rigid structure would not be able to adapt to new business opportunities in such troubled times as the second half of the 16th century. However, he combined novelty with solid trust in his chosen partners.

Table 2. Network Characteristics[19]

	Merchant Period (1553–1573)	Banker Period (1574–1598)	Entire Period (1553–1598)
Number of Nodes	648	3675	4160
Number of Edges	2132	17,282	19,226
Diameter	6	6	6
Max Degree (Simón Ruiz)	376	2060	2336
Average Degree	6.20	9.41	9.24
Cluster Coefficient	0.326	0.440	0.426
Average Path Length	2.68	2.72	2.997

The table above reveals the physical characteristics of the Simón Ruiz network, comparing the same features between the periods in which the

[19] Co-authored with Flávio Pinheiro – ATP Group.

attention of the merchant was divided between two different spheres of activity. As described before, in a similar period of 20 years, the network saw considerable enlargement after 1574. However, measuring the distance from individuals in the centre of the network to agents on its fringes – the network diameter – it remains the same. Each individual is at a maximum six steps apart from any other connected individual within this population, no matter what the number of the agents involved. It not only occurs in both trading and financing periods, but also in an overall perspective of the network. This is a small distance. In this sense, agents easily connected with each other, directly or through the contacts of someone else's neighbour, even though the network density was never too high (Barabasi, 2003: 25–53).

This feature compensates to some extent for the average number of neighbours per node being not very high. However, it rises during the financial period, as financial affairs usually resulted from a chain of credit and partnerships consisting of several agents. In the late 1570s, *asientos* made by Simón Ruiz involved partnerships with a large number of agents: the Bonvisi, Francisco de la Presa, Luis Gomes de Elvas, Tomás Ximenes, Rui Lopes de Évora, António Gomes de Évora, Manuel Gomes de Évora, António Fernandes de Elvas, Carlos Nunes.[20] A greater overall connectivity helps to divide risk, strengthens the structure, and enhances flexibility to respond rapidly to constant changes in business conditions.

Cluster coefficient measures the probability of one's neighbour also being somebody else's neighbour. This characteristic is important to evaluate the way in which people are connected, whether in small interconnected groups or in a random and disconnected way. In both periods, the Simón Ruiz network showed a high cluster coefficient, but a short average path length. On average, people are closely linked to each other at a distance of two or three steps. This combination is characteristic of a small world network. In this model, there is always a relative proximity of agents, no matter whether there are few or many nodes (Watts, 2004, 2003). This local structure, where people are close to each other and well interconnected within smaller groups, was maintained despite the main activity of Ruiz's business.

Besides the high variability of agents entering and leaving the network, the maintenance and stability of this business social structure favours a strong adaptability which allowed Ruiz's business activity to survive adverse circumstances and partners' defaults.

[20] Letter sent by António Fernandes de Elvas to Simon Ruiz 30 March 1576 – AHPV, ASR-034.1576.014.

Figure 6. Cumulative Degree Distributions[21]

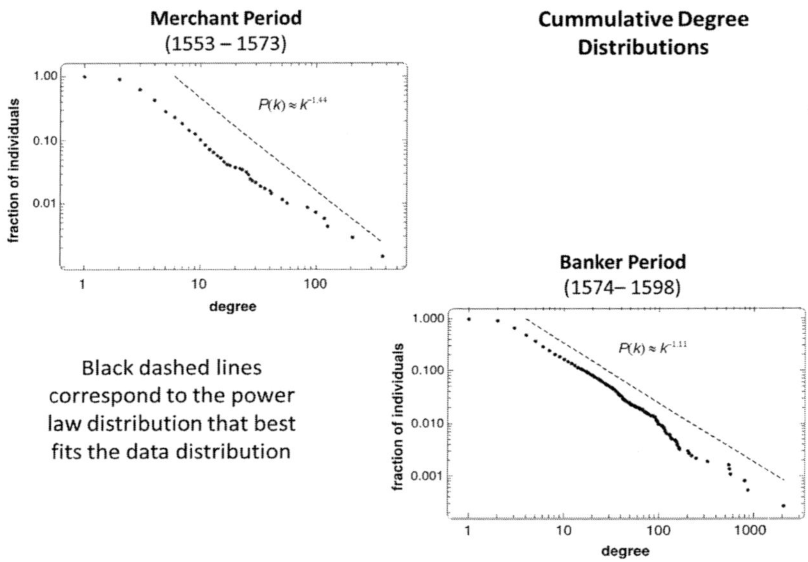

Before, we described this network's connectivity in detail and with an emphasis on the agents. However, a quantitative approach to this feature can also contribute to characterizing how this population interacted. Figure 6 shows the frequency with which an individual is connected to one or more agents.

Almost everyone in this population was connected to at least two neighbours, but most of the agents had few neighbours and only a small number had many. These individuals were the same who were the centre of the smaller stars in the network snapshots visualization in Figure 3.

Again this characteristic is independent of the preferential business activity of Simón Ruiz and of the different individuals who were involved in it. The degree distribution of Simón Ruiz's network reveals a scale-free structure, in which the connection of the individuals is organized not randomly but through an attraction mechanism. The number of future partners increased with the number of connections he already had. This is what is called "preferential attachment" (Newman, 2010: 487–490), a magnetic effect which affects agents' popularity. It is a strongly hierarchized structure and the power-law distribution points to this

[21] Co-authored with Flávio Pinheiro – ATP Group.

heterogeneity. These features characterize this past network as a "scale-free" network: heterogeneous, hierarchical and adaptive since the structure and connectivity patterns are maintained, despite the number of nodes (Barabási, 1999; Ravasz, 2003).

In a turbulent and demanding economic and financial scenario Simón Ruiz and his partners chose to vary as much as possible their contacts. However, agents with a higher connectivity became small hubs which were changing over time, according to important strategic functions displayed in specific circumstances, in a combination of financial solvency and strategic geographic locations. This flexibility allowed for quick mutations. This feature was the key mechanism to respond to demanding times. The system balances stability and flexibility, which also prevails in other biological and economic adaptive complex systems (Dopfer, 2004; Kauffman, 1993).

Nevertheless, if agents and key characters changed so rapidly, the structure and the way agents' interactions were organized became stable throughout the second half of the 16th century. This is characteristic of an emergent phenomenon (Holland, 1998, 225–231). The interaction between individuals creates structures which persist when the agents who produced them disappear and are replaced by others. Even if they owe their existence to a certain group of individuals, once they are created they determine the actions of the new individuals (Carvalho, 1999: 639).

This resilience was based on a quantitative and qualitative heterogeneity and hierarchy, where most of the agents were linked to a short number of neighbours while the main part of information, goods, and credit exchanges were concentrated in a few agents. Those were the same hubs of other small stars, other than Ruiz's himself, which means that main interactions were concentrated on the links between these star-shaped sub-graphs hubs. They controlled information, goods and credit from within their contacts, which was then spread to the other few hubs in this population. This strong hierarchy was also sustained by the short social distance between individuals fostered by a strong interconnectivity, which made the structure local to its agents. Merchants tried to be connected to important individuals so they could reduce distance to future partners. In long-distance trade this strategy enhanced the performance of the business. Whatever the dimension or the people belonging to this population, the structure of the network remains the same.

Even if these association strategies were not randomly executed, they were not imposed by a central authority or institution. The Simón Ruiz network was self-organized. Even if deeply embedded in the Iberian trading monopolies and in the policies of Spanish public debt, the connections of business partners in it were not conditioned by the states or guilds but by the emergence of an internal organization.

3. Spatial Analysis

A spatial analysis of Simón Ruiz's trade network was developed in order to evaluate the role of space in the dynamic complexity of commercial networks, in the First Global Age. This research was *(presumably)* organized pre-eminently within the theoretical framework of spatial history, aiming to highlight the role of space in the assessment of historical phenomena.

Figure 7. Simón Ruiz's business geography: a "space-movement"

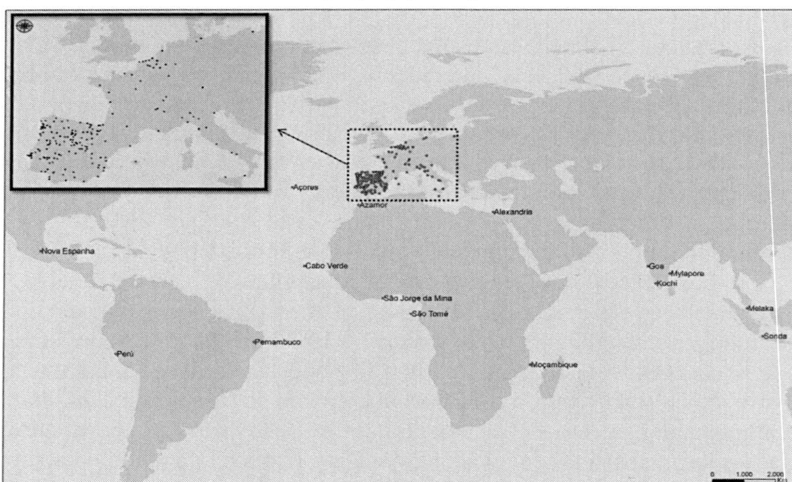

"The merchant's catchment area was a section of one or several national territories at any given period. In a period of growth, the merchant's trading surface might quickly expand, especially if he had access to the big business of the time – bills of exchange, currency, precious metals, 'royal merchandise' (like pepper, spices and silk)." (Braudel, 1983: 186).

The map in Figure 5, showing all locations mentioned in the company's documentation, perfectly illustrates Braudel's assumption. In fact, even if we did not know the kind of business that implies such geography, or even its volume, its world dimension illustrates by itself the importance of this trade company, including the way it is incorporated into the global economy of the 16th century. The geographic locations are scattered worldwide, with prominence given to the three economic spaces that early modern economic historiography has defined as the most important: the Atlantic, the Mediterranean, and overseas European settlements. In Europe, the map shows the most important fairs: Poligny, Namur, Mons, Besançon, Lyon, Cologne, Antwerp, and Piacenza. The bills of exchange flow among these financial centres, as a result of the connections of Simón Ruiz with foreign commercial houses. We can also observe the Italian cities of Salerno, Naples, Rome, Pisa, and Milan, testifying to the business relations established with the powerful Italian bankers.

The commercial correspondence completes this geography with references to the colonial space. Its analysis reveals how the capital flows were deeply engaged, and moreover were nourished, by the overseas trade. This is particularly important considering that Simón Ruiz's company operated within the Iberian empires, having privileged access to both Spanish and Portuguese colonies. That explains the presence of the Azores, the Americas with Mexico, Brazil and Peru; Africa with Cape Verde, Mina, São Tomé and Mozambique; and Asia, with India and Southeast Asia. Moreover, the commercial correspondence also attests to the European maritime connections with overseas territories – from seaports such as Dantzig, Hamburg, Dunkerque, Le Havre, La Rochelle, Cádiz, Barcelona, or Leghorn.

We can also identify a dense cluster of places in the Iberian peninsula, especially around Castile. This gathering is explained not only by the flows of bills of exchange between Iberian cities, but particularly by the fact that many of the agents mentioned in the bills – payers, beneficiaries, or even givers – are located here. To comprehend this, we must realize that the value expressed in a bill of exchange issued, for instance, in Antwerp and paid in Lyon, can in fact be a debt of a Castilian merchant. This allows us to deeply understand the origin of the money and credit circulation in the main European centres.

Figure 8 presents the ten places most often referred in the data sources, the big nodes of the Simón Ruiz spatial network. The timeline of their appearance in the data source reveals that nearly all of them, except for Florence and Rome, are present from the beginning of the company's activities. Moreover, these places, although suffering variations over time in the density or typology of their functions, continued to perform some kind of role during the fifty years of the company's operation. One of the strongest relations was established between Medina del Campo and Lisbon. Although very fragile in the first years, this connection becomes, especially in the 1570s, the main power source of the company's business. The associations between Simón Ruiz and the Portuguese merchants evolved from mercantile partnerships to limited companies for financial business, e.g. the *asientos*.

Figure 8. Functions performed by the most *(I think)* important places

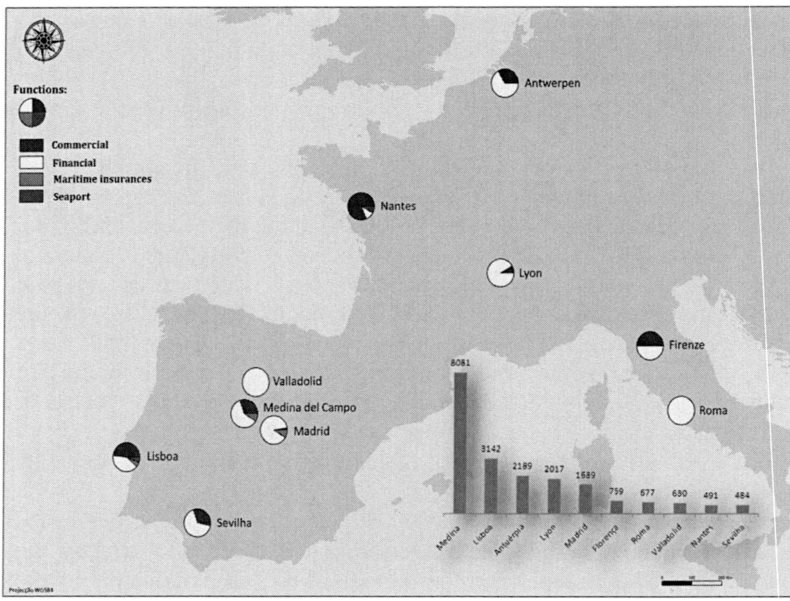

Financially, Lisbon is the payment location of many of the bills of exchange issued in Medina. Many of them represent payment for merchandise, but they also result from interest games played by bankers with the rates of exchange. Lisbon frequently suffered from the irregularities of the financial market, balancing between money overflows (*largueza*) and periods of metal scarcity (*estreiteza*). This situation frequently blocked business opportunities, and the complaints of Portuguese correspondents are frequent.[22] For example, in January 1575, Hernando de Morales complains about the one-year delay of the India and Mina armadas, and warned Lisbon of dire times if the gold from Mina did not arrive soon.[23]

[22] AHPV, ASR-020.1573.215 (29 December 1573).
[23] AHPV, ASR-028.1575.047 (14 January 1575).

However, reciprocity plays an important role in this relationship. Simón Ruiz represents, to the Portuguese merchants, the opportunity to take part in the world of high finance, namely in the *asientos*. In fact, from the 1580s onwards, Simón Ruiz starts to be truly harassed by Portuguese merchants offering their money to form short-term partnerships.

In this geography, a third place performs an important role in the dynamics of the goods and capital circuits – Antwerp. Here, the most important merchant families have their agents, like the Bonvisi, Spinola, Balbani, Maluenda, or Ximenes. The role of Lyon is also defined by the flow of bills of exchange. However, although Lisbon and Antwerp were the main centres where the bills were issued, Lyon was the place where they were sent to be paid. What we believe is that the Lyon fairs were mainly used as a middle place to assist the circulation of credit, benefiting from the direct correspondence with the Medina fairs. This means that Lyon allowed merchants to respond to the payment of a bill of exchange by issuing a new bill, maintaining the credit flow. In these circuits, the main agents were the Bonvisi family members, which frequently acted as Andrés Ruiz's commissionaires. Madrid appears in this scenario during the 1570s. After 1566 the Genoese *asientistas* arrived in Madrid, financing the crown and receiving *licenças de saca* (royal authorizations for exporting coin). This is the time of the creation of Madrid as political capital. By this time, another place assumed its role in the company's activities – Florence, establishing close connections with Medina and Madrid. Its role was essentially financial and related to the issuing of bills of exchange. In this Italian city there were no relevant relations with the Portuguese merchants who could receive the payments of their pepper cargoes sent to Italy through Medina. Later on, Rome takes its place in this geography. Bills of exchange were issued there from 1574, and testify to the financing of the church institutions, especially concerning payment of rents in the Spanish kingdom. Simón Ruiz also acts as a middle man between Rome and Lisbon, transferring funds.

After Medina, the place that receives more bills of exchange from Rome is Valladolid. The relation between these places, both privileged markets for merchant-bankers, proves the creditor action of Simón Ruiz. Valladolid, where Vitores Ruiz, Simón´s brother, operated and where Simón lived after 1580, was the place of the power, the bureaucracy, the elite, the people that supplied the market of *juros* (interests) and *censos* (rents). An individual identification of all the payers reveals that Valladolid presents the higher results for agents with honorary titles and important roles in administration and justice. It even overtakes Medina del Campo, which has the biggest number of identified agents. The presence of ecclesiastics seems also relevant, especially in examples that also include Rome. In fact, the majority of the bills of exchange paid in Valladolid were sent from Rome. They refer to loans made by Simón Ruiz's correspondents in this Italian city. As to the management of the company, the most important place, almost equal to Medina, is Nantes. Despite the geography provided by the bills of exchange, Nantes is almost irrelevant in the commercial correspondence. But it appears as the main centre in the former data source, managing the goods that circulate between the north (Antwerp) and the south (Seville). From the very beginning of his activities, Simón Ruiz benefits from the advantageous position, in Nantes, of his brother Andrés Ruiz who managed to have a royal privilege to trade between Nantes and Castile. To the south, in the Mediterranean, Seville was an open gate to the New World. Being an extremely unstable market, Seville required constant supervision and control. Here, the bankers played with their clients' funds which, together with the delays of the fleets, caused numerous bankruptcies. In 1567, after severe bank crises, Simón Ruiz was advised by his partners to abandon business in Seville.

Figure 9. Credit flows (1553–1606)

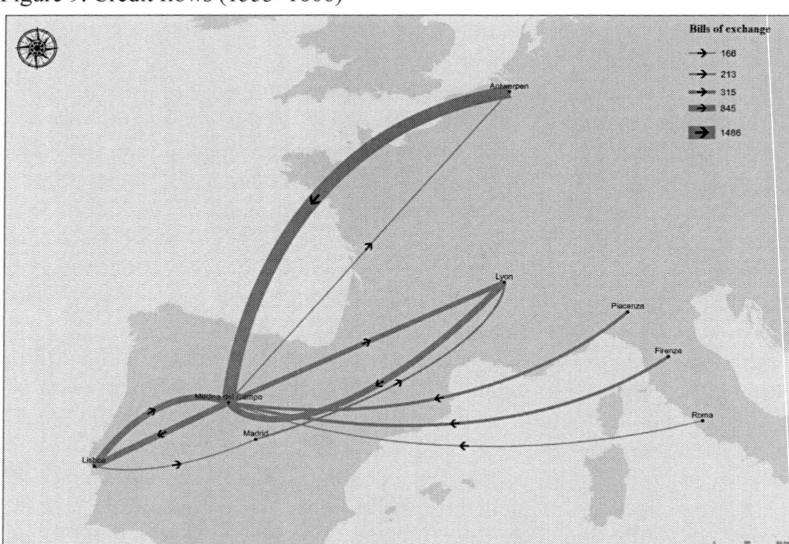

In the early 16th century an international banking system had emerged which linked the Spanish monarchy with the main European cities. This system was supported by the involvement of private agents, merchants, and bankers, organized in networks, in which private and public interests were merged (Alonso García, 2006: 98–101). Until 1551, German bankers, namely the Fugger, were the main creditors of Charles V. However, during the following years, Genoese businessmen reinforced their position, being responsible for half of the financial contracts made between 1552 and 1556 (Carlos Morales, 2006: 330).

By this time, Castile was the main stage of the international financial scene, nourished by the treasures that continually arrived in Seville. The businessmen established here, especially Genoese, had as their primary goal to obtain the precious metal and export it. Even before the 1566 royal permissions to export money (the *licenças de saca*), the Genoese used a mercantile strategy, shipping high value goods and transferring the profits. These were golden times for the Castilian fairs – Medina del Campo, Rioseco and Villalón – and the time of growth of Burgos, Valladolid, Segóvia, and Seville (Ruiz Martin, 1970: 16–17).

The business activities of the Simón Ruiz company developed against this background, so that the first partnerships were mainly mercantile. In fact, until 1566, the main credit circuits were those connecting the Castilian fairs to Lyon and Seville. The inside circuit, Medina-Seville, had operated regularly until the Sevillian market showed its fragilities. Lyon had also assumed the role performed until then by Besançon, and was the middle place between Medina and Nantes (Lapeyre, 1953: 22). As we have seen, the concession of the *licenças de saca* to the Genoese caused structural changes, converting the Italians into specialized bankers gathered in the new capital of the kingdom, Madrid. A new financial centre arose and became the big promoter of the *asientos*. At this time, most *asientos* served to provide Flanders with money funds. This task was particularly difficult from 1568, because of the privateering menaces, especially by the English and then the Dutch. Therefore, apart from some periods of financial liquidity following the arrival of American treasure, the Spanish crown was repeatedly forced to appeal to the *asientistas* (Lapeyre, 1953: 15–16).

From 1575, the connection between Antwerp and Medina intensified. Actually, the connections between Antwerp and the Castilian fairs had always been very close. On the one hand, as the balance of trade remained in favour of Spain, Antwerp merchants were forced to send bills of exchange to Castilian fairs. On the other hand, the prominent Portuguese merchant group settled in Flanders regularly used this connection. As a consequence, bills of exchange circulating directly between Lisbon and Antwerp were very rare, and transfers of funds between Portugal and

Castile were very common. This meant that the axis formed by these three centres – Lisbon, Medina, and Antwerp – was the most vital in the geography of the Simón Ruiz company. This becomes especially true following the 1580s, when credit activity became the power source of the company. Through partnerships involving Portuguese merchants, Simón Ruiz acted as creditor of the Spanish crown. Often he would not even participate directly in the *asientos*, acting instead as a fund raiser collecting money to provide the large amounts necessary to supply the crown. The returns of these loans, paid in the Genoese exchange fairs, or in the cities of Rome and Florence, were delivered to Simón Ruiz, increased by interest and capitalizing the merchant banker (Ruiz Martin, 1986: 294).

After 1595 the times were hard for the Medina fairs, which suffered several interruptions. For example, news of the Indian fleet's delay in October 1594 caused disruptions in the following May fairs. This began a time of scarcity in Castile compromising the payments to the army fighting in Flanders. Once more, the Spanish king appealed to the bankers in Madrid (Silva, 1956: 31–40). In 1596, failure to repay their loans resulted in a new disruption in the fairs. The last decades of the 16th century were harsh for the businessmen. The decreased supply of precious metal, the disturbances in the manufacture of some colonial products such as sugar, the disruption of the exchange fairs system, the closing of many commercial houses, all together blocked the credit market (Silva, 1956: 5–9). By this time, the company itself had suffered a setback with the death of Simón Ruiz, in 1597. To be correct, the management of the company had been in the hands of Simón's nephew, Cosme Ruiz Embito. His action was mainly based on the partnerships established with the Portuguese *asientistas*, such as the black slave merchant Pedro Gomes Reinel. In 1604, Cosme committed himself to a payment to Reinel of 300,000 ducats, and the next year he offered to provide galleys for Philip III. At the March fair in 1606, his bills of exchange were all refused, which caused the company to go bankrupt (Lorenzo Sanz, 1986b: 420).

Figure 10. Commercial flows: places selling goods

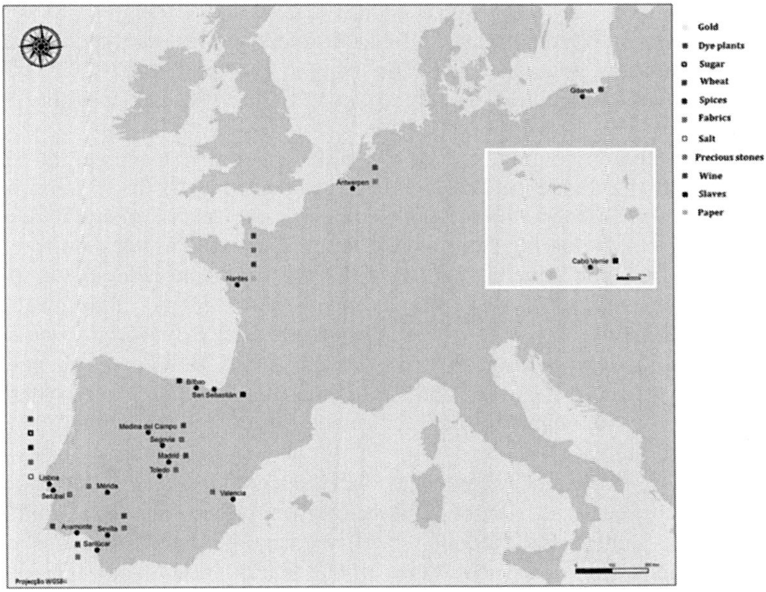

Figure 11. Commercial flows: places buying goods

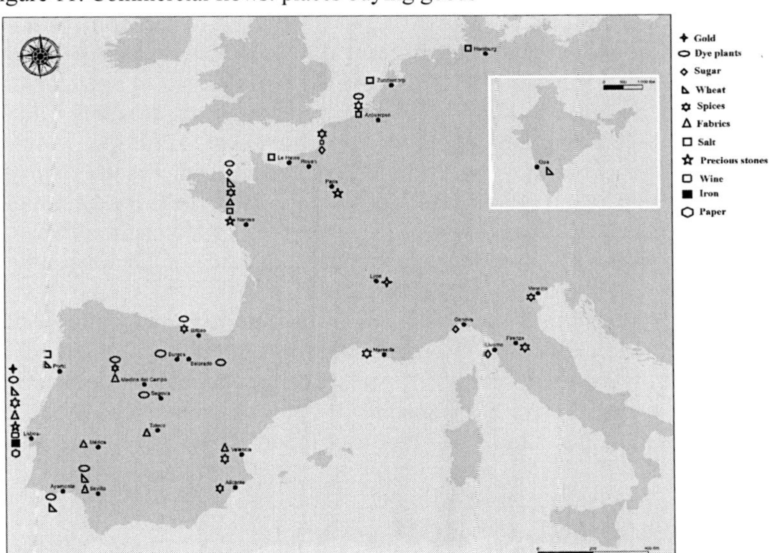

The commercial correspondence exchanged between the Portuguese agents and Simón Ruiz allows us to identify goods flows and trade routes, and to obtain some qualitative information concerning the mercantile activity, like prices and capacities. The maps in Figures 8 and 9 represent the places mentioned in the data source concerning the buying or selling of goods.

From its very beginning, one of the most important businesses of the company was the textiles trade. This had two features: the distribution of the French and Spanish fabrics and supplying the textile manufactures with the dye plants coming from the overseas. The main circuit connected Nantes, Medina, and Lisbon. Two routes were used: the maritime route linking the seaports of Nantes, Bilbao, and Lisbon; and the inland route, passing through the dry port of Badajoz. Both served important textile markets, among them Segóvia, Toledo, Valência, and Medina del Campo. Other locations such as Mérida and Ayamonte were used as places where the Portuguese merchants exchanged their cargoes of dye plants for manufactured fabrics.

In equally extreme demand was salt especially that produced in Setúbal, Alcochete, and Aveiro. The bigger importers were the French seaports like Nantes and Rouen, where the Portuguese merchants sent regular cargoes. Also very common in the letters were the orders for cereals, particularly wheat, for which there was always a demand in the Portuguese market. The main suppliers were France, Flanders, and Germany. For example, in March 1567, António Gomes sold Simón Ruiz's wheat in Lisbon and even rationed it to increase prices. Not being warned by Andrés Ruiz in Nantes, he was deeply affected by the sudden arrival of Breton boats loaded with wheat.[24] The high price of this cereal allowed it to be the bargaining chip for the pepper sent to Europe. In both markets – salt and wheat – the Portuguese suffered from French rivalry. The cereals arriving in Lisbon were sent to other Portuguese markets such as Porto, and even overseas, as in 1571, when a cargo was sent to Goa.[25] Regarding suppliers we also identify a purchase of wheat in Dantzig in the Baltic.

In their letters, the Portuguese merchants reported the coming and going of the India fleets, and informed about its sightings in the Azores, or, in bad scenarios, their delays or losses. Correspondents also informed Simón Ruiz about the price of spices like pepper, cinnamon, or even ginger, referred to as "the new drug from São Tomé".[26] The pepper

[24] AHPV, ASR-006.1567.94 (22 March 1567).
[25] AHPV, ASR-015.1571.27 (16 April 1571).
[26] AHPV, ASR-028.1575.47 (14 January 1575).

business was, however, restricted to some commercial groups, into which Simón Ruiz agents were eager to gain entry. In 1574, one of them reported a plentiful amount of pepper arriving in Lisbon, which had, however, already been chartered. He also complained that some spices, such as cinnamon, could only be bought for cash, so that he failed to do any business for the upcoming fair.[27] Another problem in this trade was its high transportation risk which led to a rise in insurance prices. In 1570, the Turkish threat put an end to the Portuguese merchants' attempts to send pepper to France.[28] However, the French market was always clamouring for this commodity and in 1575 the Portuguese agents successfully sent a cargo, using the hulks (*urcas*) usually employed in the salt trade.[29] Other consumer markets were Antwerp and the Italian cities. Although we found a cargo sent to Venice, in Italy the Portuguese faced strong competition from the Alexandria route. The partnerships between Simón Ruiz and the Portuguese were also an opportunity for the Castilians to gain access to other highly valuable commodities, among them silk, precious stones, sugar, and even black slaves. The connections between key places like Lisbon, Antwerp, or Medina, and the financial and mercantile routes, were strongly supported by another, secondary but equally essential, set of places. These included the seaports, the exchange fairs, or even the inland itineraries. They functioned as network systems promoting the circulation of money, goods, and men.

In Figure 12 we can see the most important exchange fairs where bills were sent. Medina del Campo exchange fairs operated twice a year, once in May, and again in October, each lasting 50 days. In the 16th century these fairs had connections with the ones held in Medina de Rioseco at Easter and in August, and with that of Villalón in Lent. The bills of exchange issued in the Medina May fair could also be paid during the September fair in Antwerp. There was also a correlation between the payments of August in Rioseco, and the Flemish Christmas fair; as also between the October fair in Medina and the Resurrection fair (Rodríguez González, 1995: 21). By the end of the 15th century, Medina del Campo fairs already had a privileged position among the Castilian fairs. During the 16th century their importance increased, and they absorbed the other Castilian fairs that became restricted to the commodities trade (Ruiz Martin, 1986: 290). During the reign of Charles V, the Medina fairs underwent a structural change, reinforcing their influence on state and

---

27 AHPV, ASR-023.1574.215 (25 November 1574).

28 AHPV, ASR-012.1570.238 (1 August 1570).

29 AHPV, ARS-028.1575.29 (30 October 1575).

public funding (Rodriguez Gonzalez, 1995: 23). These exchange fairs provided for the merger between private money and state.

Figure 12. Exchange fairs

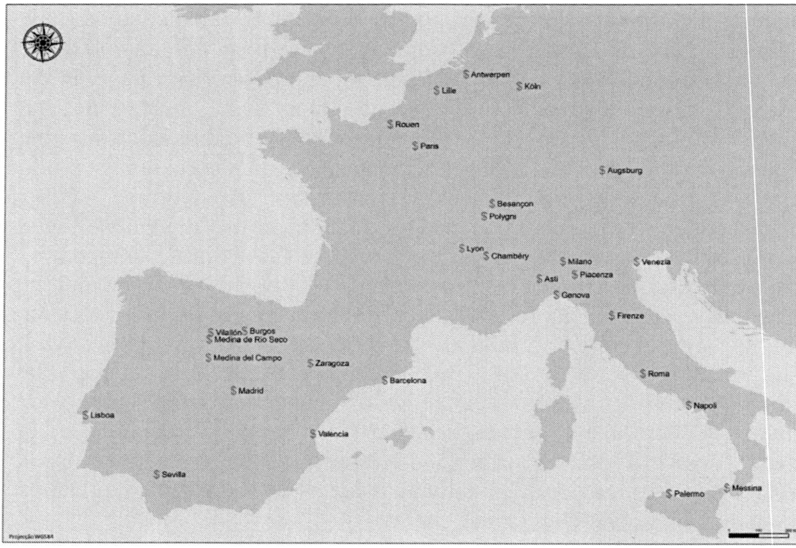

The Genoese fairs reached their peak in the very early 17th century, at which time all the large international transfers were conducted in Piaceza. Here, the Genoese bankers raised funds to supply the vast loans demanded by the Spanish crown, and appealed to various investors to diminish the associated risk. In fact, the volume of transactions produced by the state public debt, as well as by the main Genoese and Florentine groups' financial speculation, exceeded that produced by the international commercial trade (Marsilio, 2009).

In Figure 13 we spotted all the seaports mentioned in the data source in a portolan (a chart where seaports are represented).[30] Those seaports were mentioned in three different contexts: trade routes, privateering activity, or boat location. In the portolan we can observe the strong presence of the ports that assembled the two main movement spaces in the modern ages: the North Atlantic and the Mediterranean.

---

[30] Extract from the portolan of Joan Rizo Oliva, [1555]. *Yale University. General Collection, Beinecke Rare Book and Manuscript Library.*

Figure 13. Seaports

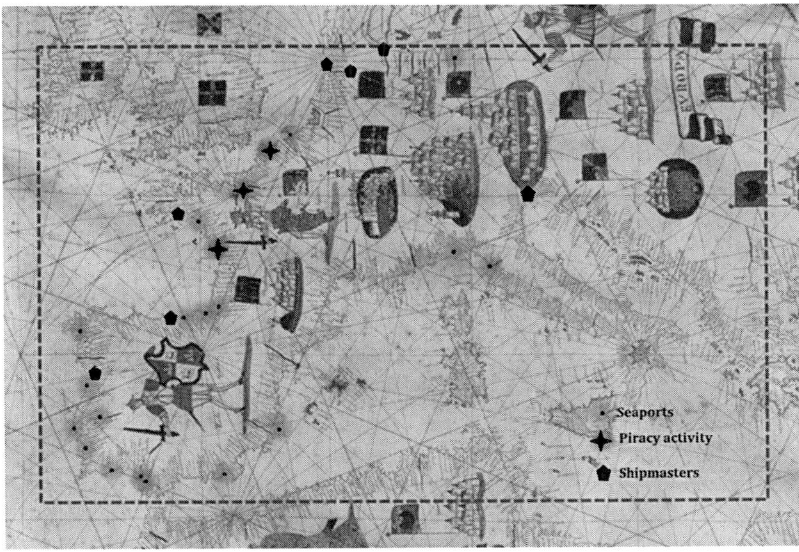

Among the various fields of historical research, port history has contributed most to the development of concepts such as complementarity, hierarchies, systems, and networking (Polonia, 2010b, 2011), as well as to the use of the time feature as an analytical tool. In fact, nowadays even the works regarding a small port or a specific maritime community explore its integration and context against a larger and profound regional, national, or even global background (Mauro, 1997; Jackson, 2001). Accordingly, the research about the Spanish empire and its overseas colonies acknowledges the Atlantic as a networked system whose nodal points were the seaports. In this perspective, the port cities represented the connections for the unification of a maritime-based "spatial economy" (Crespo Solana, 2012).

The time period we are discussing, the second half of the 16th century, is especially rich in political and religious conflicts, strains, and tensions that often materialized in the open seas, particularly through privateering. The Turkish menace, the Flemish wars, the religious wars in France, as well as the unceasing attacks between Spain and England, resulted in continuous international commercial blockades. This was the challenging scene in which Simón Ruiz developed his business activities, suffering many times from the failure of freight boats, the rise of maritime insurance, or even the capture of cargoes. The routes that we have described, after analysing the documentation produced by the Simón Ruiz company, overlap the maritime spaces considered in Michel Mollat's works (Jourdin, 1995). The European circuits discussed are those connecting the French coast to the Iberian Peninsula, and those linking the big ports in the north, such as Rouen and Nantes, with the ports in the south, such as Lisbon and Cadiz. However, this framework of commercial centres, the seaports, the privateering activities, and maritime insurance, reveals a larger geography that reaches out to the Baltic in the north, to the Mediterranean in the south, and overseas both in the east and the west. The search for commodities and markets drove the company to transcend the Atlantic circuit trade. Salt and wheat took Simón Ruiz's company to the Baltic, just as spices and sugar took it to the Mediterranean, to Alicante, Valencia, Marseilles, Genoa, and Leghorn.

In addition to drawing conclusions about business trends, or classifying the merchant activities of the company, the spatial framework outlined reveals a deep sense of opportunity. In fact, we observe that ships freighted were Spanish, French, or even Flemish, according merely to freight opportunities, ship availability, sea danger scenarios, etc. Even so, the merchant strategies did not depend on their attachment to a specific market or commodity, but, instead, on the use of the various markets, places, and conjunctures, with a single purpose: to capitalize the operations and investments of the company.

The commodity circuits were composed, not only of the origin and destination locations, but also of intermediate places such as custom houses and stopovers. Often located between the manufacturer and the consumer market, these places were the backup structures of the commercial flows. In Figure 14 we can see how the commodities, after arriving at the seaports, were conducted to the peninsula hinterland. Regarding these itineraries, the commercial correspondence has plenty of episodes of tax fraud and corruption, especially regarding customs officers.

Although we have outlined Simón Ruiz's business geography in a very large spatial range, with locations scattered worldwide, we must acknowledge that most of the action took place within a triangular spatial framework: Antwerp – Medina del Campo – Lisbon. We have also shown that the places more relevant to the activities of the company were those with a multifunctional profile and the ones that established more connections with other places. Another crucial feature was the complementarity between locations. This can be exemplified by the relation between Medina and Lisbon. Medina exchange fairs assisted and supported the capital flows between Lisbon and the Portuguese merchants settled in Antwerp. In our research work we comprehend how the spatial dynamics were a result of the interaction between the different functions of each place, as well as of the flows created and promoted by the spatial network. These dynamics were an important feature in the strategies of Simón Ruiz's business network.

Figure 14. Iberian Peninsula inland itineraries

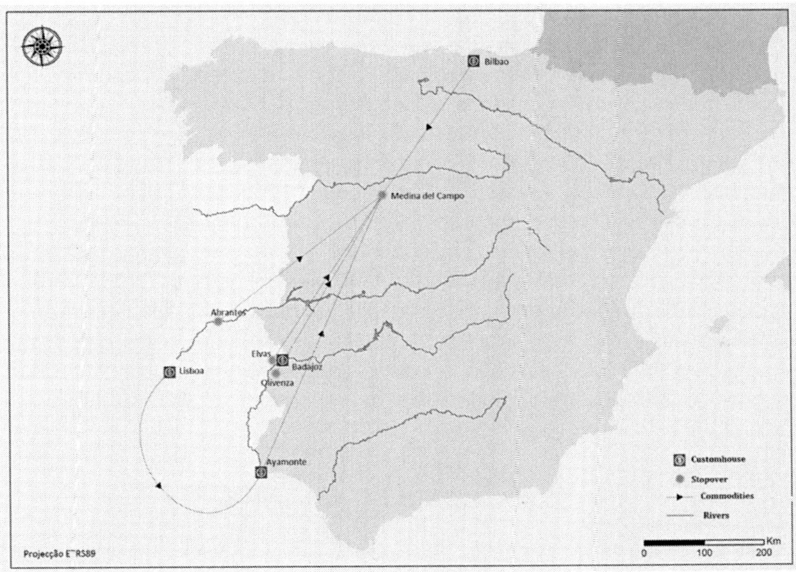

However, the operation of this system can only be verified in the framework of this network's interaction with other merchant networks. An accurate analysis of the Simón Ruiz business allows us to depict geography, its own geography, which may or may not be a path applying to other networks. The history and geography of Simón Ruiz's business also show how the conditions and variations suffered by the cities over time must be explained accordingly to flexibility, reorganization and complementarity phenomena, rather than failure, decay, or submission scenarios. E.g., Fernand Braudel interprets the redirection of the capital flows to Paris after 1570 as the end of the Lyon exchange fairs. However, studies by Henri Lapeyre and Françoise Bayard verified the vitality of the French fairs until the late 16th century. Similarly, Felipe Ruiz Martin attributed the decay of the Medina fairs to the rise of Madrid as the new financial centre. Still, as reported in our dissertation, Medina del Campo remained one of the main payment locations for bills of exchange throughout the 1580s, with some difficulties arising only in the 1590s. Even Seville resisted the banking crises around 1565, maintaining its financial centre status. In fact, Seville is the paradigm in the dichotomy of risk and opportunity that places represent to a mercantile company business strategy. It is remarkable how the risks and the restrictions presented to the trade activity are often simply surpassed and even ignored. We could easily point out privateering, war or high customs duties as mandatory reasons to exclude markets. These factors were undeniably considered by the merchants, but it can be shown that they did not overrule other considerations. The analysis of Simón Ruiz data source has also shown how the Atlantic trade between the French and Spanish coast was maintained by means of arrangements with the privateers. The connections between seaports and inland routes also allowed reorganizations of the shipments in order to avoid piracy risk (Pinto, 2012: 211–215). Finally, mercantile and financial flows were supported by a spatial network structured by central and border places. These border places include inland ports and seaports, resource centres, stopovers, custom houses – backup structures essential to the company activities. They also include cities like Rome and Florence that even if they did not constitute big nodes of capital flows, were crucial to the gathering of funds.

4. Summary and Conclusion

In conclusion, the functions performed by a specific place can be enhanced through its co-operation with another place, and by creating a relation of complementarity, resulting in an outcome to the functioning and success of the network. The spatial analysis and visualization of Simón Ruiz's spatial network turned out to be essential in order to understand how the placement of agents in specific locations and their mobility over time, how the building-up of places as economic, financial and political entities interacted with the dynamic evolution of a business company and contributed to the success or failure of a business network, and vice versa, as the use of Social Network Analysis (SNA) enabled us to understand how the network functioned when facing the challenges of an overall historical context. Using Social Network Analysis methodology historians are able to understand the action of individual agents within a certain structure which is much more than the sum of individual agents' choices and actions. In this case study, the way early modern merchants' relationships and organization were displayed illustrates strategies for making the exchange of goods, money, or information more efficient.

Even if early modern historiography has been attentive to the constitution of early modern trade networks, the social characteristics of its actors and preferential partner choice attributes,[31] it has focused far less on the mechanisms of connection and association of individuals from a complex perspective, in which the system is much more than the sum of individuals' actions. Unfortunately, the absence of similar case studies, based on the same methodologies and questions, for the same period, but also for previous periods, does not allow a comparative and evolutionary approach. Trends in business linking strategies help to quantify hierarchical characteristics of the individuals, as well as to identify key individuals in business circuits. It permits us to relativize an individual's position rather than just describe his or her actions. Understanding the internal organization of a merchant population can explain unexpected partner choice. In this case study, not only the social attributes of the agents are relevant for linking individuals. The possibility of increasing a person's social efficiency by being well connected was no less relevant than choices based on the acquaintance of someone's good reputation or wealth.

On the whole, the DynCoopNet-Pt experience gave us the opportunity of sharing historical epistemological focus and methodologies with other disciplines and sciences, as it happens with GIS and the spatial history approach. In terms of analytical focus our research went from the analysis of events and the weaving of individual biographies to the analysis of historical dynamics; from methodologies based on source criticism to those based on statistical calculations, categorizing and modelling, including mathematic modelling;[32] from monodisciplinarity to interdisciplinarity (or even transdisciplinarity). The DynCoopNet-Pt experience proved that crossing frontiers of knowledge is not only possible, but may be the only way to build new paths of knowledge, without artificial borders or unacceptable hierarchies. As historians, we learned that multidisciplinary and multimethodological approaches can improve the understanding of history, and that history can provide data and frameworks of analysis which in turn can present, by their complexity and awareness, challenges to other sciences, even those considered as hard sciences. Crossing frontiers is, in fact, not only possible, but essential to building a new paradigm of knowledge.

---

[31] Subrahmanyan, 1996; Curto, 2002; Priotti, 2004; Schulte-Beerbühl, 2007; Roitman, 2008; Antunes, 2009; Trivellato, 2009; Lamikiz, 2010.

[32] A contribution from Jorge Pacheco and the ATP Group has to be acknowledged as an important input to DynCoopNet-Pt. For more information on this team and its research see: http://www.ciul.ul.pt/~ATP/.

CHAPTER SEVEN

COMPLEXITY AND MERCHANT NETWORKS: EVIDENCES FROM EARLY MODERN CASTILE[1]

DAVID ALONSO GARCÍA[2]

1. Theoretical Introduction

The last few decades have seen an abundance of studies on merchant networks. It was these networks that formed the basis of commercial and financial expansion in the modern era. Cities, regions, countries, and even continents became connected in different ways through global trade, including the transmission of information and the development of means of payment between places that were at great distances from each other. Different scholars have shown that the end of the 15th century marked the beginning of a world economy as a first stage of global history (Mazlish, 2006: 16). Trading networks, in which a key factor was the informal element, became essential for promoting trust and establishing new contacts to reduce transaction costs.

Current literature on the mercantile world is full of terms like network, transnational communities, diaspora, and cross-cultural trade. This is nothing strange. Historians go to great lengths to recreate contact networks, identify actors (nodes) and ultimately define as precisely as possible the links through which money, information, and power flowed (Subramanyam, 1996). Therefore, in any evaluation of the mechanisms used to generate networks of trust, elements such as family, neighbourhood, and religion are always present. It would be no exaggeration to say that at the present time all serious studies use the concept of *network* in their analyses. Everything appears to derive from

---

[1] This publication has been sponsored by the Research Project: *Geografía Fiscal y Poder financiero en Castilla en un siglo de transición. Análisis mediante Sistemas de Información Geográfica* Ref. HAR2010-15168, MICINN, Spain.

[2] Universidad Complutense, Madrid.

the idea of close relationships, which include family members, neighbours, friends, and friends of friends.

With the proliferation of studies on trade networks there is a risk that the theoretical apparatus for explaining rather than simply describing them will become obscured. This is, first, because the use of the term *network* seems to justify any reality that includes "relations" between people. Perhaps a greater cause for concern, however, is that in *recreating* networks and describing them as mercantile, political, or familial, we are actually reflecting a partial vision of reality. In other words, if we intuited the existence of an interconnected world from elements that already existed in modern times, then focusing our attention on a single network or a specific group, in isolation from other parts of the system (or another part of a same piece of the network) could cause us to treat the part as the whole. Also, most analyses individualize the "merchant" with respect to other realities (the "politician" or the "noble") and this has an impact on the image of partiality frequently transmitted by literature on merchant groups.

Using Castile as an example, this study sets out to show that discussing networks is not as straightforward as simply discussing merchants. This is logical if we consider that in the modern era, there was no clear distinction between *public, professional,* and *private* (Chittolini, 1995). The family was therefore an extension of the company, and vice versa. The same could be said for neighbours, friends, acquaintances, and those who were close in a religious capacity, with whom, by the same criteria, relations were personal and socio-economic at the same time (Trivellato, 2009). We therefore propose an analysis from different perspectives, not only from those of economic groups but also from a spatial point of view based on the interrelation of different social spheres, in line with the idea of *cross-cultural trade* that was so brilliantly explained by Philip D. Curtin (Curtin, 1986). Thus, a merchant network may be regarded as part of a wider whole in which different spheres come into play. As a result, it is more than just an economic reality.

With respect to approaches to transnational networks, the theoretical contributions of the "sciences of complexity" hold particular attraction. J. B. Owens has defined modern times in terms of complex systems, in which *self-organized networks* and the mutual interrelation and influence between different places have special prominence (Owens, 2012). This proposal sounds frankly interesting in that it provides a theoretical framework for trying to make sense of the rise of networks ("emergence" in sciences of complexity terms) in the early Middle Ages and late modern eras. Indeed, historical development can be measured in terms of *self-*

organized networks. These formed a single reality with different sections, where geographical or social sectoralization could not be detached from the whole. Historiography faces a considerable challenge when it comes to demonstrating whether the sciences of complexity really offer a theoretical framework that would explain the properties and evolution of work networks. In this respect, the present study examines whether financial and commercial networks in Castile at the beginning of the modern era included any of the basic conditions for complex systems. To carry out the study, we have done a graph analysis with data gathered on almost 5000 individuals from different archives in Spain and Rome, analysed with GIS[3] and the Gephi graph analysis programme.[4] Graph analysis measures both their relations and their densities. Our database therefore includes different weightings depending on the greater or lesser proximity to an individual. This way we have tried to discriminate between strong and weak ties (Imízcoz, Oliveri Korta, 2010). The closest family members in the line of descendancy have been given more importance than more distant ones. The same goes for business associates: contact with those who are closer will be more intense and generally longer-lasting than with those who collaborate occasionally. Similarly, negative weightings have been assigned to relations resulting from enmity or conflict as these are also factors that lead to contact between people.

The sciences of complexity define a series of characteristics for non-linear systems which are otherwise fulfilled in a large number of phenomena found in nature and the universe, or in the international financial system itself (Johnson, 2007: 113–127). We intend to define here the principal characteristics of complex systems before moving on to analyse whether it is feasible to apply them to Castilian financial networks at the beginning of the modern era (García, 2006; Solé, 2009).

1. A non-linear system is defined as a reality where the sum of the parts is not equal to the whole. In other words, its elements cannot be studied in isolation, or at least without taking into account that they belong to a wider whole in which only interaction between the parts can give us the key to how the system operates. This approach presents historiography with two huge challenges: i) to map the vast amount of interpersonal relations making up the network, or at least attempt to delimit it in order to analyse its characteristics; ii) to establish research teams to integrate specific analyses of the parts comprising a non-linear system with analysis of the system as a whole, given the impossibility of a single researcher

[3] http://www.qgis.org/.

[4] https://gephi.org/. I wish to express my gratitude to S. F. Mesa for his help in drawing up the different analyses.

recreating and studying all the relations spread through the networks. This would make it possible to analyse networks from the perspectives of the sciences of complexity (R. García, 2006: 68).

2. A non-linear system is a representation of reality composed of multiple subsystems. It is therefore essential to define each subsystem (node) and see how it is integrated in the system as a whole. The first step, even before the subsystem is analysed, is the creation of types of subsystems since the networks can be observed from different angles. In our opinion, networks can be seen and examined from at least three analytical perspectives; *social, geographical,* and *political.* The first of these is related to agents and enables the most influential people (nodes) to be identified, together with their links to other individuals. This approach is probably the one most commonly used in current studies on merchant communities and social networks. A geographical approach would be based on the hypotheses of *Spatial History* (White, 2010), which enable studies to be carried out from a spatial perspective in order to determine the hierarchy of places. In this respect, it would be interesting to study the multi-faceted relations that were established between the many places found in the network, bearing in mind that without an analysis of these relations it would be difficult to understand the system as a whole. Finally, in third place, the political perspective refers to the many centres of power (including institutions) that were present in the formation and development of networks. Each centre of power or institution could be considered as a node that was related to other organizations at the same time, with networks forming between them. Current historiography persists in defining networks as social, mercantile, or political, even though they were all interrelated and had an influence on the system as a whole (Recio Morales, 2012). In other words, there was no emergence or expansion of merchant networks; this only applied to networks that involved different areas of power, which were susceptible to being identified as subsystems. In short, the networks would form a single entity but, almost like Rubik's cube, they could be approached from different angles, all of them complementary and able to contribute to a more complex understanding of the networks themselves.

3. The concept of *emergence*, which will not be dealt with in this study, constitutes an absolutely crucial element in the sciences of complexity. Emergence is related to the appearance of complex systems and the variations they experience. Such systems are not static and neither were commercial networks in the modern era, for obvious reasons. Non-linear systems are defined as being self-organized entities, where small changes and alterations do not affect the system as a whole. In contrast, a

variation or loss in one of the more important centres may well have consequences for the entire system. Mutations may be caused by *exogenous* or *endogenous* factors; the former is related to historical events, such as wars, discoveries and mass migration, while endogenous mutations refer to changes within the network, i.e., the conditions in which trust between agents is built up and maintained, variations in each of the geographical centres and changes in politico-institutional matters. Although this aspect is not fully explored in the present article, it would be interesting to identify which internal and external changes make transformations possible in the system as a whole, and which are absorbed without any variation in network morphology resulting solely from their action.

2. Contacts and Relations: At the Limits of the System

Some very clear conclusions can be drawn from the morphology of the network of links between Castilian agents at the beginning of the modern era. Most of the almost 5000 records comprising the database show some type of contact with other members, for example, personal relations through trade, politics or enmity. While it should be remembered that the database only shows a sample of the real situation, this does not preclude an evident conclusion: this was a sector that was closely interconnected. In fact, no less than 4000 relations were detected between individuals belonging to commercial, financial, and political spheres in Castile and other areas. This is a clear indication of the magnitude and importance of personal ties. At the same time, these are dispersed over the whole diagram, which is significant. Although there were people who accumulated more relations, these were distributed throughout the entire network.

A quick look at the previous network (which is still only partial) raises the question of whether we are dealing with a set of relations that would give shape to a non-linear system. It certainly appears to fulfil some of the characteristics: on the one hand, it was an open system presenting a large number of relations in lattice formation, which in practice gave rise to different formulas for collaboration between agents, both at fiscal and financial level (Alonso García, 2012) and at a strictly commercial level (Casado Alonso, 2012). The possibility of relying on people who were close and on other people close to their families made for easier access to information and reduced transaction costs. This was essential if we take various factors into consideration: the world was a very uncertain place where there was no regulated access to "official" or "institutional" information and where communications were much more dangerous than they are today. At the same time, numerous practices of a capitalist nature had already been established. The most significant of these were the use of new accounting techniques, the existence of coordinated exchange fairs and, in particular, a generalized use of bills of exchange (Casado Alonso, 1995). This all led to the consolidation of the networking practices that formed the basis of mercantile relations in that period. As a result, *trust* and *reputation* became the key elements of a merchant's potential success.

Figure 1. Castilian financial and political networks (1504–1525)

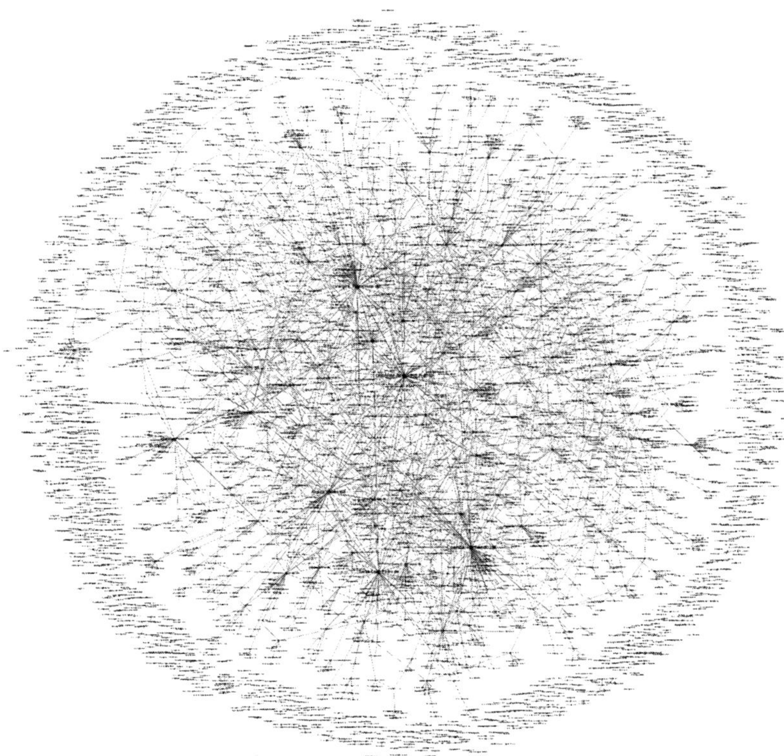

Castile was fully integrated in the economic circuits of the times. Moreover, the funding of the royal exchequer depended to a large extent on leases from local financiers, including the *obligados a guardas* who financed the army. Although present, international banking still had only a limited influence on Castilian finance in general at the beginning of the

16th century (Alonso García, 2007a). Consequently, it comes as no surprise that a whole system existed – or subsystem of a much wider network – that connected the different nodes. As many as 70 direct relations and collaborators have been detected by Erdös number for Pedro del Alcázar, governor of Seville and one of its main financiers (Alonso García, 2012). As collaborations were established, distances between different points of the network were shortened, an element that also characterizes complex systems, a good example being the Internet. Therefore, the object of study is the whole network, which cannot be explained in terms of the sciences of complexity if the aim is only to analyse each of its parts.

Another element that should be noted is the density of the different relations.

Figure 2. Density of relations

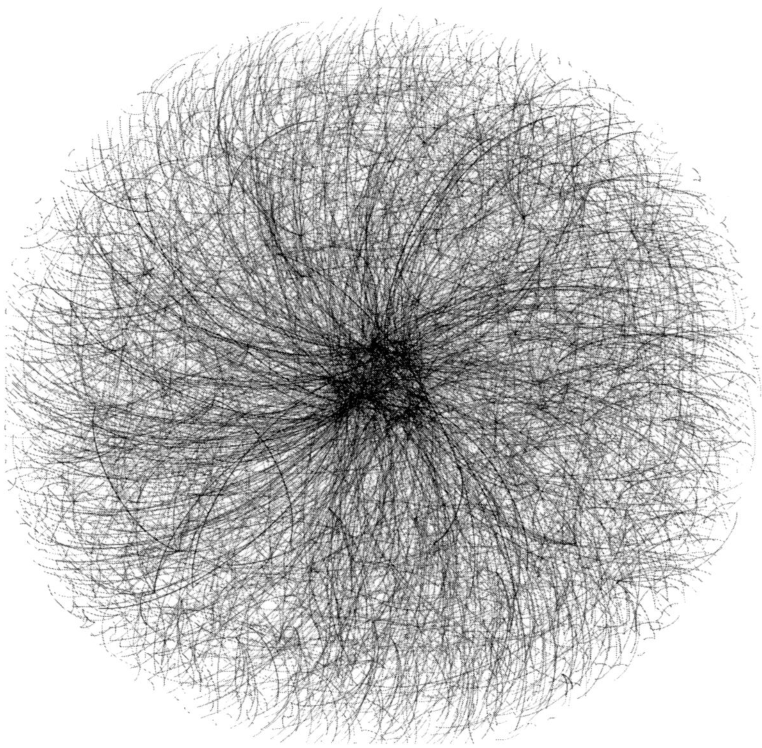

In Figure 2, two elements stand out with respect to the density of relations. First, most of the relations are grouped around the centre, where the main financiers of Castile were found. This would have an immediate consequence that may explain (this is only a hypothesis) the potential success of these agents: the greater the number of connections and the nearer the actors were to each other, the greater the possibility of having access to information. Moreover, such information was likely to be of a high quality, a factor that was just as important as the actual act of acquiring news and knowledge (Casado Alonso, 2008). Secondly, it is again clear that relations were present over the length and breadth of the whole system, not only at one of its centres. Although one part of the network had more branches and connections, the reality was that anyone involved was within the network to a greater or lesser extent

3. Visions of Networks: Agents, Places and Spaces

We shall now analyse the subsystems that made up the network of Castilian financiers at the beginning of the 16th century. Figure 1 shows the presence of modularity, in other words, certain nodes that accumulate a greater number of interconnections. Seen from this perspective, the basic element of analysis would be the agent, with a higher number of relations showing greater influence over the network as a whole. In non-linear systems it is essential to define the degree of modularity and the principal actors involved because any substantial variation affecting one of the centres of connection with most relations could cause the whole system to collapse, or at least undergo considerable changes. It is therefore appropriate to define a hierarchy of people and identify particularly those who occupied a higher position or who were more interconnected with other individuals.

Figure 3 shows some of the main Castilian financiers at the beginning of the 16th century, based on their familial, social, and commercial relations. The most prominent is Fernando de Cuenca, one of the best known financiers of the time. His decisive intervention in tax farming during 1519 gave him a subsequent importance that surpassed any fame that others in this particular world might have (Pérez, 1997: 135; Carretero Zamora, 1999), although the presence of Fernando de Cuenca as one of the main players in this auction would certainly have come as no surprise to his contemporaries as he had already become one of the most important players in the taxation business some years before.

Figure 3. Tag cloud (detail)

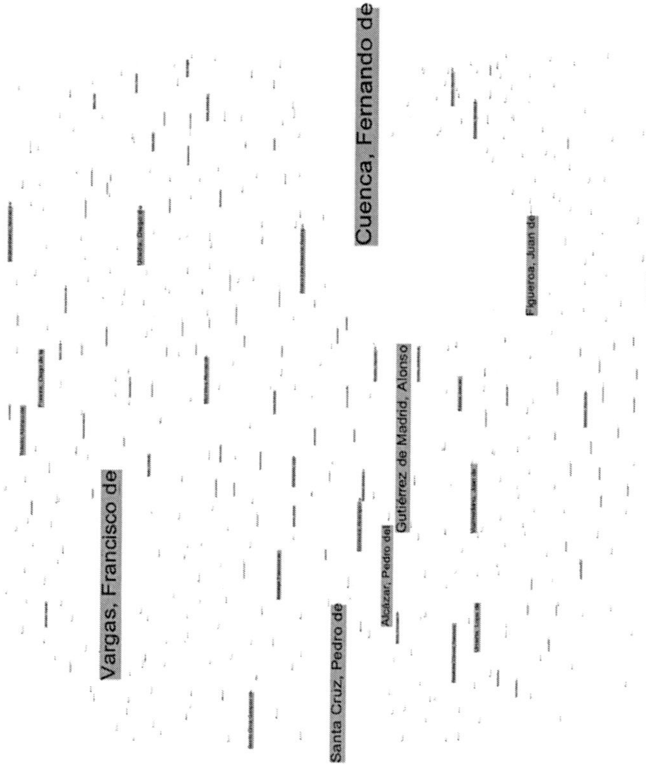

A glance at the list of guarantors he presented in 1520 will reveal some of the sources of his support (Table 1):

Being able to count on Don Pedro López de Ayala as a guarantor was something that was not within everyone's reach. It should be remembered that he belonged to one of the most prestigious factions in Toledo, which is why Cisneros would be more than satisfied to have him as a faithful servant (Giménez Fernández, 1984: 242). Diego Hurtado de Mendoza headed one of the factions in Cuenca, where he was governor (Diago Hernando, 1994). Doña Isabel of Bobadilla, the wife of Hurtado de Mendoza, also figured as guarantor in different lease agreements.[5] In other words, among the guarantors from Cuenca were some of the most

[5] AGS, EMR, legs. 131-2, 142, 144-1.

prominent members of the world of the Castilian oligarchy, although there is no doubt that the essence of this faction lay not only in the presence of the powerful but in the different geographical locations of its members, which tells us the extent of Fernando de Cuenca's personal relations. In this respect, two basic nuclei stand out: Aranda de Duero and the Cuenca area. Although the social class of the guarantors recruited in Aranda de Duero is not known, we do know that in this town Pedro de Salazar was known as a very rich and trustworthy man (*"muy más rico y abonado"*). His assets not only included properties, kilns, a vegetable garden, a store and a "very good cellar" but also a selection of tapestries and an arms collection.[6] We also know that Juan Sánchez de Quemada owned different houses, vineyards, arable land and a tavern.[7] Apart from Don Diego Hurtado de Mendoza, other prominent figures at the heart of Cuenca were the lawyer Carlos de Molina and the notary Miguel de Villanueva. The link between Fernando de Cuenca and Huete was supplied by Pedro Afán de Ribera and Francisco de Ribera. This was no ordinary contact as Afán de Ribera married Doña Isabel Sánchez, who was not only a guarantor but also Fernando de Cuenca's daughter.[8] The lawyer Luis de Soto from Valladolid owned properties in both the towns of Pisuerga and Santiago de Compostela.[9] Finally, the presence of the family component was inevitable in such a large organization: as well as the ties between Afán de Ribera and Fernando de Cuenca, Bernardo de Pisa, whose family was well known within the financial oligarchy of Almagro, was also Fernando de Cuenca's brother-in-law[10] (Parello, 1999).

Most of Fernando de Cuenca's financial business was linked to leases in Galicia and financing the army, by virtue of which he also controlled various *receptorías de encabezados* or tax collection offices. It is worth mentioning that his businesses did not date from before the turn of the century but emerged suddenly in 1508 when he took control of leasing and tax collection in Santiago de Compostela. This was the beginning of a period in which there were many years when sums in excess of ten, fifteen and up to twenty million *maravedíes* were brought in from activities related to leases and tax collection.

---

[6] AGS, EMR, leg. 169-1.
[7] AGS, EMR, leg. 164.
[8] AGS, EMR, leg. 168.
[9] AGS, EMR, leg. 169-1.
[10] AGS, EMR, leg. 164.

Table 1. Guarantors of Fernando de Cuenca (1520)

Name	**Neighbourhood**	**Amount of Surety Offered (in *Maravedíes*)**
Luis de Soto	Valladolid	500,000
Alonso de Valencia	Aranda de Duero	500,000
Don Martín de Acuña and doña Isabel de Avellaneda	Aranda de Duero	600,000
Pedro de Salazar	Aranda de Duero	900,000
Don Pedro López de Ayala, Count of Fuensalida		1,500,000
Juan del Pozo		
Pedro Afán de Ribera	Huete	6,000 *maravedíes de juro*
Bernardo de Pisa and doña Catalina Valle de Anaya	Santiago de Compostela	650,000
Sebastián de Bedoya	Calahorra	
Don Diego Hurtado de Mendoza		Towns of La Parrilla and Belmontejo
Rui López Coronel	Ágreda	
Carlos de Molina	Cuenca	
Antonio de Quemada	Aranda de Duero	
Juan Sánchez de Quemada	Aranda de Duero	
Francisco de Ribera	Huete	
Miguel de Villanueva	Cuenca	

Source: AGS, EMR, legs., 164 and 171

Another example of someone with close links to the commercial, financial and political circles of the kingdom was Pedro de Santa Cruz. This agent was the most outstanding of a group of *arrendadores de rentas* or tax farmers from Aranda de Duero (Diago Hernando, 1991a). Apart from links with his own relatives, in particular his son and partner Gaspar de Santa Cruz, his contacts included people from Segovia, Avila and Seville, as the following image shows:

Figure 4. Links of Pedro de Santa Cruz

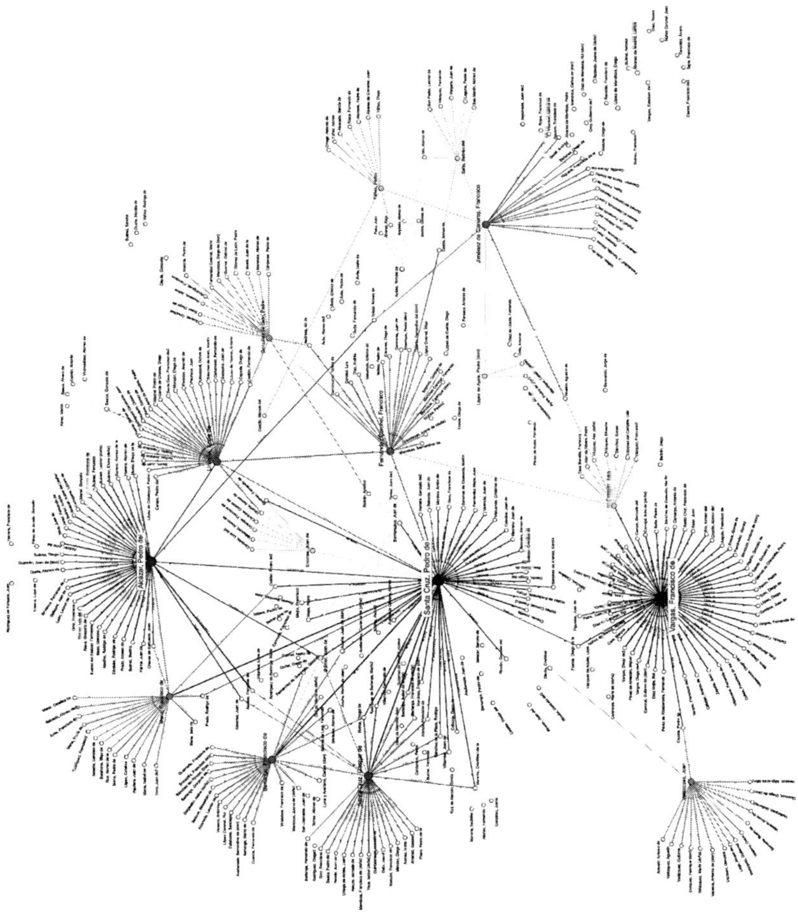

The chart shows three levels of relationships stemming from Pedro de Santa Cruz. The modularity of the system can be seen perfectly in the sense that there were some people who brought together a greater number of contacts. Those who stand out are Gaspar de Santa Cruz and the wealthy and influential Sevillian trader Pedro del Alcázar (Gil, 2000–2003, vol. III: 194–205). In close proximity to them, and with direct relations, is Francisco Fernández Coronel, the most important tax farmer in Castile during the second decade of the 16th century. It comes as no surprise that although Pedro de Santa Cruz lived in Aranda del Duero, he was able to farm taxes in areas as far away as the Canary Islands.

Each person who brought together the greatest number of relations can be considered in terms of a subsystem. These "subsystems", in other words, the most important people, did not only belong to the world of finance or trade. They also included notable figures like Francisco de Vargas, who was general treasurer of Castile until the defeat of the Castilian Communities (Carlos Morales, 2000), the chief accountant Juan Velázquez, and the exchequer Alonso Gutiérrez de Madrid. As can be seen, these people held great power and it was thanks to their contacts that the economic machinery of the monarchy was kept working (Alonso García, 2007a).

Figure 5. Network of Treasurer Francisco de Vargas' neighbours (on 2 relation levels)

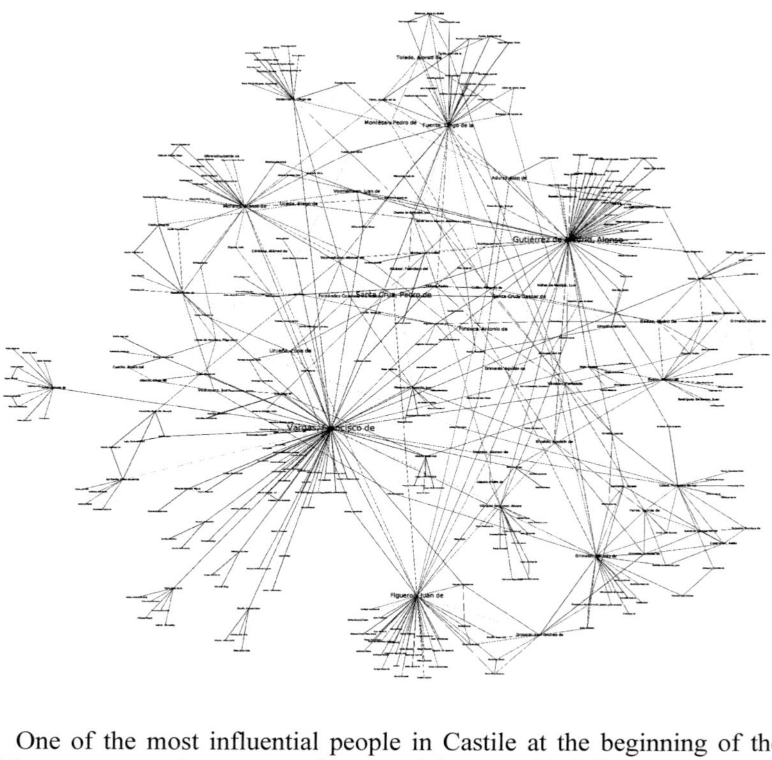

One of the most influential people in Castile at the beginning of the 16th century was the treasurer Vargas, right up to his fall from grace and death at the beginning of the third decade. His position as councillor and general treasurer, with business dealings with the army in Africa and relatives belonging to the Madrid oligarchy, meant that much of Castile's credit passed through his hands, particularly when external money transfers were involved. In close proximity to him, with first and second level relations, were prominent financiers like Juan de Figueroa, the Cazalla family, who were merchants involved in military administration (Alonso García, 2007b), the Baeza family and Pedro de Santa Cruz himself. Seen from the perspective of Francisco de Vargas, there would be a multitude of political relations on a par with economic ones, so that the financiers and merchants in these circles could be considered as a political network. Any changes in the set of relations they had in Vargas, the leading actor, had a considerable influence on the creation of the *Consejo de Hacienda* and reforms in General Treasury. In this sense, changes in the set of Vargas' links coincided with cities and international banks strengthening their positions in Castilian tax administration (Alonso García, 2007a: 332–344). For Vargas, this simply meant his fall from grace and, after that, his death.

It seems strange that in Vargas' network of relations his rival, Alonso Gutiérrez de Madrid, also appears. As from the end of the 15th century, Gutiérrez de Madrid had an impressive register of tax farming operations and was accompanied by other renowned financiers such as Fernán Núñez and Fernando de Villarreal (Ortego Rico, 2012). He was also governor of Toledo, formed part of the Madrid oligarchy – like Francisco de Vargas – and was accused of leading various movements to gain control of the main economic institutions of Castile. What is perhaps most surprising is that when he was appointed Receiver General of Castile in 1524, his access to credit was done through some of the agents who had also collaborated with the treasurer, Vargas. This happened in the case of Pedro Afán de Ribera, to whom he ceded an aldermanry in Seville. Ribera had collaborated with Juan Vázquez del Campillo, one of Francisco de Vargas' closest supporters (Alonso García, 2007a: 198). Something similar had happened at the height of the dynastic crises between 1504 and 1507, when Italian financiers could be found working for both Ferdinand the Catholic and Philip I at the same time. In terms of complex systems, this behaviour is quite normal. Because of the existence of so many connections throughout the system, networks show an extraordinary capacity for adapting to possible loss of nodes, even important ones, where some of their previous links become connected to other people. This way, the systems continue to work despite the loss of one of their parts. In the case of Castile, the fall from grace of Vargas implied the decline of some of the important financiers in his circle; in contrast, others who had also collaborated with him, continued to play a leading role, albeit in a different institutional environment.

Figure 6. Origin of tax farmers in different districts of Toledo (1509)

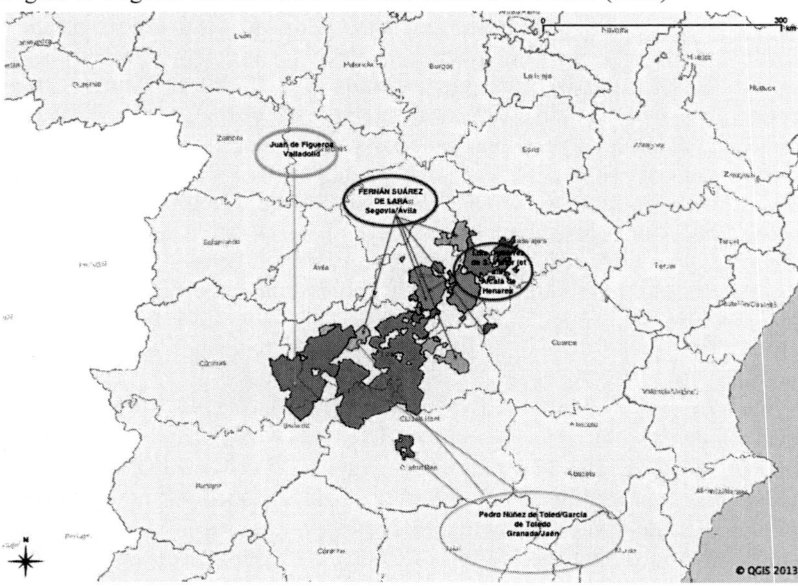

As understood in terms of complex systems, networks can also be studied using a geographical approach. Geographically-integrated history understands that every act or process takes place in a specific place, which in turn is related to the rest, so that any event in a specific space has – or is liable to have – an influence on other places. In this respect, the system created a hierarchy of spaces and cities that were also interrelated (Costa Pinto, 2013). Thus, from this perspective, the dynamic system becomes a network of centres with their own links; the network is the same but it is understood in a spatial sense, with nodes in the shape of a city rather than a person (who would ultimately be in a certain place at a certain time).

In the case of networks of financiers in Castile, this can be observed perfectly. Fiscal jurisdictions had a component that was both geographical and personal, understood as the relation between the Royal Exchequer and a specific tax farmer (Alonso García, 2011).This tax farmer would collaborate with others, so that spatial relations were established between places where they were in proximity. This way, for example, tax farming from different districts in the Toledo area in 1509 could be considered from the logic of people (of relations between them) at the same time as from a geographical point of view; in reality, the different financiers involved in managing it were concentrated in specific places or locations.

In Figure 6 two factors stand out. Firstly, different financiers reproduced groups or subsystems with a different location, and secondly, it was possible for agents in different parts of Castile to manage taxes even when the fiscal districts were at a distance from their places of origin or residence. So it is that the network we have previously seen expressed in terms of individual agents is susceptible to analysis in geographical terms.

Figure 7. Cities and Financial Networks (detail)

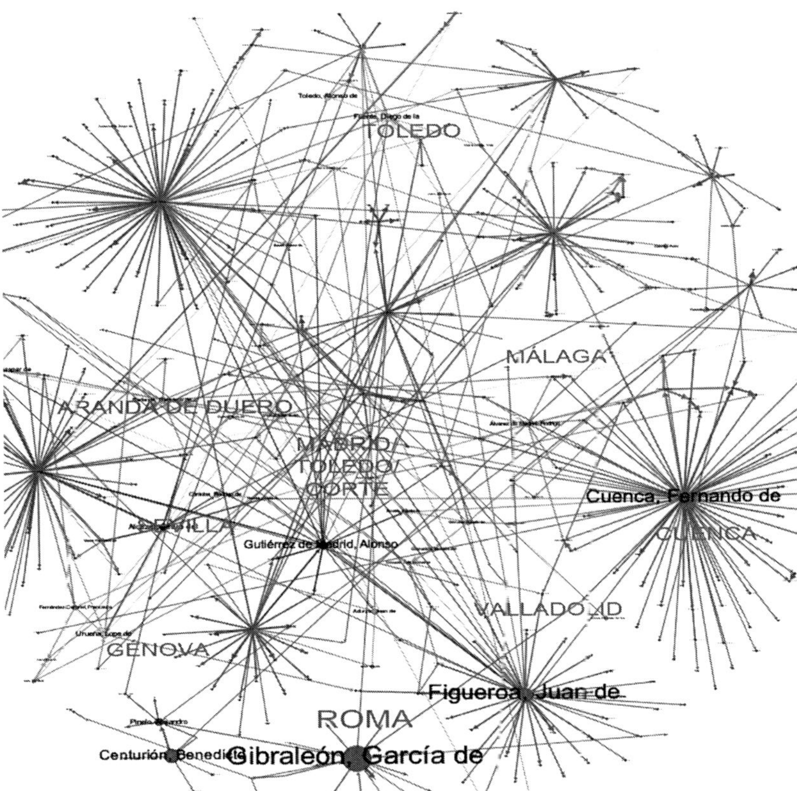

Also, from this perspective it is possible to detect subsystems that demonstrate the growing internationalization that networks were undergoing. It is well known that trade relations between the Iberian peninsula and other European countries had existed since the end of the Middle Ages (Casado Alonso, 1995; Marcos Martín, 2000). Thus, Vargas, the treasurer, and Alonso de Morales, his predecessor in the post, kept in contact with García de Gibraleón, one of the most influential Spaniards in Rome (Vaquero Piñeiro, 1999). Gibraleón, whose origins in Rome went back to the expulsion of the Jews in 1492, maintained contacts with his family in Seville and with others who, directly or indirectly, linked him to the Bank of Genoa, the general treasurers in Castile and Fernando de Cuenca. Despite making most of his fortune in Rome, Gibraleón was transferred to Seville Cathedral (Gil, 2000–2003, vol. II: 37–50).

The list of places that appear in the network is extensive. Note how this figure represents only a very small part of the total amount of relations that might develop between people from very different places and therefore between the places themselves. Sometimes these centres were at great distances, which, seen from this angle, acquired a *small world* perspective. In short, links between people were also links between different spaces in many locations, at the very moment the first steps in the First Global Age were being taken (Costa Pinto, 2013).

The interrelation of spaces should not only be understood in geographical terms. Another perspective for analysis, which complements the previous ones, is the one in which institutions or spheres of power were established as subsystems. Each institution maintained a considerable number of personal relations in the form of clienteles. People working in an organization also maintained links with agents who acted in other organizations. Thus, we find the treasurer Vargas and, in close proximity to him, Juan Velázquez de Cuéllar, one of two accountants in the Exchequer at the beginning of the 16th century (Diago Hernando, 1991b) and Cardinal Cisneros, regent of Castile between 1516 and 1517 (Alonso García, 2012). We also find lessors and traders who worked for nobles or aristocrats who, directly or indirectly, participated in or supported financial companies. It is impossible to ignore that part of García de Gibraleón's social success derived from his post as apostolic protonotary and the fact he was related to Pope Leo X. As a result, his personal relations were strongly influenced by work he carried out in papal circles. So it is that this potential dynamic system can be understood in terms of spaces of power (or institutions) as a complementary reality to actual relations between people. Thus, distances between the "economic" and the "political" or the "religious" would be considerably reduced, as part of the same scenario approached from different viewpoints. The Torres coat-of-arms can still be seen today at a fine palace in the Piazza Navona in Rome. This coat-of-arms belonged to Luis de Torres, first secretary to the Pope and later bishop of Salerno. Luis de Torres was the son of Fernando de Córdoba, a rich merchant from Murcia, who was well known in commercial and financial circles in Spain, thanks, among other things, to his dealings with the Royal Exchequer and Francisco de Vargas (López Beltrán, 2012: 71).

4. Conclusion

A common feature of dynamic systems is the presence of an *unstable equilibrium.* The system is always in motion with the rise and fall of people and the creation and demise of relations. It tends towards stability; however, as most changes of this type occur in sectors that are not significant enough to bring about a systemic change, instability, until this instability finds its own (unstable) harmony. In contrast, at other times changes may be so great or affect nodes of such importance that the system as a whole is profoundly altered and has the potential to disappear.

It is therefore a question of identifying changes that are capable of making a subsystem emerge or disappear, as well as the possible influence this may have on the system as a whole. Changes may be divided into two types. On the one hand are the changes known in terms of the sciences of complexity as changes in the surrounding conditions (García: 128), in other words, transformations external to the system or not directly produced within it, and to which it has to adapt. In our case, changes in the surrounding conditions may be related to historical events which, even if transitory, were highly significant in the early stages of the modern world. The most notable of these events were the arrival of Europeans in America, the existence of a favourable economic framework as from the second third of the 16th century, greater market integration at international level, thanks to fairs and bills of exchange and, in a Hispanic context, the expulsion of the Jews (we have seen the repercussions this had on the networks through the figure of García de Gibraleón), political instability in Castile between the death of Isabella I (1504) and the end of the Revolt of the 'Comuneros' (1522). All these elements have been extensively covered by historians and what perhaps remains is to study the influence they might have had on the financial networks of Castile at the beginning of the 16th century. The second type of change has an internal origin, which is precisely where historiography has been focused in recent years as concepts such as reputation; trust and family come under analysis (Subrahmanyam, 1996; Trivellato, 2009).

The sciences of complexity may provide an innovative interpretative framework for explaining the emergence of trade networks during the initial stages of globalization, thereby making sense of this phenomenon in accordance with properties used to explain many other realities in the world of natural, medical and social sciences. Complexity is all around us and is present in a multitude of everyday processes. It pervades the world and the universe and its principles are being applied in multiple fields of knowledge to explain very different dynamics. Non-linear complex systems have been defined as systems that are open, closely interrelated and self-organizing, with subsystems where the sum of the parts does not explain the system as a whole and with properties that explain their origin, stability and change. Applying this theoretical framework to historical research presents a real challenge that has not yet been confronted, but taking it on would be perfectly feasible with the help of new technologies (databases, Geographic Information Systems, statistics programs and graphical analysis.) and new forms of organizing research. Although the application of complexity to the field of history will be slow and difficult, it is a viable proposition that is potentially susceptible to generating a great change in the way we tackle historical research.

CHAPTER EIGHT

SPATIAL CONCENTRATION IN EARLY MODERN MARITIME TRADE AND TRANSPORT: TOWARDS A MODEL FOR ECONOMIC-HISTORICAL ANALYSIS USING GIS

WERNER SCHELTJENS[1]

1. Introduction

Among the many challenges that maritime-economic historians have to face when dealing with the international distribution of commodities in the early modern period, the problems arising from the questions "what goes where" and "who did what" have great impact on the visual representation of research results and on their potential use in comparative or time-series analyses.

The first question – what goes where? – is most commonly dealt with. Indeed, the primary focus of studies in early modern maritime history is on patterns of international trade, which are often reconstructed on the basis of tax registrations at port of departure, port of arrival or at intermediary points such as the Sound, the narrow strait between the North Sea and the Baltic, where from as early as 1429 the Danish king levied a tax on all ships passing by (Gøbel, 2010: 305–308). Often, the frequency of ship movements is used as a proxy for the volumes traded. In most cases, time series are generated to indicate the development of certain ports and the decline of others. Sometimes, more detailed accounts of the volumes of goods shipped are given. Cross-tables listing ports and calculations of the volumes transported are generated with relative ease. However, since weights and measures are not usually dealt with, a general assessment of

[1] University of Leipzig.

the volume of trade becomes almost impossible. The production of time series based on shipping frequencies comes at the cost of losing the spatial perspective. The production of crosstabs based on non-standardized volumes comes at the cost of losing the temporal perspective.

The second question – who did what? – is not commonly posed, even though several sources provide detailed information about individual maritime transporters, in addition to information about the goods they carried and the routes they sailed. Registered domiciles of shipmasters are sometimes used to provide some general information about individual shipmasters and their activities. Again, time series using frequency data and crosstabs based on non-standardized volumes are the most common forms of visualization when the question "who did what?" is discussed. Interestingly, the relation between a shipmaster and the routes he sailed is often treated in a descriptive manner, without underlying quantification. Perhaps, the complexity of dealing simultaneously with four variables – time, domiciles, transport routes and volumes shipped – can explain this. As a consequence, knowledge of early modern transport systems falls short of the advances made in the study of international trade, even though both have to face similar methodological problems.

This exploratory paper is devoted to the questions "what goes where?" and "who did what?" in early modern maritime trade and transport, but rather than separating them from each other, they are treated in connection. The goal of the paper is to put forth a number of technical and methodological procedures that may bring the development of a model for economic-historical analysis of early modern trade and transport data using GIS a step closer. In particular, the issue of spatial concentration and its visual representation will be discussed. Hence, changes in the spatial concentration of the distribution of goods (what goes where?) and the distribution of services (who did what?) are indicative of fundamental changes in the patterns of international trade and the structure of transport systems during the early modern period. Moreover, answers to these key questions may provide valuable insights into changes in regional economic structures, the emergence of regional services economies and the role of specialization in the economic exchange of goods and services. Therefore, it seems worthwhile to discuss what a potential method for analyzing changes in spatial concentration in early modern trade and transport might look like and what the contribution of GIS tools may be.

The paper is structured as follows. In the following section, I briefly discuss recent development in the use of GIS for the study of early modern trade and transport and I evaluate how advances may contribute to the development of a method for the study of changes in spatial concentration.

In section 3, I introduce the Danish Sound toll registers, which are the source used to support my argument, and I briefly describe the several steps that were taken to homogenize the raw Sound Toll data and prepare them for further use. Without going into detail, section 4 contains a short overview of the various steps necessary to convert the homogenized data from the Danish Sound toll registers into volumetric estimates measured in ton-km. It should be noted that the conversion results presented as examples in this paper are based on work in progress and have to be used with caution. In section 5, I explain how to derive data series for domiciles, routes and goods from the converted Sound Toll data and I use these data series to calculate market shares and market concentration indices for domiciles and routes in the distribution of goods and services. It will be discussed how these calculations contribute to the analysis of change in the spatial concentration of the distribution of goods and services. In section 6, I discuss how the results generated in section 5 may be adequately incorporated in historical GIS systems dealing with early modern trade and transport.

2. Historiography

In this section, I briefly discuss recent developments in the use of GIS for the study of early modern trade and transport and I evaluate how advances may contribute to the development of a method for the study of changes in spatial concentration. It should be noticed that recent historical research on early modern trade and transport has been influenced by two related developments. The first is the emergence of historical databases of maritime shipping; the second is the use of GIS software to visualize the contents of these databases.

A pioneering effort in the creation of historical databases of maritime shipping was executed by a research group headed by Pierre Chaunu in the early 1950s. His group of French historians made an attempt to compute the period from 1784 to 1793 of the Danish Sound toll registers. Two journal articles notwithstanding, the project has not resulted in the publication of an electronic database (Desfeuilles, 1959: 492–520; Desfeuilles, 1961: 56–74). In the 1970s, the Danish historian Hans Christian Johansen undertook a similar effort, which eventually led to the release of an electronic database of the Danish Sound toll registers for the years 1784 to 1795, also published on microfiche as an appendix to Johansen's dissertation (Johansen, 1983). This database has recently been converted to contemporary standards for historical databases by the Dutch historian George Welling,[2] who has also published a database of part of the so-called *Paalgeldregisters*, in which all ships arriving at the port of Amsterdam between 1771 and 1787 are recorded (Welling, 1998). The examples given here represent the earlier stage of the computation of large historical datasets. Despite the assistance of students in all above-mentioned projects, the creation of historical databases of maritime shipping was (and still is) mostly an individual endeavour, which explains the rather limited spatial and temporal scope of the results.[3]

Recently, a number of large-scale collaborative efforts have significantly advanced the availability of historical databases of maritime shipping and trade in the age of sail. Representative examples are the Slave Voyages database, Amphitrete, Navigocorpus, DynCoopNet and Sound Toll Registers Online (Eltis, 2012: 257–264; Scheltjens & Veluwenkamp, 2012: 301–330; Dedieu, Marzagalli, Pourchasse & Scheltjens, 2012: 331–359 and Crespo Solana & Alonso García, 2012). In conjunction with these collaborative database projects, an increased interest can be witnessed in data visualization techniques and in the issues involving the application of GIS in maritime-historical research in the early modern period. In most cases, the use of GIS does not go beyond the level of static map illustrations of frequency data to support descriptive analyses of changing patterns of trade and transport. Of course, there is nothing wrong with visualizing straightforward projections of frequencies to points on two-dimensional maps. Visual representations of early modern maritime-historical data using GIS do provide a much clearer view of what the data may tell us about the questions "what goes where" and "who did what" than what could be achieved with crosstabs or graphs. But the full potential of GIS is not exploited. Moreover, transport routes and trade routes are dissected and the use of points involves losing insight into their evolution in time and space. As a consequence, GIS tools do not bring us closer to identifying and interpreting fundamental changes in the patterns of international trade and the structure of transport systems during the early modern period.

While the issues involved with the use of "messy" historical data are generally acknowledged (Owens, 2012: 51), it is nevertheless beneficial to

---

[2] http://www.let.rug.nl/welling/sont/johansen.htm.

[3] Of course, databases are still very often created as parts of individual research projects (such as dissertation projects, for example). The limitations mentioned in the text – limited spatial and temporal scope – apply to most of these databases as well. Additionally, it must be noted databases prepared for individual research project often provide selections of the data provided by the original source. The formulation of specific research questions plays a decisive role in this regard.

pursue their use, especially when the goal is to incorporate space and time in historical GIS. Recent collaboration in the creation of large-scale electronic databases of early modern maritime shipping has devoted much attention to data preparation, standardization and conversion. Thanks to advances made in the standardization and interpretation of names of products, the geo-referencing of historical names of places and the standardization, interpretation and conversion of pre-modern weights and measures, it is now possible to homogenize early modern data sets better than before. Obviously, certain margins of error, loss of data due to standardization and simplification need to be taken into account. They do not outweigh the advantages offered by the homogenization of historical data sets, especially when the conversion of weights and measures to metric equivalents is pursued. The main benefit of such conversion is that the complex relationship between the distribution of goods (what goes where?) and services (who did what?) can now essentially be captured in the expression of "what?" as a ton-km estimate. Non-weighted frequencies such as the number of arrivals are no longer necessary to estimate spatial concentration and the disconnection of the distribution of goods from the distribution of services is no longer inevitable.

In the following section, I will briefly describe the main features of a conversion to ton-km estimates as executed using data from the Sound Toll Registers Online (further referred to as STRO). This description constitutes the basis for the further elaboration of my argument in sections 4, 5, and 6.

3. Data Preparation

In this section, STRO, which is the source used to support my argument, and I briefly describe the several steps that were taken to homogenize the raw data from STRO and prepare them for further use. The section contains an overview of the steps necessary to convert the raw data from STRO into volumetric estimates measured in ton-km. The conversion results used in this paper are based on work in progress and should be interpreted as preliminary, indicative results.

The STRO are the records of the toll levied by the king of Denmark on the passage of ships through the Sound, the strait between Denmark and Sweden connecting the North and Baltic Seas. The Sound Toll Registers Online project, or STRO, has been introduced at several conferences and in a number of publications. They are stored by the Danish National Archives. The more than 700 volumes of the STRO that have been preserved cover include a practically uninterrupted series from 1574 to

1857 and some scattered records for the period between 1497 and 1574. They hold information on about 1.8 million passages. For each individual passage, both westward and eastward, the STRO contain the passage date, the name of the shipmaster, his domicile, his port of departure and – from the mid-1660s – his port of destination, the composition of the cargo and the toll paid. STRO is a relational database set up to make the STRO data instantaneously available to all via www.soundtoll.nl. In many ways, STRO is the delayed continuation of Johansen's pioneering effort (Scheltjens & Veluwenkamp, 2012: 301–302).

The conversion method used in this paper is derived from previous work, in which a method was developed to estimate the volume of Dutch Baltic shipping at the end of the eighteenth century (Scheltjens, 2009: 74–102). Standardization of geographical data, product data and weights and measures data were necessary preparatory steps before the actual conversion of volumes expressed in pre-modern weights and measures could start. Standardized geographical data was provided by George Welling, who has created a georeferenced data set of all places mentioned in STRO, which includes both the original spelling variations, a standardized place name and additional geographical categories.[4] Some of these categories refer to historical regions; others refer to present-day provinces and countries. The larger geographical categories in Welling's thesaurus of place names correspond to the "small regions" and "big regions" elaborated earlier by Hans Christian Johansen.

Data items referring to products were standardized in the following way. First, the raw data about products in STRO, which sometimes takes the form of complex and detailed descriptions of cargo items, was simplified. All elements of the description were separated so as to obtain a cargo description divided among several data fields. For example, the cargo description "5 a 6 All. ord. deehler" (*Eng*. Ordinary deals with a size between 5 and 6 *alen*) was divided among 4 columns. "deehler" was put in the main data field; the other items – ord., 5 a 6, and All. – were put in data fields describing the additional features of the cargo item. As a result, the information provided in the main column corresponds to a general description of the cargo, excluding its additional features. Upon completion of this data manipulation process, spelling variations were eliminated from each data field so as to obtain a cargo description that is relatively easy to use.

The conversion procedure for weights and measures, as well as the procedures followed to estimate distances between ports have been

[4] The thesaurus of place names is incorporated in the online search engine of STRO. See www.soundtoll.nl.

described in more detailed fashion (Scheltjens, 2009: 74–102). The conversion procedure employed in this paper is somewhat simplified and the level of detail and variation that is taken into account during the conversion process is lower. Distances between ports, for example, were estimated more generally as distances between the regions to which these ports belong, which resulted in a significant reduction of the number of combinations that had to be computed. Similarly, rather than converting unique combinations of product, measure, and (port of) origin to their metric equivalent, as was the case in the previous method, product data was left out of the procedure. Again, this resulted in a significant reduction of the number of combinations that had to be computed. The consequence of these simplifications is that the degree of precision is somewhat lower than in the previous method. However, the time period that could be covered thanks to these simplifications is significantly bigger: with the previous method, 12 years of Sound Toll data from the Johansen database were converted to ton-km estimates; with the new method, a preliminary conversion for the entire period between 1670 and 1800 could be produced.

4. Aggregation

In the previous section it was suggested how the largely qualitative content of a complex and non-standardized data set for maritime shipping in the early modern period, namely: STRO, may be converted to ton-km estimates. One of the main assets of the conversion method described above is its inclusive structure. Even though the conversion of STRO data to ton-km estimates and their subsequent aggregation into annual series can only capture part of the information in the original registers, geographical and product data are not left out *a priori*. As we will see in sections 5 and 6, their incorporation in the aggregated data sets allows generating detailed indices and the production of maps, whereas abridged quantification formulas inevitably lead to loss of geographical detail and severe limitations to the potential use of the data.

In order to prepare the analysis of market concentration in the distribution of goods and services, the converted STRO data is aggregated to annual data and stored in one table with the following fields: year, domicile *d*, port of departure and port of destination, which together constitutes route *r*, good *g* and its volume (expressed in ton-km). This table serves as the basis for all further calculations.

Four basic time series – (1) *D*, or the total volume transported annually by shipmasters from domicile *d*, (2) *R*, or the total volume shipped annually on route *r*, (3) *G*, or the total volume of good *g* shipped annually through the Sound and (4) *T*, which indicates total volumes shipped annually – can be easily derived from the converted STRO table. Time series (1) to (4) will be referred to as *D*, *R*, *G*, and *T* respectively. The corresponding variables are domicile *d*, good *g* and route *r*. The unit of measurement is ton-km.

Six subseries can now be calculated, namely: (5) *Dg*, or the share of good *g* in the total volume transported annually by shipmasters from domicile *d*, (6) *Dr*, or the share of route *r* in the total volume transported annually by shipmasters from domicile *d*, (7) *Rg*, or the share of good *g* in the total volume shipped annually on route *r*, (8) *Rd*, or the share of domicile *d* in the total volume shipped annually on route *r*, (9) *Gd*, or the share of domicile *d* in the total annual volume of good *g*, and (10) *Gr*, or the share of route *r* in the total annual volume of good *g*. These shares can be expressed in absolute numbers, using the ton-km estimates found in the converted STRO data, or in percentages of the total volumes found in the basic time series.

The basic time series were derived from the general table, which contains all converted STRO data in aggregated annual data series. The six subseries were based on three preliminary select queries so as to structure and speed up the calculation process. The first query allows us to specify time and domicile, in the second time and route are specified and in the third time and product are preselected. The subseries *Dg* and *Dr* are based on the results of the first select query, the subseries *Rg* and *Rd* are based on the results of the second select query; and the subseries *Gr* and *Gd* are based on the results of the third select query. The structure of these select queries corresponds to that of the general table containing all converted STRO data in aggregated annual data series.

To sum up, the database now contains the following elements:

Table 1. Overview of the database structure derived from the table containing all converted STRO data

	General table containing all converted STRO data in aggregated annual data series				
	Basic Series		Subseries		
Tool:	Totals query		Select query based on the general table	Select query based on the general table	Select query based on the general table

Selection criteria:			Good	Domicile	Route
			Time	Time	Time
Resulting series	*T*				
	G		*Gd*		
			Gr		
	D			*Dg*	
				Dr	
	R				*Rg*
					Rd

In short, the following time series are derived from the general table containing all converted and aggregated data:

Table 2. Overview of basic time series and subseries derived from the table containing all converted STRO data

Time series	Description
T	Total overall volume shipped annually
G	Total volume of good *g* shipped annually
D	Total volume transported annually by shipmasters from domicile *d*
R	Total volume shipped annually on route *r*
Gd	Share of domicile *d* in the total annual volume of good *g*
Gr	Share of route *r* in the total annual volume of good *g*
Dg	Share of good *g* in the total volume transported annually by domicile *d*
Dr	Share of route *r* in the total volume transported annually by domicile *d*
Rg	Share of good *g* in the total volume shipped annually on route *r*
Rd	Share of domicile *d* in the total volume shipped annually on route *r*

5. Market Concentration

In this section, I discuss the application of ton-km estimates of shipping through the Sound, analysing spatial concentration in the distribution of goods and services. Data series for domiciles, routes and goods have been derived from the converted STRO data. In this section, they are used to calculate market concentration in the distribution of goods and services between the Baltic and North Seas.

Market concentration can be measured in absolute and relative terms (Lippe, 1993: 141–142). In this paper, we limit the analysis to the measurement of absolute market concentration and more precisely to the so-called Herfindahl-Hirschman Index (or HHI). The HHI is commonly employed in (maritime) economics to estimate absolute market concentration. HHI is generally accepted as being superior to more basic concentration ratios, since it takes all participants in the market into account, rather than describing the share of the largest 3, 5 or 10 participants in the market. Moreover, the HHI also takes variations in the number of participants into account, whereas relative market concentration indices, such as the Gini-coefficient, only comprise the disparity aspect of market concentration. Usually, the HHI returns low values. A commonly used contemporary interpretation of HHI values finds that a Herfindahl-Hirschman Index value below 0.1 means a low degree of market concentration, a value between 0.10 and 0.18 indicates a moderate degree of market concentration and all values higher than 0.18 are signs of a high degree of market concentration. Theoretically, an index value of 1 implicates the existence of an absolute monopoly.

Index values of this kind can be calculated for all data series described above. Since each of these series offers a different perspective on the complex issues of market concentration and specialization in early modern maritime trade and transport, the corresponding HHI values all represent different aspects of market concentration and changes therein.

In the following paragraphs, I will explain how the indices are calculated and what insights they provide about trade patterns and transport systems. Two examples will be used to explain the various steps. In the first example, ton-km estimates for the volumes of grain and timber shipped through the Sound between 1670 and 1800 are used to calculate absolute market concentration indices for the domiciles and routes involved in the distribution of these product groups. This example allows highlighting the contribution of the data series *Gd* and *Gr* to the study of Early Modern maritime trade and transport. In the second example, ton-km estimates for the volumes shipped by shipmasters from Hindeloopen and Rotterdam are used to calculate absolute market concentration indices for the goods they transported and the routes on which shipmasters from these domiciles sailed. The second example helps to explain the contribution of the data series *Dg* and *Dr* to the study of Early Modern maritime trade and transport. Ton-km estimates for the volumes shipped along specific routes could also be used to calculate absolute market concentration indices for the domiciles active on this route and for the goods transported between the port of origin and the port of destination. This would allow us to highlight the contribution of the data series *Rd* and *Rg* to the study of Early Modern maritime trade and transport. Its elaboration, however, is left out of the present article since it follows the same logic as the first two examples.

5.1. Grain and Timber

In the first example, ton-km estimates for the volume of grain and timber shipped through the Sound between 1670 and 1800 are used to calculate absolute market concentration indices for the domiciles and routes involved in their distribution. This example allows highlighting the contribution of the data series *Gd* and *Gr* to the study of Early Modern maritime trade and transport. Market shares for domiciles *d* and routes *r* in the transportation of good *g* were calculated as *Gd* and *Gr*.

Graph 1. HHI values for market concentration of domiciles in grain and timber transports through the Sound. Source: STRO

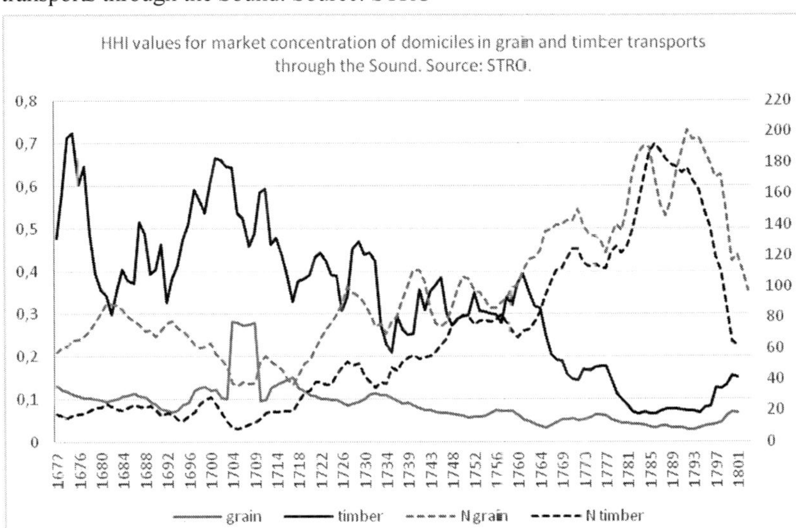

HHI can be interpreted a measure of the degree to which domiciles specialized in the transportation of timber and grain through the Sound between 1670 and 1800.The respective markets that are captured in this graph are the grain transport market and the timber transport market. The

graph shows that until the 1770s timber transport was the business of specialists, while grain transport was the exact opposite. An index value of 0,1 on average shows that the transportation of grain was an activity pursued by many, leading to a low degree of market concentration. In modern terms, the index values indicate a market structure for grain transport varying between not concentrated (< 0.1) and moderately concentrated (> 0.1 and < 0.18). The period between 1700 and 1710 is a remarkable exception, which can be explained as the effect of the severe disturbances that the Great Northern War (1700–1721) caused to shipping through the Sound. Low volumes of grain were transported by a limited number of shipmasters from relatively few domiciles (see Graph 1). Only a few of these domiciles had a significant share in the grain transport market, as can be established by calculating the concentration ratio for domiciles in the grain transport market (see Table 3).

Table 3. Concentration ratio for domiciles active in the grain transport market between 1701 and 1730. Source: STRO

	1701–1710	1711–1720	1721–1730
Top 3	40%	46%	45%
Top 5	59%	64%	58%
Top 10	76%	82%	72%

During this war period, grain transport was mainly carried out by shipmasters from the domiciles Vlieland, Ameland, and Hoorn, who accounted for about 40% of the grain transport market between 1700 and 1710. Hoorn was replaced by Amsterdam in the top 3 domiciles in the second decade of the 18th century. Between 1720 and 1730, the top 3 were Ameland, Amsterdam, and Terschelling. None of this information, however, can be immediately derived from Graph 1 or Table 3. To get this information, we need to return to the source. Similarly, changes in the market position of these domiciles during, before, or after the war cannot be captured in the graphical representation of the HHI. The use of GIS tools, however, makes it possible to visualize the data series underlying the HHI values of specific time periods differently, so as to obtain a geographical insight into the composition of HHI values. This topic will be elaborated further in section 6.

5.2. Hindeloopen and Rotterdam

In the second example, ton-km estimates for the volumes shipped by shipmasters from Hindeloopen and Rotterdam are used to calculate absolute market concentration indices for the routes on which shipmasters from these domiciles sailed. Alternatively, indices for the goods transported by shipmasters from these domiciles could also be generated, but these are not further explored in this paper. The second example helps to explain the contribution of the data series *Dr* (and alternatively *Dg*) to the study of Early Modern maritime trade and transport. In the second example, time and domicile are preselected in order to generate data series concerning the shares of routes in total transportation of domicile *d*. As in the previous paragraphs, rather than visualizing market shares directly as concentration ratios, it is more attractive to focus on the absolute market shares of each route in the overall transportation realized by a specific domicile *d*. The resulting HHI values can be interpreted as indices for the spatial concentration of shipmasters from pre-selected domiciles in the transportation of goods through the Sound.

Graph 2. Spatial concentration of Rotterdam in Sound shipping. Source: STRO

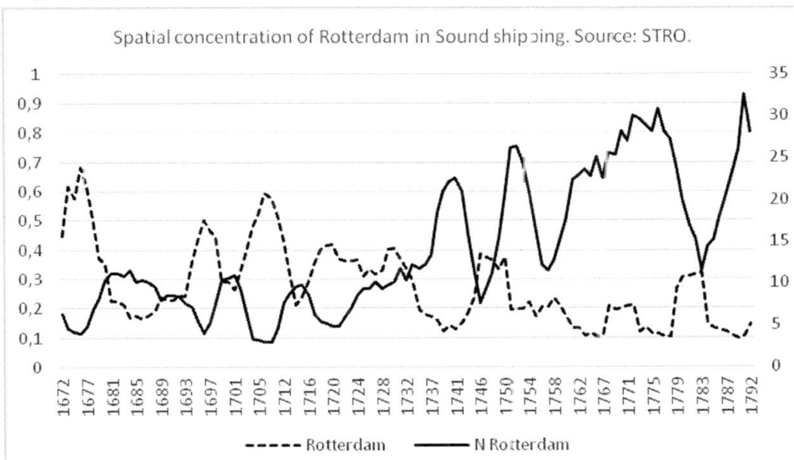

N Rotterdam indicates the number of different routes sailed by shipmasters from domicile Rotterdam. Its value is scaled on the right axis.

Graph 3. Spatial concentration of Hindeloopen in Sound shipping. Source STRO

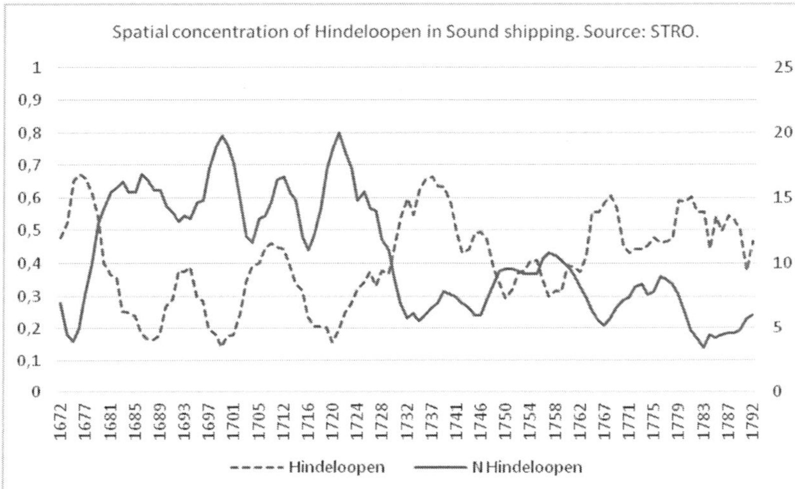

N Hindeloopen indicates the number of different routes sailed by shipmasters from domicile Hindeloopen. Its value is scaled on the right axis.

Graphs 2 and 3 show that the spatial concentration of the transport activities of shipmasters from Hindeloopen and Rotterdam developed quite differently between 1670 and 1800. Especially after 1720 it becomes clear that the transport patterns of shipmasters from Hindeloopen were highly concentrated spatially. A few routes clearly dominated their transport patterns and degree of concentration grew over time. In the same period, the transport patterns of shipmasters from Rotterdam became more and more diffuse, although a moderate concentration is maintained for most of the period (see Graph 2).

Like in the previous example, the graphs presented here do not provide any information about changes in actual transport routes or in goods distribution patterns that might explain this development. Again, the use of GIS may be beneficial in this case. Since the underlying data series do contain all necessary details about the annual volumes shipped by shipmasters from Hindeloopen on each route, it is possible to visualize these routes and to explain their role in the transport patterns of shipmasters from Hindeloopen. A more general description of the contribution of GIS to the study of market concentration in early modern trade and transport will be the topic of section 6.

6. Use of GIS

In this section, I describe briefly how GIS tools may contribute to the study of market concentration in early modern trade and transport. GIS software provides powerful tools for the visual analysis of the complex relations between domiciles, routes and goods in Early Modern trade and transport. GIS software can bring the geographical data on routes and domiciles, which are reduced to mere numbers in the Herfindahl-Hirschman Index, back into the picture. Technically, it should then be sufficient to highlight a specific year on a graph in order to execute the geographical visualization of the underlying data. There is only one important condition that needs to be fulfilled before such connection between graphs and GIS can successfully be established: the quantitative data used to calculate HHI values must be the result of a procedure in which both numerical and alphanumerical data items are preserved for the duration of the entire process calculation process. This is the case in the method described in this paper. For example, the tables underlying Graph 1 are composed of a list of all domiciles that were involved in the transportation of grain or timber through the Sound between 1670 and 1800. For each domicile, the annual estimated transported volume, expressed in ton-km, is given. This information may serve perfectly as a basis for the visualization of the relative position of each domicile in grain transport through the Sound in a given year. Similarly, the tables underlying Graphs 2 and 3 consist of lists of all routes – represented as combinations of port of departure and port of destination – and the volumes transported annually on each route. Rather than using points as the basis for visualization in GIS, segments linking the port of departure and destination of each route seem to be the better choice here.

Connecting graphs and GIS visualizations in this way is a fairly easy and straightforward procedure, but it has its limitations. Most importantly, GIS visualizations of this type are not dynamic, even though generating (annual) sequences of static GIS visualizations may be considered as a step in the right direction. Moreover, the procedures described in this paper essentially involve generating stand-alone Herfindahl-Hirschman indices for the analysis of actor behaviour on specific markets (e.g. the grain transport market). Visual analysis of more complex relations between the variables represented in the four basic time series and the six subseries described in this paper remains problematic. It may, for example, be interesting to visualize how the relation between *Gd* and *Gr* changes over time and which factors determine these changes. This, however, would require taking into account several parameters simultaneously, which greatly complicates generating manageable statistics. Perhaps, GIS tools may have to be included earlier in the calculation process in order to overcome these difficulties.

CHAPTER NINE

FROM REGION TO LOCALE AND FROM SPACE TO PLACE: LOCATING SRIHATTA/SYLHET IN THE MEDIEVAL BAY OF BENGAL WORLD

RILA MUKHERJEE[1]

1. Introduction: Space and Space Production

Space and place are dynamic concepts and, with the new academic interest in spatial humanities, increasingly deserve our attention. Ever since Edward Casey wrote about "getting back into place", a whole generation of scholars has been looking at the historical production of space (Casey, 1993; 1997). Why are we looking at the production of space? The new "spatial turn" in the academy emphasizes that space is neither a neutral category nor a passive container of human actions and social relations; it becomes an instrument of social and economic change, space and space production being intimately connected with territoriality, means of production and control (Lefebvre, 1991; Zhang, 2001: 179–205). Historians are increasingly aware that the history of a place cannot be understood properly without taking into account its relations with other places. "In the new century", Owens notes, "historians increasingly argue that the history of a geographical place, no matter how large or small, cannot be adequately understood without taking into account how it has been connected to other locations" (Owens, 2007: 1). These connections may be social, political, religious, economic, or just plain cultural, and within them are various sub-categories of relations and links. Once

[1] University of Hyderabad.

connections become structural – in other words once they transform into connectivities – they are furthered through networks. We also find that relations between places are contingent on the relations people build at these places. Relations are based on social networks: networks of trade, trust, justice, finance, culture, political allegiances, religious exchange, etc. Short, self-organizing and informal networks are more sustainable than official networks. The third premise is that people build relations not only with one another and with places but also with space. Space is partly produced by the "connections (of a given place) to the rest of the world" and different connections lie at the heart of the production of different kinds of spaces. Here spaces become "social locations" embedded in 'webs of cultural, social, economic, and political relationships' (Radil, Flint and Tita, 2010: 309). The last premise is that globality rather than globalization is the issue here. Globality is used here as a condition rather than process – the latter is evident as globalization – where crucial *structural* shifts in pre-modern commerce are discernible, where power shifts from traditional economic centers of power and influence – the shift from Asia to Europe in the period between 1400 and 1800 is one case. In the case of Srihatta, the subject of this paper, globality is observable from much earlier.

2. Srihatta/Sylhet and its Myriad Identities

2.1. Spatial Identities: Puranic and Ptolemaic

Let me introduce to you the landlocked region of Srihatta, today known as Sylhet – the name deriving from "Sarkar Silhat" of Mughal Bengal – and located in Bangladesh.[2] This region was initially beyond the reach of Hinduism and the South Asian world, lying in what the *Puranas* termed *Bhadrasvavarsa* as opposed to *Bharatavarsa*. The *Markandeya Purana* divided *Bharatavarsha* (India) into two units: *Bharatavarsha* or

[2] Two gazetteers have been used in this research: Austin, Robert F., *A Historical Gazetteer of Southeast Asia* [a preliminary edition] (U. Michigan, U.S., 1983) www-personal.umich.edu/~copyrght/image/monog04/, 118 pp. Department of Geography University of Missouri-Columbia, Columbia, Missouri 65211, USA. Library of Congress Identification Number DCLC 8622956-B. Published online as Monograph Four by the Institute of Mathematical Geography, Ann Arbor, MI 48104; and *Wa Area [Burma/Myanmar, Yunnan] Online Gazetteer*, available at mercury.soas.ac.uk/wadict/wa_placenames.html [A gazetteer of towns, villages, geographical features; together with their various current and past names (in Chinese, Burmese, Wa, Lao, etc), and their lat/long coordinates].

India proper and *Bhadrasvavarsa*, i.e. the Region Beyond India. The Karatoya River was the boundary between the two units. The first unit (that is, India proper) lay on the west coast of the *Lauhitya Sagar* (the Brahmaputra) and *Bhadrasvavarsa* lay on its eastern side. It is worth noting that *Bhadrasvavarsha* is also one of the island continents in the Puranic classification of the cosmos (Ghosh, 1984: 84–86; Sircar, 1990: 18–20). It is also worth noting that Ptolemy too divided South Asia into India Intra Gangem and India Extra Gangem. These were therefore accepted spatial categories in the ancient world. Srihatta is located within India Extra Gangem. See Map 1 to see its present location in Bangladesh, relative to India. The 15th-century *Ming Shi Lu* affirms that this whole region was well known to the Chinese from quite early times and most definitely from the Tang period.

Map 1. Map of Sylhet[3]

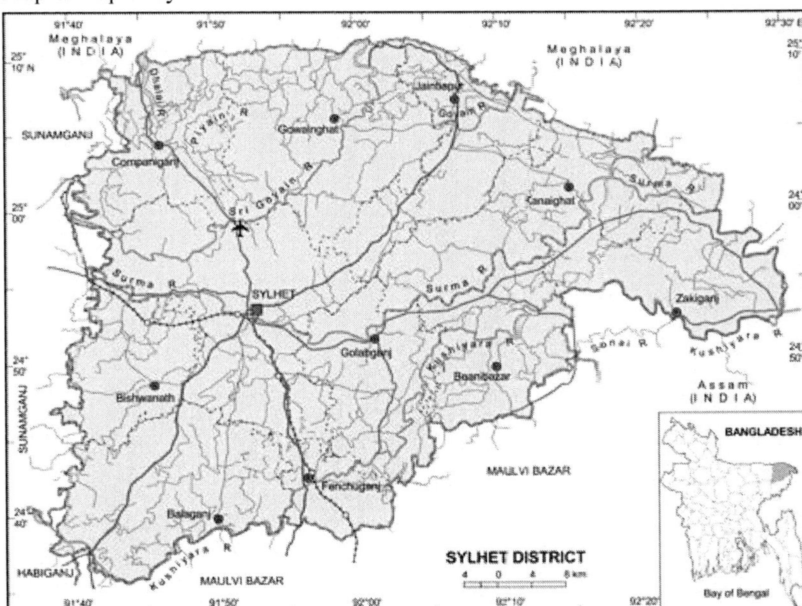

[3] http://www.google.com/imgres?imgurl=http://www.kamalbazar.net/sylhetdistrict %2520map.gif – accessed 14 January 2013.

2.2. Spatial Identities: Classical

Rather than simply revealing spatial identities from political integration and disintegration I would like to introduce here first the spatial categories used in early and medieval Bengal. These categories, forming well-contoured geographic units, were known as Vanga, Samatata, Varendra and Harikela, corresponding to west, south, north and north-east Bengal. To locate them today, we may argue that Kolkata (India) lies in Vanga, and Dhaka (Bangladesh) in Samatata while the early Pundravardhana or Jalpaiguri (both in India) lie in Varendra and Chittagong (Bangladesh) in situated in Harikela. See Map 2 for these units.

Map 2. Bengal: Ancient Political Divisions[4]

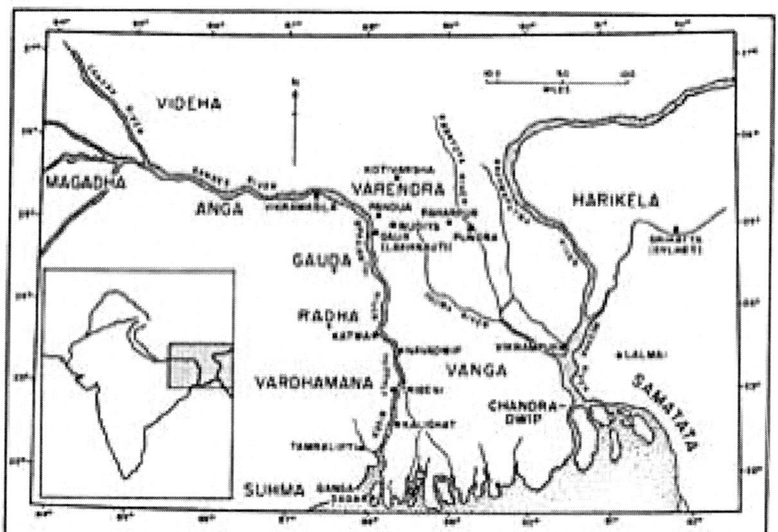

These spatial categories were not only geographic but cultural categories as well, again forming reasonably well-defined spatial units. The easternmost unit, Harikela, for example, was the unregulated no-man's frontier, and a markedly Sinitic one at that. It was regarded as lying

---

[4] From R.M. Eaton, *The Rise of Islam and the Bengal Frontier, 1204–1760*, Berkeley; Los Angeles; London: University of California Press, 1993.

beyond the reach of Indic civilization and history, being situated on the corridor from India to China.

Moreover these were dynamic units whose boundaries fluctuated over time. Vanga and Samatata, for example, exhibited particular mobility during the 7th and 8th centuries. This mobility is particularly observable in the case of Srihatta. The 7th-century Kalapur copper plate of Marundanatha suggests that Srihatta was part of Samatata. However, in the 8th and 9th centuries Srihatta formed part of Harikela, as the large numbers of Harikela coins found tell us. In the 10th century the Paschimbhag copper plate (found in 1958 at Srihatta) of the Chandra monarch Srichandra noted Srihatta as part of *Pundravardhanabhukti* or province, that is, of Varendra. The two Bhatera copper plates of the 11th/12th centuries mention Srihatta as an independent kingdom, with a southern outlet to the sea, indicating its southward shift from the earlier *Pundravardhanabhukti*.[5]

Although Srihatta initially formed the frontier between the Indic and Sinic worlds, after the Muslim conquest of north Bengal in the early 13th century, the frontiers of the Muslim kingdom in the north gradually extended as far as the Karatoya River. Srihatta therefore now exhibited a dual frontier. And with the conquest of Srihatta by Shamsu'd-Din Firuz in 1301, it became an Islamic frontier area, warding off the effects of Sinitic beliefs, Buddhist vestiges and an emergent Hinduism. What factors lay behind these spatial swings in Srihatta's case? Were they caused only by political factors? Let us take a look at its political history.

2.3. A Political Identity within Empires and Kingdoms

At inception there was really nothing called Srihatta; it lay in a fuzzy zone in the extreme north-east of South Asia. The region within which Srihatta lay, conveniently positioned on the Surma-Meghna river system which connected Srihatta to Bagan and Yunnan, was supposedly part of the ancient kingdoms of Pragjyotisa and Kamrupa (both in present Assam in India); and after these successive kingdoms declined Srihatta may have been an autonomous state for a while as well, because Hiuen Tsang noted it as Silichatalo (actually Shih-li-ch'a-to-lo) in the 7th century (Bhattacharya, 1920: 1–19).

But it may have been fragmented again because in the same century the Nidhanpur copper plate of the king of Kamrupa, Bhaskarvarman, mentioned its Chandrapur section as part of Kamrup. By virtue of his

[5] K.L. Barua's reading of the Bhatera plates in D.C. Sircar, *Studies,* p. 164.

control over the land and river route into China that passed through the region – the Surma-Meghna system around Srihatta that fed into the Brahmaputra and then into Yunnan via North Burma – Bhaskaravarman styled himself "King of Eastern India" (Barua Bahadur, 1966). Significantly, his contemporary in Nanzhao, the kingdom in Yunnan, used a similar title, *utibwa*, or "King of the East" (Yang, 2011: 317–342). The chronicles titled *Sri Sri Rajmala* of Tripura refer to the 16th-century ruler of Mong Mao as "the great King of the East" – this being a clear case of the former titles of Bhaskarvarman and the Nanzhao ruler continuing to exist and being passed on to yet another monarch of the region (Nath, 1999; Arambam Parrat, 2005). The persistence of this title in chronicles and historical memory needs further investigation. Was this a title bestowed in popular perception or did it have actual, sacral moorings? How, and on what grounds, was the greatest ruler of the east perceived? How were the boundaries of the "East" determined? It is very difficult to discern the geographies of this region.

Srihatta certainly became an autonomous state from around the 10th century as the sovereign *Srihatta-rajya* (or kingdom), although its monarch never achieved the status of the King of the East. The Barak Valley and the Cachar area of present Assam was included within it (Bhattacharyya, 2003). The 10th-century Paschimbhag plaque of the Chandras, in 65 lines of Sanskrit in *nagari* script, emphasized the importance of the fluvial network of the région around Srihatta, namely the Meghna, and, more importantly, mentioned the arrival of the Cambodians at the Chandra capital at Vikrampur. The plaque also indicated that the Chandras (both Srichandra and his father) had invaded countries up to the Himalayan foothills (Boisselier, 1971: 90–94). We can therefore definitely establish, from the 10th century at least, a network of communications between Srihatta and the eastern Himalayan foothills, and a circulation of peoples and goods between Srihatta and the minor hill kingdoms.

Around the 12th century Srihatta was probably included within the Tripuri kingdom, and in the 14th century Afghan Muslims conquered Srihatta – possibly as early as 1301 – but this conquest was short-lived. Ludden notes that "when Ibn Batuta traveled to meet Shah Jalal, in 1346, the region comprised diverse localities held by Khasias, Garos, Hindus, Muslims, and others. Thus, in Shah Jalal's time, a definite 'Sylhet region' does not appear to have existed. Instead, unbounded geographical spaces spanning mountains and plains seem to have housed a great many small territories loosely connected to one another and to distant places, including

Sultanate capitals" (Ludden, 2003: 1–54).[6] Yet Srihatta became part of the Islamic cultural world and of the diverse networks spawned by it. A list of *sufis*, *darbeshs*, *pirs* and *faqirs* in this region reads like a who's who of some 158 personages, displaying networks between south-eastern Bengal, Yemen, Turkestan and Baghdad (Bhattacharyya, 2008). When Ibn Batutta went down the Meghna River, at the time when the Afghans ruled Srihatta, he saw water wheels, gardens and villages such as those along the banks of the Nile in Egypt. Was this a case of some long forgotten technological link with Arabia? He wrote that "for 15 days we sailed down the river passing through villages and orchards as though we were going through a mart". He did not describe market towns, as he did routinely elsewhere. Ludden notes that "fourteenth century Sylhet seems to have no big market towns, but its saint from Turkestan and visitor from Morocco show it did have wide connections, and these connections, networks and mobility changed the land in centuries to come" (Ludden, 2003: 5082). This mixture prevailed in the 18th century, Ludden noting in another article that ethnic spaces here held mixtures of Persian, Sanskrit, Bengali, and disparate tribal cultures (Ludden, 2003: 21).

In the 15th century Srihatta was included in the Tibeto-Burman Khen dynasty's territory and served as a sort of buffer state between Assam and the Islamic state of Bengal. As its contours shrank in the 16th century, for some time Srihatta also formed part of the larger Koch kingdom of c. 1560 in north-eastern India and it was later also a part of the Ahom kingdom of Assam. It is clear that "little" Srihatta was becoming the prey of large states in the region. At this time, Srihatta established connections with the Hindu kingdom of Tirhut or Mithila, now in Bihar, India and welcomed some of the earliest Brahman settlers of Srihatta from Mithila (Choudhury, n.d.: 78). Between the 16th and 19th centuries, at least 22 scholars from Srihatta wrote various editions of the *Manasa Mangal* and *Padma Purana* (Datta, 2004; Choudhury, 2007). Was Srihatta redefining itself once again, by tapping into other cultural networks? It became so famous as a scholarly centre that the reformist Bengali monk Sri Chaitanya visited it in the 16th century.

Its many histories show that Srihatta lay in an area that witnessed the rise and decline of regional empires and local states.[7] In terms of its

---

[6] http://www.sas.upenn.edu/~dludden/JASB-Boundary.pdf – accessed 14 January 2013, different pagination. Page nos. refer to the second version available online. See p. 2.

[7] Rila Mukherjee, 'Between Land and Sea: The Integration of 'Small' Polities into a Northern Bay of Bengal Economic System Between the 9th and 19th centuries', forthcoming Nalanda Sriwijaya Centre/ISEAS, Singapore.

integration into large state forms, Srihatta therefore formed a part of Kamrupa (7th century), then a part of a succession of Islamic states centred on Bengal – Sultani (13th century), Afghan (14th century), then Mughal (17th century), later still Nawabi (18th century) – and finally it became a part of the British province of Assam in the 19th century. Srihatta's bearing and networks were therefore reoriented from east to west over some 800 years and its 1000-year history shows various spatial identities, as well as many political characteristics. The historical process evinced in the making and unmaking of Srihatta: its incorporation into the Pragjyotisha and Kamrupa kingdoms and its markedly Sinitic character, its subsequent break-away from these kingdoms as they declined, its re-emergence as a Hindu/Buddhist and later an Afghan Muslim state, saw various networks jostling for space in what was a fairly marginal region with access to the Bay of Bengal via its port of Indresvara on the Meghna River.

2.4. Into the Bay of Bengal

So, Srihatta had yet another identity, that of a Bay of Bengal port. References to Indresvar are few. I have not been able to find it on any of the maps of the region that I have seen but the medieval Bengali ballad poem, the *Manasa Mangal*, mentions such a port: Chand Saudagar's ship, the *Madhukara*, on its way to the Incomparable Port or Anupama Patan in Sri Lanka, passes Indresvar but this location can be contested as this may be another Indresvar situated in western Bengal (Bhattacharya, 2007: 197). It is more likely the Indresvara of Srihatta was a riverine port, situated on the floodplain of the Meghna River. In the ancient inscriptions of Srihatta there are references to large bodies of water surrounding the area; and the 10th-century Paschimbhag copper plate mentions a port or anchorage called Indresvara Naubandha. The Bhatera plaques of the kings Kesavadeva and Isanadeva in the 11th and 12th centuries, also found at Srihatta, show maritime associations eastward. They refer to Indresvara Naubandha as being the mooring point of innumerable war boats (*nauvatakas*) which presupposes that in ancient times Srihatta was a place that was surrounded by large tracts of water (*haor* in the local dialect, itself a corruption of *sagar* or sea) being a place that saw much maritime activity.

An interesting feature of the inscription in the Bhatera plaque no. 1 is that many houses and structures including outhouses, kitchens, cowsheds etc. were devoted to the use of the various helping hands of the temple establishment like *gapa* (milkman), *rajaka* (washerman), *vaniya* (goldsmith), *navika* (boatman) and *malo* (fisherman). Note again the marine element in the inscription – grants given to boatmen and fishermen. The second plaque, containing 32 lines in Sanskrit in a proto-Bengali script, has been regarded as showing a style similar to Khmer plaques from Cambodia; there was an attempt to strip away the esoteric associations and the power of a literate language and to make its meaning accessible to all (Boisselier, 1971). This changeover in linguistic style is significant, testifying to the region's deep associations with Cambodia. Not just in mainland south-east Asia but in maritime south-east Asia as well, networks linking the various ports in the Bay became more visible at this time and these are reflected in the plaques found at Srihatta. Indresvara is today located in Maulvi Bazar district of Sylhet in Bangladesh and is called Indresvara. The Hakaluki Haor in the Maulvi Bazar district is a complex of more than 80 inter-connecting *beels* (large lakes). During the dry season, the *beels* cover an area of approximately 4,400 ha. However, during the rainy season, the entire area gets flooded, and the *beels* are united as one large lake, or *haor*, with an area of approximately 18,000 ha. This is also referred to as *sagar* or the sea. Maulvi Bazar lies on the banks of the Manu River. The Manu is the sacred river of Tripura, yet another important silver area to the east, and the river provided connectivity between the two regions.

2.5. An Economic Space: Silver, Money, and Coinage in Srihatta

Being situated in and around the Brahmaputra valley, on the Surma-Meghna river system that connected to the Brahmaputra, Srihatta had easy access to two important networks that spanned the valley: one was the crucial corridor linking it to the coast and the Bay of Bengal, the other was the metal corridor linking it to the gold mines of Assam and Tripura, the silver mines in north Burma (Bagan) and the gold and silver mines of Yunnan. Thus Srihatta had an entrée to metals from inception, important for both technological development and monetisation. However, as we see later on, Srihatta largely avoided silver coinage, preferring to use *kauri* shells (*cypria moneta*) instead, the only exception being a short period under Sultani rule in the 14th century. Kuroda notes that the beginnings of Mongol rule in China coincided, in part, with change in eastern Bengal. From 1339 CE (AH 740) numismatic evidence from the Srihatta find confirmed the continuous issue of silver rupees. Those bearing the inscription of Mubarak Shah were issued in consecutive years from 1339 to 1349 CE (AH 740–50), Mubarak Shah being the ruler of the Bengal Sultanate who expanded his territory towards the southeast and the coast.

Thus, political and military developments in Yunnan and Bengal seem to have caused a flow of silver from China to India, via Bagan to Srihatta, from 1339 (Kuroda, 2009: 255–256).

But Srihatta still lay within a *kauri* economy zone and not within a silver currency area, although we associate only the latter with the Indian Ocean trade. According to Heimann however the lack of silver does not indicate absence of trade and for him the *kauri* economy formed the basis of what he calls the Indian Ocean Trade System, a belief also shared by Bin Yang (Heimann, 1980: 48–69; Yang, 2004: 2011: 1–26). This connection continued into the 19th century, Srihatta being situated in a region rich in products that were much coveted in the pre-colonial commerce of the Bay. We noted that some of the items mentioned in the *Periplus* might have orginated from this region. Aloe wood and China-root appear as Srihatta's products in the 16th-century Mughal census *Ain-i-Akbari* (Allami, 1873).

2.6. An Agrarian and Commercial Frontier

Ludden tells us that elephants and ivory travelled down the mountains, and salt and rice moved up the river. Srihatta's first resident Collector, Robert Lindsay, in 1782–1783, underlined the commercial value of these eastern mountains by buying limestone and burnt lime to sell in Dhaka, Calcutta, and elsewhere. He even once had built four sea-going ships on mountain tops, using local forest timber; and then had the ships filled with mountain products and to sail on the flood to Bakarganj, the old Chandra port of Chandradwip, where Lindsay's agents sold forest products and bought rice to sell in Madras (Ludden, 2003: 5082).

This agrarian frontier was as linked to the *kauri* zone as Bin Yang's Yunnan (Yang, 2004). Ludden writes that Lindsay's ships perhaps returned to Calcutta or Chattagrama with *kauri* shells, Srihatta's only coin (Ludden, 2003: 5082–5083). Srihatta depended entirely on *kauris*, tiny shells from the Maldives that served as the cheapest coin all around the Indian Ocean and in South East Asia and Africa.

Srihatta also connected vertically with the overland Silk Routes by supplying eunuchs to both east and west – to the courts of China and Byzantium – and thus welcomed merchants from afar. Through its overland connections with Byzantium Srihatta became known to the Arab world, as its medieval history affirms. Srihatta's varied history, its almost *mandala* like political and economic contours, therefore reveal various networks scattered over time and space from very early times. Just as the history of Srihatta reveals a trajectory from regional empires to local kingdoms, so too does its economy reveal a trajectory from the regional to the local.

3. From Region to Locale

Part Hindu, part Buddhist and markedly Tantric and part Muslim, Srihatta offers a fascinating case study of a crossroads region. From very early times Srihatta was situated in a region that had considerable commercial and diplomatic relations with China, and was indeed one of the important corridors into China from eastern India. It was famed for the production of malabothrum, i.e. bay leaf and was probably part of the ancient kingdom of the Kiratas or Kirradoi referred to in the *Periplus of the Erythraean Sea* (McCrindle, 1879). At the end of the first century CE the Kirata kingdom formed an intermediate political and economic unit between China and India proper and at its borders was a large fair where the Chinese sold silks and bought bay leaf (Das, 2009: 37–38). It has been speculated that the shadowy Thina or Thinae of the *Periplus* (Schoff edition), usually taken to be China, may in actual fact incorporate India's north-east and might therefore be Pragjyotisa, the first kingdom of Assam, containing within it Srihatta (Choudhury, 1988; Schoff, 1912). Casson's commentary on the *Periplus* supports the geographic location of this eastern production area by noting that most commercial items originated from the area around Dhaka and Tripura, and that those such as malabathron and tortoiseshell came from the shadowy area between present Assam and Myanmar, notably Kamrupa and Srihatta (Casson, 1989).

It is now clear that ancient Srihatta lay in this shadowy region. From this region Srihatta was transformed into an independent little state in the early medieval period, as we noted, and it lost its independent status when absorbed into the Tripuri state between the 10th and 12th centuries. It was again conquered by the Muslim Sikandar Shah Gazi in 1303, with the blessings of the famous medieval Muslim saint Hazarat Shah Jalal from Turkestan, and after that its political and cultural networks bifurcated: one set continued eastward, but a new set of networks was now directed westward. Srihatta became a truly multicultural area, housing the saint until his death in 1384, and welcomed Ibn Batuta from Morocco who came to visit him in 1346.

But as Srihatta remained a border region, its eastward associations continued to exist. In the 15th century the region was mentioned in the *Ming Shih Lu*, as we read. In the 16th-century Arakan (an independent kingdom, absorbed in 1785 into Burma) claimed it as a historic possession ceded to the Bengal sultans under the name of "Silak", another name for Sylhet (Leider, 2002: 145). As late as the 16th and 17th centuries it lay in a markedly Sinitic cultural zone, although nominally it remained a Muslim kingdom. At the beginning of the 17th century it was conquered by the Mughals and European cartographers referred to this space variously as "Cirote", "Sirote", "Salhat" and "Silhet" and marked it accordingly on most maps of India and the Indian Ocean, thereby revealing its importance as region and place(see Map 3)[8].

Map 3. Srihatta as Sirote in Nicholas Sanson d'Abbeville's *India Vetus intra et extra Gangem,* Paris, 1702

8 Map 3 is reproduced with the kind permission of Prof. Frances Pritchett of Columbia University, N.Y. Most of them are available at http://www.columbia.edu/itc/mealac/pritchett/00maplinks/ and are a testimony to Fran Pritchett's love of maps and her commitment to visual pedagogic aids. We acknowledge here her generosity in sharing this resource.

4. The Shrinking of Sylhet and the Dwindling of Spatial Boundaries

Map 4. Sylhet in Bangladesh and its Riverine Access to the Bay of Bengal

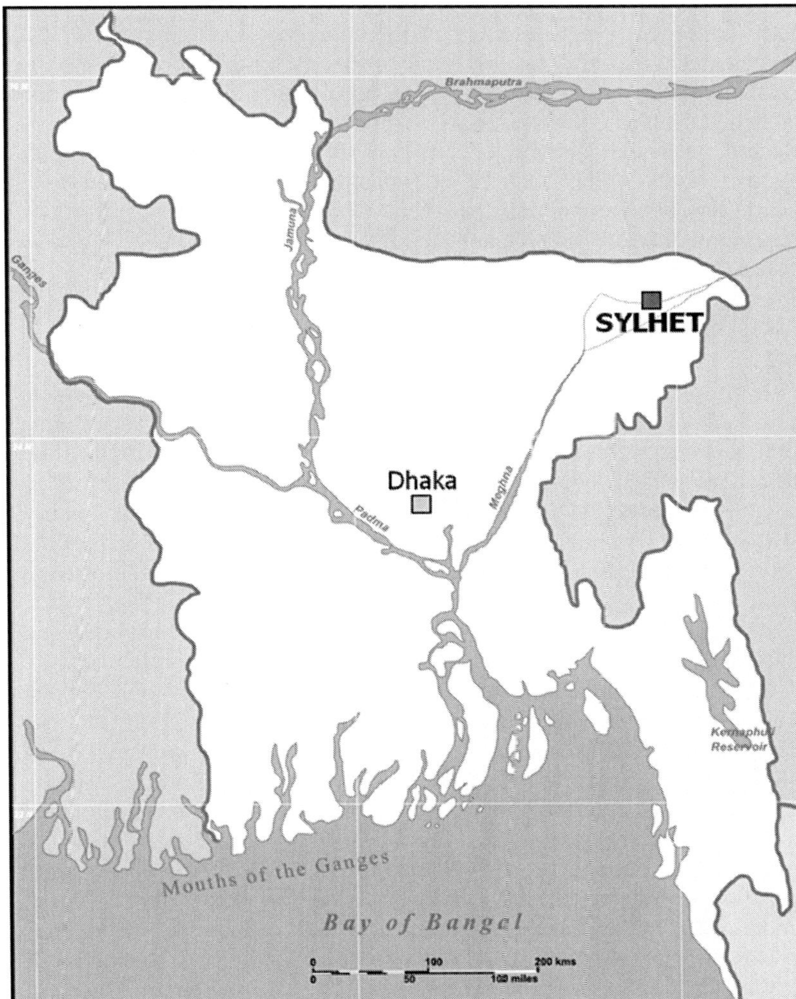

But Srihatta's economic decline was visible soon after. Srihatta still depended on *kauris* for its economic transactions although neighbouring

kingdoms such as Koch Bihar, Jaintiapur, Manipur and Tripura, were issuing silver coinage from the seventeenth century, thereby indicating a certain amount of commercialization and monetization in contiguous areas. As yet, we know of no such coins issued from Srihatta in the medieval period except those of Mubarak Shah commemorating his victory, although a large number of the so-called Harikela coins of the early medieval period are atrributed to the Srihatta region. From the second half of the 18th century, as the new political dispensation established itself at Calcutta, the whole north-eastern region, and Sylhet as a place in particular, became an economic wasteland. By the 19th century a vastly shrunk Sylhet became an upland appendage of the ports of Calcutta and Dhaka under British rule, thereby demonstrating again new spatial networks. See the location of Sylhet in Bangladesh and its Riverine access to the Bay of Bengal in Map 4.[9]

5. Conclusion: Sylhet: From Space to Place

I have tried, in this chapter, to show the tensions between Srihatta as region (space) and Sylhet as locale (place) over a thousand years as this region sought to remake and define itself. The final shift occurred in the 17th century when Srihatta disappeared and Sylhet rose in its place. This changeover echoed Sylhet's history in the 17th century; in that century Sylhet formed a frontier town as the Mughals moved in to conquer the Khasis, Dimasas, Ahoms and other "hill folk". Thus new boundaries were forged.

As Ludden notes, "geographies of nature, land use, markets, and culture had no strict boundaries; they remained open to limitless mixing and mobility. Inside them, however, permanent farm cultivation defined more strictly bounded local geographies of social investment on the land, where families endeavored to keep land free of forest and safe from animals and marauders, to accumulate wealth to pass on to their children. Agrarian states also defined geography in more strictly bounded territorial terms, and drew their boundaries around central sites of state authority. After 1303, the geography of Sylhet governance revolved around Sylhet town, but in the eighteenth century, Sylhet's internal and external boundaries did not appear on maps or visibly on the land. State boundaries were instead defined by transactions among individuals in the ranks of state authority. The spatial boundaries of governance emerged in transactions among the men of rank who participated in the ongoing

[9] From http://en.wikipedia.org/wiki/Sylhet – accessed 13 January 2013.

process of producing and maintaining state authority, in places connected to one another by relationships among elites that focused on state headquarters in Sylhet town" (Ludden, 2003: 21).

What this chapter also underlined is that like most border regions, Srihatta cannot be studied with reference to national frames. The modern nation state erases distinctions and cloaks them under a spurious homogeneity. The history of Srihatta shows, even today, the persistence of older networks in its language, food and customs. I will end with a line from Ludden: "the mobility that characterises globalisation today operates simultaneously in many spatial and temporal registers, and forms many geographies, which coexist, conflict, and complicate one another, and have done for a very long time" (Ludden, 2003: 5086). This is so true of Srihatta.

CHAPTER TEN

MAPPING AND VISUALIZATION OF COMMERCIAL NETWORKS IN THE PACIFIC OCEAN DURING THE 17TH CENTURY[1]

ANTONI PICAZO MUNTANER[2]

1. Introduction

The possibilities offered by the study of networks are quite broad and even more so if performed within an investigation that aims to analyse the complexity of a comprehensive system and variety of dynamics that emerge within it. Undoubtedly, the sum of the objectives pursued by Europeans in Asia influenced the maturation of capitalism in the old world and they made the most of the advantages the new situation gave them. They learned from each error committed and collected data (Habermas, 1995; Husserl, 1990). Carayol and Roux (2005: 307–332) stated that innovation truly meant a linear function of accumulation of information in a manner not seen until that time, all of which generated truly spectacular economic growth.

But first, and very briefly, some observations must be made that should be taken into account about the nature of networks. We must take heed of emergence theory and the need to implement detailed models of the system as proposed by Holland (1998: 17), and linked to its internal processes; a network is much more than the sum of all its parts, as well as that of its qualities. As Ricard Solé (2009: 20) states, there is still a certain invariance of everything despite modifications the parts may undergo. The networks studied, which will be explained in detail further on, tend to become "small worlds" (Milgram, 1967: 62–67) framed in their own,

---

[1] This paper is funded by the MICINN, Ref HAR 2011-27694 'Ruta Global: Análisis histórico experimental con base de datos y geovisualización espacio-temporal del comercio marítimo internacional, 1717–1850' (GlobalNet).

[2] Universitat Illes Balears.

"complex, non-linear and self-organising" (Arthur, 1999: 107–109) dynamics, in pursuit of a common goal. These complex systems co-operated and collaborated on specific tasks. Anderson (1999: 216) wrote: "Simon defined a complex system as one made up of a large number of parts that have many interactions (Simon, 1996). Thompson (1967: 6) described a complex organization as a set of interdependent parts, which together make up a whole that is interdependent with some larger environment".

One of the determining factors that clearly demonstrate their extreme versatility and complexity was that, in certain situations, they could develop unpredictable models. Anderson (1999: 217) explains: "Modem complexity theory suggests that some systems with many interactions among highly differentiated parts can produce surprisingly simple, predictable behavior, while others generate behavior that is impossible to forecast, though they feature simple laws and few actors". The network should be self-perpetuating within controllable parameters and controlled by the same structure that supported it. Its size and its anchorages (political and economic) made its own survival and effectiveness possible. In this sense, its effectiveness was basically a response to two logical premises: the larger the size, the greater the efficiency; and the greater the number of political anchors the greater will be the strength and resistance to exogenous attempts of destabilisation.

To visualize this dynamic, our objective in this paper is to weigh and report the analytical possibilities of new software developed for mapping networks. This study will focus on two major areas that will serve to exemplify the evolution of the basic structures, how its internal dynamics progressed and how that same evolution caused major collisions with rival networks. Thus, we shall delve into those that conducted their business in the Asia-Pacific area through two major trade blocs.

First, we will compare the development of systemic relationships by assessing three complex networks that will be looked at separately, namely the East India Company (EIC), the Vereenigde Oost-Indische Compagnie (VOC) and the one implemented by the Spanish in Manila. Finally, we shall assess their links and the end result: their complexity. Targeted networks were not acting alone, in fact, far from it. They were part of a much denser totality, which gradually took hold in markets, establishing connections with other groups who were conducting a similar activity, as pointed out by Kearney, Tracy, Barendse and others, (Kearney, 2004; Tracy; 1990; Mcpherson, 1993; Erikson and Bearman, 2004; Barendse, 2002) forming co-operation and collaboration agreements with one goal:

mutually to enhance each other's business activities. In short, a network is always open to new members.

Secondly, we will detect structures of some invisible networks (i.e. underground ones), designed to increase their profits through illegal actions (read contraband). Those networks circulated in environments of political power, particularly among the governors and captains-general of the Philippines. We shall observe their internal structure and evolution, but also how competition was affected. Those contractual networks functioned like galaxies rotating around each other. Only in very critical moments, such as a change in the gravitational weight of an agent, or when a sudden mutation in the density of these networks caused a collision with the immediate consequence being one of them was terminated. In that regard we shall see how two violent shocks were generated. The first among them was the network created by Governor Diego Salcedo and the coup promoted by his enemies. This involution ended in the imprisonment and death of the former and his thought out, meditated, pragmatic revenge and, finally, elimination by the colonial authorities of rivals who had risen to power, but not before creating new clashes between the groups of small networks forming that great coalition. The second major battle occurred under the government of Juan de Vargas (Picazo, 2011: 1–15); his collision with the great merchants of Manila, his fall from grace upon discovery of a batch of smuggled items and the rise to power of his adversaries, the network created around Curucelaegui, which also had at its centre a certain, though not decisive, division.

Above all, however, we should make a small methodological consideration in regard to the networks we shall visualize. To economize space, and for clarity in network mapping, we shall only observe their general structure, and the nuclei that formed them. Hence, we shall ignore peripheral centres, sacrificing them to give greater definition, but also, and not to overlook them entirely, we shall study some specific examples of connection between these two parts (Robinson, 1976: 71–86).

2. Systematic Relations and Network Structures

Visualization of the mapping of the two large global companies, the EIC and the VOC, and their development in the Asia-Pacific area clearly illustrates not only the strategy followed but also the two main areas that maintained their cohesion as a system: first, centralization, and secondly dispersion. An example will be given to clarify that. The VOC – at least throughout the 17th century – was located in the central area of business transactions: Batavia. The strategic location of this port, geographically equidistant from the two major economic regions (the Indian Ocean on one hand, and the Pacific on the other) and in close proximity to major spice producers caused it to become the headquarters for the expansion of Dutch trade.

Each of the ports of the network, of which we have only represented the core ones, maintained continuous contact with its own hinterland. Thus, for example, operations were deployed in many villages and cities in Japan (Osaka, Edo, Miyako, Akumora, etc.) from the great Hirado factory and political relations activated that allowed trade throughout the entire area of Japan. However, some external actions were also undertaken by Hirado along the nearby coasts, such as those carried out in Korea. Therefore, the relationships made with secondary ports, with an average of five ports or towns, should be added to the results from this nuclear network of ports. In addition, and as a methodological issue, it should be added that the network presented here is a basic one and limited by space, and when laid out on a printed page could not be displayed clearly. Hence, the multiple relationships established between different nuclear ports of both companies are also omitted, one can assume in a pragmatic way. Some of them had more or less constant correlations and others, undoubtedly, were totally sporadic. Another of the basic features that allowed Europeans gradually to control the market was the high capacity to generate, share and transmit information, thereby strengthening the contacts between all agents and making it subject to headquarters. However, one factor that partially limited that control was distance and, consequently, the time required for a message or an agreement to be reported, analysed and approved. Below are some examples of how that dynamic could be affected in transmission with some examples.

Figure 1. Dynamic of merchants' correspondence

Sender node	**Receiver node**	**Speed (in days)**
Xinonaxeque (Japan)	Hirado	10
Bantam	Texel	181
Bantam	Hirado	115
Osaka	Hirado	13
Patani	Hirado	31

The main factor that should be taken into account in assessing a network is its fragility. The weakness of the network of the two global companies, or its effectiveness, would be measured by two items that, in principle, appear to be at opposite poles. The weakness would be subject to the number of core sites that unite the whole system, so that the lowest number of them determines their maximum vulnerability. If a network only has one nucleus and it collapses, the entire network falls. On the other hand, if there is more than one, a total collapse will not occur and the network remains partially operational. Thus, the VOC would be much weaker than the EIC as it only had a main port connector, rather than the EIC's two.

Moreover, their potential would be measured by the total number of connectors implemented: the larger the number, the greater the efficiency. In response to these circumstances, the network efficiency rating of the EIC and VOC, and their weaknesses, leads to the conclusion that the East India Company's dynamic was far more effective than that of the Dutch and was less exposed to a total collapse. The EIC maintained two large hub ports, compared to the VOC's single one; it also had 35 nuclear ports compared to the Dutch's 32 and 180 secondary supply ports compared to 160 for the Dutch. The strategic importance of many of these ports should be mentioned, both for their location in the system and for ease of communication with the interior. Most of these large nodes, for both the VOC and the EIC, were located at the mouths of rivers (Danvers and Foster, 1899: 24–25), which facilitated the spice trade from the interior and thus increased activity, thereby reducing shipping costs very significantly (Kathirithamby-Wells, 1986: 333–351).

The use of secondary ports on different routes established by the VOC illustrates a long-term global strategy based on its presence in all areas: Africa, the Middle East, India, Japan, China, and the large platform of producer islands such as Java, Sumatra, Makassar, etc. But, undoubtedly, the great resources Sri Lanka provided, especially spices, and good relations with the local business and political oligarchy gave a significant shift to the VOC's trade policy. In National Archives Sri Lanka there are many documents about it, in "Ducht Central Governnement of Coast Ceylon", as "Treaty between Adam Westerwolt on behalf of the V.O.C. and king Raja Sinha II at Batticaloa on 1638 May 23", and the "renovations of that treaty between governor Joan Maetsuycker and the King at Galle on 1649 August 6"; "Memorandum on the relationship between the V.O.C. in Ceylon and King Raja Sinha by the 'directeur generaal' at Batavia Joan van Hoorn, to the governor-general at Batavia, Willem van Outshoorn. 1700 August 23" or "Treaty between the V.O.C. and the late king of Calicoilang, as recorded by the coopman Schoorl on an oral rendering by the 'gricol' and Brahman Wammena of Calicoilang. Certified copy made in Galle 1662 April 25". Colombo (Arasaratnam, 1958; Goonewardena, 1958 and Pieris, 1924) and Negombo were, little by little, eventually consolidated in the latter part of the 17th century as the main ports of transactions (Sluiter, 1942: 29–41).

Figure 2. Efficiency

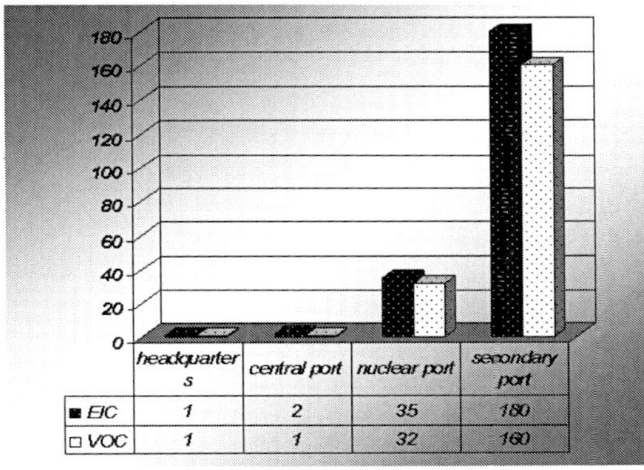

The structure of the VOC network could be visualized thus:

Figure 3. VOC network

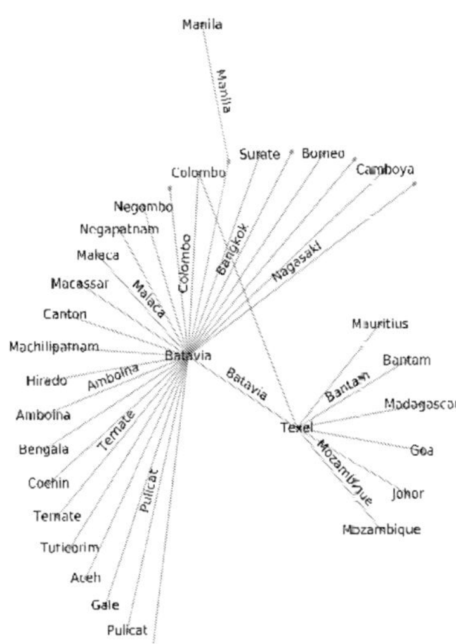

The commercial and, therefore, spatial budgets of the EIC did not differ much from those proposed by the VOC. About EIC Ron Harris explains: "This institutional innovation took place at a critical juncture in Eurasian trade. It also created a decisive transition in the history of the organization of business, from partnerships, guilds and regulated corporations, to widely-held joint stock corporations. In retrospect, it signified a turning point in the economic development of England and in the rise of Northwestern Europe to dominance within Europe and globally. Studying the formation of the EIC as a cooperation-enhancing institution may contribute to our understanding that innovative institutional design, no less than inventions in sail, navigation and gun technology, was crucial in the rise of Europe and England to global commercial dominance" (Harris, 2005: 12).

Throughout the entire period studied, the EIC's main port was, undoubtedly, Bantam (Lawson, 1993), followed by Surat, Bombay, and Bengal. That is, for the most part the majority of trips centred on an area that facilitated the connection between the two areas that made up the subsystem, the Indian and the Pacific Oceans. India eventually took on a much greater specific weight and some factories in south-east Asia were abandoned as they were unable to increase production. This was the case, among others, with the port of Aceh, abandoned in 1669 by the EIC because it was unable to exceed 600 tons of pepper annually (Basset, 1999: 625–643).

The detailed study of the nuclear ports makes evident the following models:

a. Those located in the south-east Asian area prevailed between 1600 and 1625, with Bantam being totally dominant, although Surat in the Indian Ocean was also noteworthy and second in relative weight.

b. In the second quarter of that century, until 1650, the south-east Asian area was also worth mentioning. Bantam acquired, if anything, much more weight, eventually assuming 73% of all trips. That was a significant increase compared to the first period. Surat maintained its presence, although with a lower value.

c. From 1651 to 1675, Bantam's economic dynamic remained steady and became the EIC's main site for transacting business.

d. The EIC's implementation policy changed markedly at the end of the 17th century. India was set up as the company's main area of commercial traffic and fundamental changes were made. Business in Surat (Guppta, 1979) waned and other areas of hitherto minor importance were strengthened as in the case of Madras, Bombay, and Bengal. Major ports also arose for the first time in the Pacific, especially Amoy. The network implemented by the EIC would be:

Figure 4. Network relations by the EIC

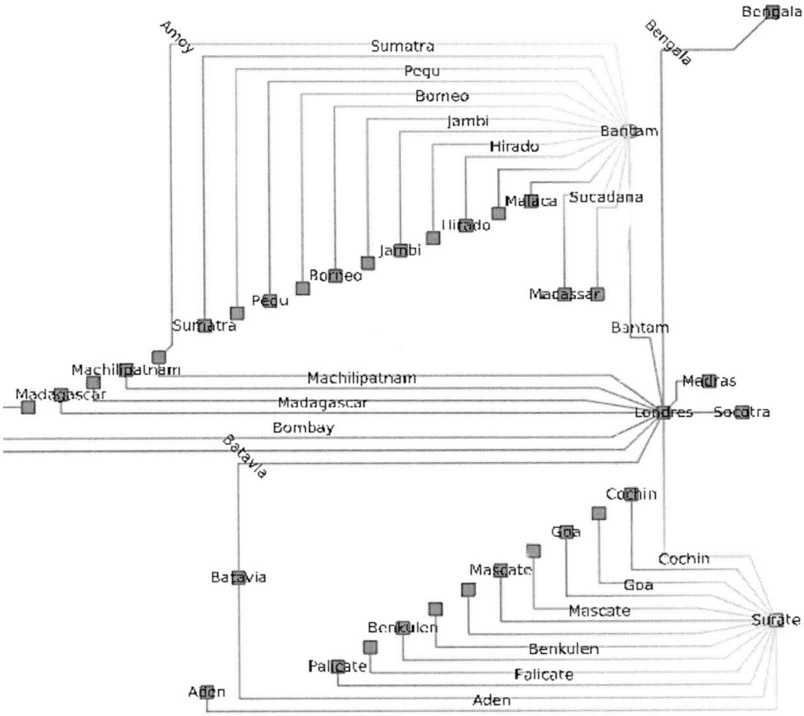

While networks implemented by the VOC and the EIC in the Asian-Pacific area were highly efficient, arising from great dispersion, the one made by the Spanish monarchy was, undoubtedly, much weaker and more precarious than the previous two. Regarding the comparison of the centrality of the network, the Spanish one differed in one very remarkable aspect. Whereas a single nuclear port remained active, remaining stable throughout the colonial period (the other two were mutating and the share of trade changed over time), the relationship with headquarters became dispersed over time and space by a long route of intermediate stops, each with its own sometimes conflicting decision making elements. Manila had a clear economic policy but it was monitored and reduced from New Spain and, in many cases, restricted by law from the court. The end result was the implementation of a highly coercive policy, greatly limiting the trade capacity of the Philippine archipelago. Another of its peculiarities was that the number of nuclear ports that supplied the Spanish merchants in Manila (whose own number scarcely changed) was around the same number that the VOC and the EIC had, in this case 34. But the remarkable thing was that some of them were controlled by their economic rivals in the area: two from Portugal, one from Holland, and another from England.

However, the crucial problem for the islands was that they did not transact business directly with any secondary port. There was no policy for provisioning through factories that stimulated the surrounding areas or other ports or production areas. The Spanish limited themselves to expecting that goods were shipped to the port of Cavite by native merchants. They even allowed, in some cases, the Portuguese, Dutch, and English to perform this trade. This, of course, greatly weakened their capacity and converted it into hostage trade, which was the case of the ports such as Canton, Ching, and Amoy, which accounted for 84% of all trade. With decreasing economic activity due to, among other factors, sudden and violent political and social changes experienced in China, the Philippines traffic plummeted to minimal levels. This situation led to a time when the crisis was so severe that authorities were forced to change their attitude; they had to make purchases of strategic items necessary for the maintenance of the islands directly from production areas. The lack of trade activity led to decisions for enhanced production in the islands of commodities considered as strategic, such as iron and wheat. However, demand for some highly specialized products made their manufacture in Manila entirely impossible, as in the case of anchors. To alleviate these needs, a boat was urgently sent to Cambodia to buy anchors worth 2,000 pesos.

Thus, after analysing the dynamic developed in the Philippines by the Spanish, totally opposite to that generated by the first global companies, we see the network mapping and more clearly perceive its weakness.

The structures of the three networks seen separately deviate from the extreme complexity emanating from their merger into a single system. That merger triggered direct relationships and also boosted secondary ports and connections to markets outside European control. Hence, we find a very high trade density in which headquarters were connected with trade bases in the Asian-Pacific area and with the whole. In addition, the relationships established with Manila and trade with Mexico, which stretched to the Caribbean and large cities, converted the area's complexity into a global system.

Figure 5. Structure of three networks

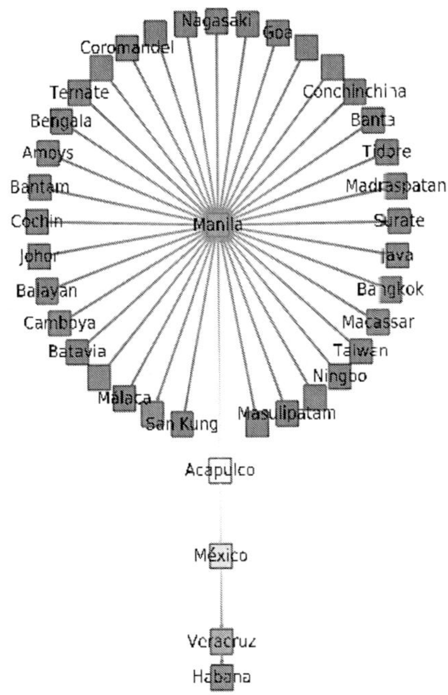

A radial mapping visualization of this system would be as follows:

Figure 6. Visualization

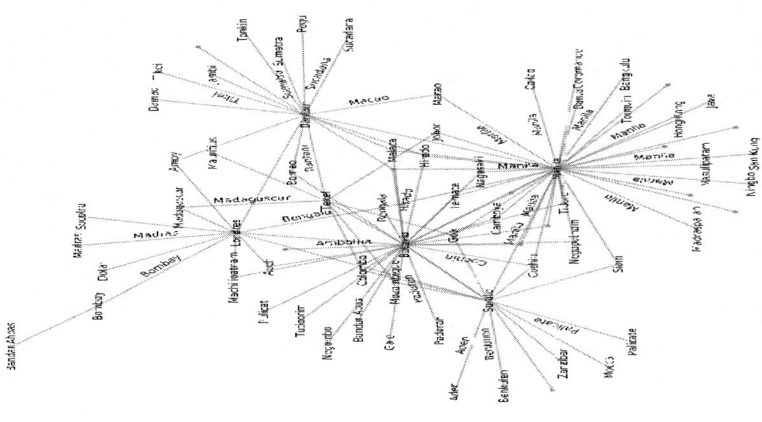

In contrast, a circular vision of it makes it easier to read in a simpler way:

Figure 7. Circular structure of networks

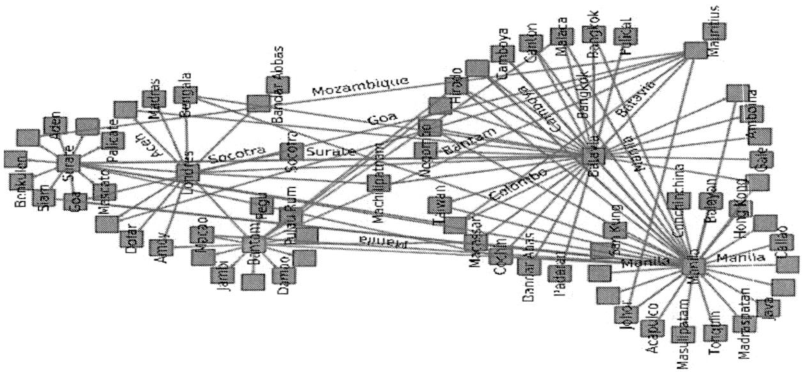

3. Invisible Networks: Structure and Collision

Manila's trade with New Spain was limited by a mercantilist policy implemented by the court that failed to understand the problems of the First Global Age. Resort to coercion and trade limitations precipitated the building of invisible networking, i.e. smuggling, generated around the archipelago's authorities. Thus, the islands' governors and captains-general such as Diego Salcedo, Juan de Vargas, Curucelaegui and others developed alliances and created true underground networks. The basic features of the structure of these networks were that forged around a key figure, a large central node that limited its operation. Other networks appeared alongside these major networks with an opposing character. The existence of a single galleon a year with connection to Acapulco and, therefore, with very limited loads caused extreme competitiveness, especially when investments were high and losing the shipment could mean ruin. Hence, in most cases the great merchants, along with council officials or judges of the court, formed other networks to try to counteract the influence of the governor and his followers. Usually, attempts were made to try to maintain a certain balance that prevented a violent clash. The inclusion of a key character in the transpacific race altered the status quo and that was when the great collisions occurred. That was the case of the aforementioned governors. A small confederation of networks finished off the government of Salcedo. He was arrested and his assets frozen and then a violent conflict erupted. The lessons of these disputes were quickly understood, so that the opposition acted differently towards Juan de Vargas. Violence was left aside and the viceroyalty justice system itself terminated the governor's smuggling.

Thus, a number of general conclusions can be reached if one looks at the mapping of the network. First, each one formed around a central nucleus that upon collision collapsed, thereby causing the entire network to collapse. Secondly, in most cases there were few connections or ties that associated them with the main court authorities, or to the monarch himself; the average was two links. Thirdly, our attention is highly drawn to the fact that each and every one of the networks studied was created by the governors of the Philippines, and had international connections, a definite reflection of global trade. Take the case of Salcedo, who achieved collaboration with Dutch agents from Batavia and Ternate, with large Chinese merchants from Canton, with the King of Siam, or with Armenian merchants based in major Indian ports (Bhattacharya, 2005: 277–322; Lang, 1981). Finally, he developed an extension on the periphery based on cooperation and collaboration from sales and production agents located in various places around the archipelago. Those agents were crucial for the network to be efficient and generate maximum profits. An example will clarify this idea. Tomás Bravo bought wax from natives from the province of Aklan for Salcedo and Miguel de Alarcón. Silvestre de Olmedo also carried out this process; he sent various amounts with a sloop to Matienzo and Salcedo. Pedro de Mondiguren, mayor of the strategic port of Ybalon, sent several wax blocks to John de Berastain, Luis de Morales and Matienzo.

Let us take a look at the network mapping of Diego de Salcedo in which links with power can be observed, in this case judges from the Mexican courts, the viceroy, or its double link with the monarch himself. At the other extreme, we find connections with the Dutch from Batavia or with the ambassadors of the King of Siam, among others.

Figure 8. Network of Diego de Salcedo

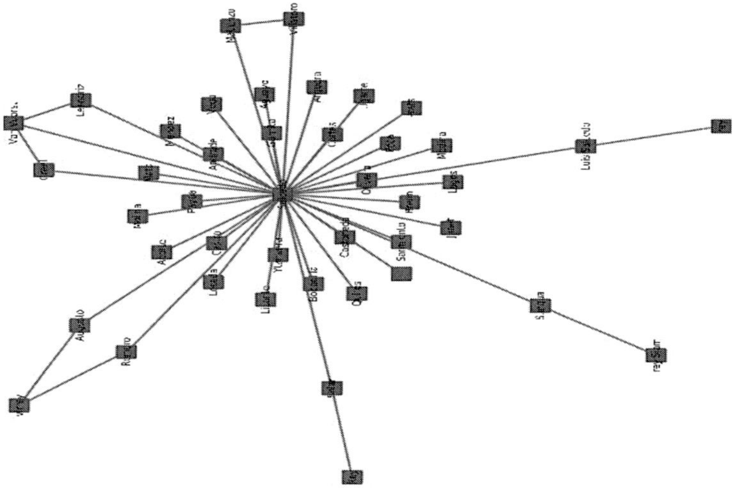

In contrast, the coalition of networks formed to combat and cancel out Salcedo was as follows.

Figure 9. Coalition of the Network of Diego de Salcedo

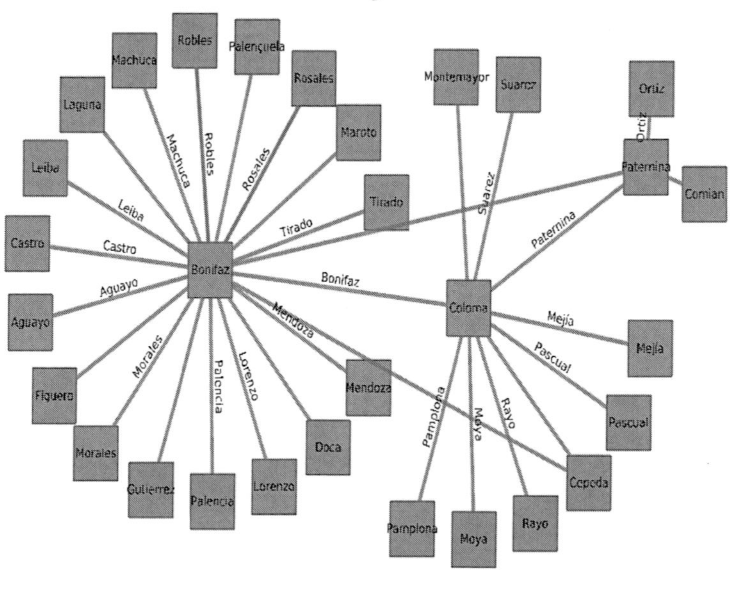

4. Conclusions

Comparison of the various network mappings studied leads to a number of general conclusions about their effectiveness, but at the same time helps clarify what their weaknesses were. First, it is worth noting the low number of steps required to move from being an agent on the periphery to the centres of decision-making power. Secondly, network efficiency was dependent on its extension, ease of transmitting information from the periphery to its centre and in the multiple roles agents executed and which resulted in profit. Only in some cases were there single nuclei the fall of which could lead to the total collapse of the network. However, in most cases, the loss of a peripheral element was automatically compensated by turning to other elements. Even in some of them, the possible collapse of one of the central nuclei could be lessened by the continued existence of a secondary one and diversion of its products to supply Europe.

The merger of all these networks into a single global system enabled a trade increase between Europe and Asia and also led to an extension of it to mainland America and the Caribbean. It was the result of the first globalization and the emergence of the first problems, including the ability of Asians to increase efficiency in packaging for manufactured goods (there was a large difference if it was made in China or in Acapulco); much higher production, and low cost, which took place in Asia and adversely impacted Atlantic trade. Other new problems also appeared arising from the big increase in production, especially in China, and the high demand for some items. The result was increased loss in quality to the point that remittances were profoundly altered and measures had to be arbitrated to reverse the precarious supply and poor quality of certain items. Finally was the disturbance created by new manufacturing dynamics by the VOC and the EIC. In the late 17th century they began a policy of outsourcing manufacturing in Asia, following the guidelines of separate companies, resulting in goods being sold worldwide as goods produced in the old European production centres.

CHAPTER ELEVEN

GEOGRAPHIC INFORMATION SYSTEMS APPLIED TO METRIC ANALYSIS OF OLD MAPS: 16TH-CENTURY MAPS

ANTONIO CRESPO SANZ[1]

"Other maps are such shapes, with their islands and capes!
But we've got our brave Captain to thank:"
(So the crew would protest) "that he's bought us the best –
A perfect and absolute blank!"

Lewis Carroll, *The Hunting of the Snark*, 1876

1. Introduction

Geographical Information Systems (GIS) are a very useful working tool for research into the history of maps and mapping. This subject is the interface between cartography, geography, history, the history of science and technology, and other related fields, and is of considerable interest to a wide range of researchers from different academic backgrounds who consider maps from different points of view and enrich and extend our knowledge overall. The use of GIS has proved extremely useful in the study of old maps, allowing them to be accessed online; facilitating metric analysis to evaluate their precision, scale, degree size, the projection used and other aspects of mathematical calculation; helping to clarify the methods used for data collection and depiction; improving the capacity for analysis of the geographic features and, above all, enabling their comparison with other modern or contemporary documents, demonstrating simply and easily every conceivable comparison and superimposition (even in 3D). GIS also provide task solutions for the

[1] Universidad de Valladolid.

automatic identification of geographical features, for texts or for storing and assigning non-spatial data related to maps.

The digitization of the collections of many archives and map libraries is under way or already completed, and their databases are available online. Although the presentation and characteristics of the material available is not yet standardized, significant advances have been made in this field. Many databases are currently being incorporated into the Spanish spatial data infrastructure (*Infraestructura de Datos Espaciales de España* – IDEE), which is responsible for online availability of the most frequently used GIS which conform to specific standards. In other words, users accessing the IDEE website (http://www.idee.es/) can consult the basic topographical maps of the *Mapa Topográfico Nacional,* and superimpose a crop, forest, or cadastral map of the relevant area or the maps produced by any official institution included in this geo-website, on a national, regional, or municipal scale. In addition to this, specific software is available to help add these maps to the GIS as an additional layer. This offers a range of possibilities for cartographic analysis, and allows statistical data to be included, associated with a specific space, or viewed in different ways, and other advantages which each user/researcher discovers and can contribute to the knowledge of the user community. This article does not explain the principles and theory of GIS and the multiple possibilities on offer; nor does it detail all the new programmes which are emerging. The aim here is to use some of the widely available existing tools and apply them to a case study comparing two old maps. We have already used these programmes for the metric analysis of a unique 16th-century map of Spain – the *Atlas de El Escorial* – to calculate the scale, degree size, and characteristics, and the working method used in the mapping with the distortions which affect it. Here we use these methods again to superimpose maps.

Three complementary systems can be used to compare two maps. First, traditional mathematical and statistical methods can be used to obtain results to evaluate the map characteristics, determine the scale, exactness, the size of the degree, position of geographical features, hydrography, territorial limits, etc. This system requires a large number of calculations and tables, making it complicated and not very instructive. The next step is to use a GIS which provides immediate visual output, is user-friendly and easily understood. Although any of the widely available programmes could be used, we chose to use Google Earth, the best-known and most widely used map viewer, which allows any cartography to be geo-referenced and superimposed on other old or modern maps. The third option is to use MapAnalyst, a very simple cartographic analysis tool which combines aspects of both the numerical and the visual options above to compare two maps. The user must define homologous or common points on the two maps and can then calculate meshes and distortion vectors, determine an approximate scale and obtain the rotation factor (Jenny, 2010). To carry out the comparative study of two 16th-century maps (the *Atlas de El Escorial* and a map of *Lusitania* drawn by Fernando Álvaro Seco) we decided to use the systems mentioned above as additional, complementary research tools. These two maps are unique and highly significant within the context of the Renaissance. The *Atlas de El Escorial* (c. 1538–1545) by Alonso de Santa Cruz is a map of the Iberian peninsula made up of an index map and twenty map pages on a scale of 1:400,000. This cartographic task was abandoned when Maestro Esquivel started work on a much more precise representation of Spain, applying the topographic methods of Gemma Frisius and Peter Apian. The *Atlas* was forgotten and stored along with other documents which had been very valuable in their day for strategic reasons, but were then considered of little interest. Later, Philip II gave it to the Library of the Monastery of El Escorial and there it still is to this day, catalogued under reference K-1-1. Looking through this cartographic document, interesting coincidences emerge between Portugal as represented in the *Atlas de El Escorial* and on the map of *Lusitania* published in 1561 and attributed to Fernando Álvaro Seco. This suggests that the *Atlas* incorporated information from some contemporary cartographies and confirms the hypothesis that this document was produced from the fusion of existing maps, travellers' accounts, itineraries, and various other sources.

2. The Atlas of El Escorial (c. 1538–1545)

The manuscript of the Atlas of El Escorial is catalogued under K-1-1 in the collection of the Library of the Royal Monastery of El Escorial, and includes an index or general map and twenty pages of maps which cover the whole of the Iberian Peninsula. These offer an unusual image of Spain during the Renaissance on account of both the scale and the volume of data contained (Crespo Sanz, 2008; Paladini Cuadrado, 1996: 35–50; Vázquez Maure, 1982: 203–215). The Atlas was received in the Monastery library in El Escorial in May 1576, with the first consignment of books Philip II sent to the monks of the Order of St Jerome, and it has been there ever since. Alonso de Santa Cruz started work on the Atlas between 1538 and 1539 – at the command of Charles V – and continued to work on it until 1554, when he went to Valladolid and discovered that a more precise and ambitious mapping project than his own was under way.

His unfinished manuscript was left in the palace library and was often consulted, judging by the wear and tear on the lower edges of the pages. Some indirect evidence helps to date the map to c 1561 and the correspondence between Santa Cruz and Charles V seems to show that the cosmographer began a map of Spain around 1538 and worked on it for several years. Further proof of the connection between Alonso de Santa Cruz and the *Atlas de El Escorial* are the tables of coordinates which appear in his *Astronómico Real*, a scientific work of astronomy written in 1556 for Charles V. A comparison of the longitudes recorded in both these documents confirms the link between the two works, and we consider that the "Corographia de España" [Chorography of Spain] from which the coordinates were obtained for the *Astronómico* can only have been the *Atlas de El Escorial*. The data in the *Atlas* were taken from existing maps, itineraries, accounts, and direct references, organized by the co-ordinates of the important cities. The points are located with an uncertainty of less than 10′, although in some areas there are larger errors, inherited from each source. To reach this conclusion, the latitude and longitude of the main towns were considered separately, compared with current data with tables of co-ordinates and using Google Earth and MapAnalyst.

The general or index map which precedes the other pages is an important innovation in the context of Renaissance mapping. The scale used is approximately 1:2,600,000, similar to other contemporary maps of the peninsula, and although it was drawn based on the other pages, the position of some towns and their symbols are different. This shows that it is not simply a reduced version, but rather an original work based on the twenty map sheets. In contrast to all the contemporary examples, it shows hardly any navigation-related features and its rounded coastline is very different from the portolan charts, making it clearly a geographical or land-based map. It shows typical features of printed cartography, such as the shaded mountains, the broken lines for the sea or the perpendicular hatching of the coast, none of which is used in the twenty detailed maps. Rivers, coastlines, and frontiers are generalized or simplified, although their outlines are approximately the same as on our maps. It is interesting to note that the Balearic Islands are not shown, nor the borders of most of the kingdoms, except Catalonia and Portugal, marked by a dotted line. The relief is depicted with shaded, schematic profiles following the main mountain axes, easily identifiable even though they are not accurately located, which affects the courses of the rivers. The border with France is marked by the Pyrenees, which are shown very wide and placed further north than they really are, to include Cerdanya, Roussillon, and part of Aquitaine within Spanish territory. Some 116 localities are shown, fewer than on other maps on a similar scale, although this is justified by the quality of the index and summary. Some of the circles used to mark the position of towns include symbols differentiating eight archbishoprics with a double cross, forty bishoprics with a single cross and six fortified or walled towns with appropriate pictograms. Some features have been exaggerated and are not shown to scale (e.g. the Tagus gorge in Toledo or its estuary in Lisbon) and there are some obvious deformations in Catalonia, Portugal, or on the Basque coast, but it is more exact than its contemporaries. In this article we use the index map in the comparative studies, since its scale and characteristics facilitate the interpretation of the images obtained with graphic methods. Nevertheless, any of the proposed systems can be applied to the twenty pages of the *Atlas* and to any other historical map.

The general or index map and the twenty map pages are bound together. They contain an extraordinary amount of information; this includes more than 9000 geographical features with their corresponding toponyms, making this one of the most detailed and complete Renaissance maps for the volume of data and the scale used (c. 1:400,000). Each sheet is numbered with its ordinal at the top edge, with the rectangular outline divided into whole degrees of equal longitude and latitude, showing that there is no projection and that the coordinates are drawn according to a flat squared grid. The first page has two plain graphic scales, one for "long leagues" and the other for "common leagues", from which we discover that the degree used is 16⅔ leagues. The scale was determined in different ways; first, measuring the longitude on the maps and comparing it with real magnitudes, giving a mean of 1:430,000 oscillating between pages and sectors, reflecting the variable precision of the map. Another way of establishing the map scale starts with the traditional equivalences between the measurements known as *pulgada* (inch), *pie de Burgos* (Burgos foot) and *legua común* (common league), giving a value rounded to 1:400,000.

The *Atlas* was a draft at an advanced stage which had to include or change some features, confirmed by the existing bald patches, crossings out, corrections, and ink stains. A final "fair copy" would have been made before it was presented but work on it came to a halt and it was forgotten, and a complete map of Spain did not become available until the late 18th century. While it has unique features, it has some similarities with three other contemporary maps. The first is a well-known map of the dioceses of Andalucia included in the *Theatrum* of Ortelius entitled "Hispalenses Conventus" (1579) where many of the same names can be found as in the *Atlas*, although their positions do not coincide exactly. Its author was Jerónimo de Chaves, connected with the Casa de Contratación, who may have used the maps by Alonso de Santa Cruz as his source. The second map to consider here is also from the *Theatrum*, called the "Carpetaniae Pars Descriptio" (1584), which represents the winding course of the Tagus between Aranjuez and Toledo at a scale of 1:250,000. This map has important coincidences with the *Atlas de El Escorial* in the towns shown and their relative positions. The third map considered here is the most unusual. This is a map of Portugal drawn by Fernando Álvaro Seco, which coincides with the *Atlas* in almost all aspects: number of towns, hydrography, coastal outline and frontiers, and with only slight differences in the toponymy. The El Escorial manuscript contains original features (bridges, towns, inlets), which suggests that both the Portuguese and the Spanish maps were drawn from the same original model, which has now disappeared.

3. The Map of Portugal by Fernando Álvaro Seco (1561)

Just as with the *Atlas de El Escorial*, the map by Fernando Álvaro Seco poses many unanswered questions, including those related to its author, a geographer and mathematician who is credited with this representation of Portugal published in Rome in 1561.[2] His intervention was apparently limited to selecting and designing a map (based on an older source) which was to be printed in Italy and given as a gift to Cardinal Sforza for his work as ambassador.[3] The text of the cartouche reveals that the promoter was Aquiles Estaco, secretary to Cardinal Guido Sforza, who dedicated this map to his master. Another text on the lower part identifies the printer as Michaelis Tramezini,[4] one of the cartographers who formed part of the school of Lafreri, the author of numerous maps printed in Venice and Rome between 1544 and 1585 (Tooley, 1982). This map of Portugal is a careful and very detailed representation, which was reprinted for over a century with only slight (and not always accurate) changes in the most famous Renaissance atlases (Cartografía, 1994). Some 20 examples of the

---

[2] Azevedo e Silva, 1998: 317-342; *Tesouros da Cartografía Portuguesa*, 1997: 143; *Cartografía impresa dos Seculos XVI e XVII. Imagens de Portugal e Ilhas Atlánticas. VI centenario do nascimento do infante D. Henrique*. Ed. Comissao nacional para as comemoraçoes dos descubrimientos portugueses, 1994; Cortesâo and Teixeira Mota, 1960: Volumes I and IV; Alegria, Daveau, García, Relaño, 2007: 1034-1061.

[3] The promoter of the primitive map may have been Francisco de Melo, between 1521 and 1524, as suggested in Daveau, 2010: 146.

[4] Michaelis Tramezini formis. Cum Summi Pontificis/ac Veneti Senatus priuilegio. Sebastianus a Regibus Clodiensis in aereicidebat.

1561 map still exist, two of which are conserved in the Library of the Royal Palace in Madrid. The best-known version of the map of Portugal was made by Abraham Ortelius: it was included in the *Theatrum Orbis Terrarum* of 1570 and mentions Fernando Álvaro Seco as the author of the original work. Efforts to confirm the author of the Portuguese map offer a particular detail in the presence on the map of a village named as *quinta dos Secos*, which was where the family of the cartographer lived; this and other small details have allowed the map to be dated to around 1557 (Alegria, Daveau, García and Relaño, 2007: 1040).

Figure 1. Map of Portugal by Fernando Álvaro Seco (1561), published in Rome. Source: Biblioteca Nacional de Portugal (BNP)

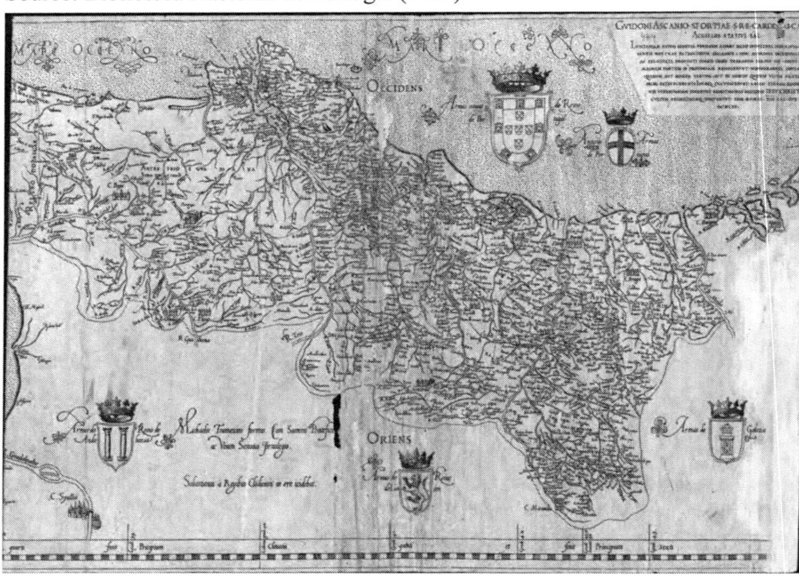

4. Comparison between the *Atlas de El Escorial* and the Map of Portugal

The surprising similarity between the *Atlas de El Escorial* and the map by Álvaro Seco led us to carry out a more exact geographical and metric study. Until recently it was difficult to find detailed reproductions of the Portuguese map and we started working with an map by Ortelius conserved in the National Library of Portugal (BNP)[5] and with a copy – obtained from microfilm – of Lafreri's maps from the collection of the Royal Palace in Madrid. We recently acquired the digital file of this map, with images complementing the examples we already had, and almost at the same time we discovered a new image in the BNP of the 1561 map (cutting off the Algarve and Galicia), in a free digital format with excellent resolution.

4.1. Towns

Once we had obtained the maps, the difference in scale between the Portuguese map (1:1,300,000) and the Spanish map had to be resolved, but this was easily done by enlarging the former – although without distorting it – until equivalent images were obtained.[6] As the *Atlas* has a general map (1:2,600,000) and twenty pages (1:400,000), we carried out two different studies. We began with the index map which although it contains little information shows certain features, traits, and cartographic techniques similar to those of the printed works. It shows the ten most important towns in the kingdom of Portugal, seven of which are marked with a conventional sign for bishopric and archbishopric, while on the Portuguese map these were drawn as a cluster of buildings. The remaining three are located by a circle on both maps. There are minor differences in the texts (*Silues* for *Siluis*, or *Lisboa* for *Lisbona*), due to the absence of a definitive spelling, so that each author transcribed the texts as he saw fit, and above all due to the Italian engraver, who made some errors which were not revised or corrected.

The second comparison was between the pages of the *Atlas* (which form a mosaic covering the whole Peninsula[7]) and the map by Álvaro Seco. Both maps had to be adapted to a similar scale to facilitate the comparative analysis, which was made even more complicated by their different orientation (the Portuguese map to the east and the Spanish map to the north). At first sight the Portuguese map appears to contain more towns, but a count of the localities on both maps reveals a similar total of

[5] *Portugalliae que olim Lusitania, novissima y exactissima descriptio.* Auctore Vernando Alvaro Secco. Antuerpiae: s.n., [157-].

[6] The index card for the map in the BNP (1575) notes the scale of 1:1,300,000 and dimensions 33.80 x 51.10 cm on a sheet 36.30 x 53.20 cm. http://purl.pt/5901. These dimensions are similar to those of the Ortelius map (37 x 68 cm) and the Lafreri map (35 x 66 cm).

[7] The page numbers of the *Atlas de El Escorial* containing Portuguese territory are: 1, 4, 8, 9, 12 and 13.

around 1200 on both. The reason for this apparent visual discrepancy is due to the scale chosen for each of the maps. On the Álvaro Seco map the scale is smaller but the lettering size is larger, so that the information appears more compact, with fewer spaces left blank. In contrast, the *Atlas* uses a larger scale, with more separation between the geographical features, giving the impression that there are fewer localities shown; in fact the number is very similar on both.

Figure 2. Representation of towns on the maps in the *Atlas de El Escorial* and on the map of Portugal

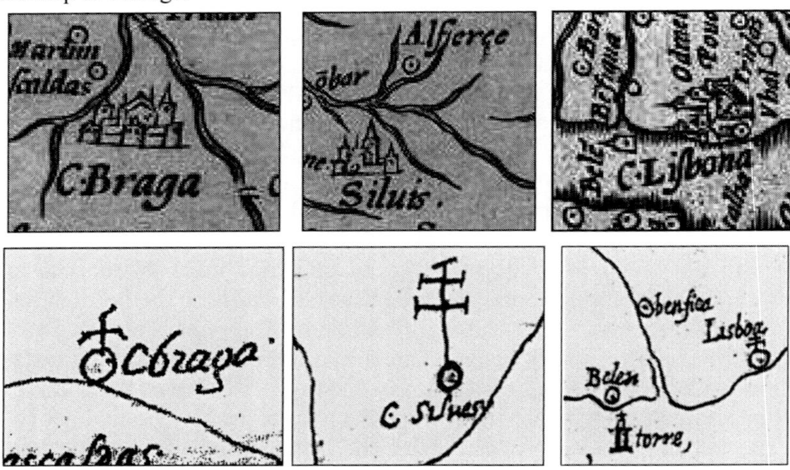

The *Atlas* uses conventional signs to indicate hierarchy: the most important cities include a single cross (bishopric) or double cross (archbishopric) in the circle and pictograms for walls or fortifications. On the map of Portugal drawings of buildings are used. One of the most significant coincidences is the *Torre de Belém* near Lisbon, which on the Portuguese map is shown on the left bank of the river. On both maps the main Portuguese cities are preceded by the letter *C*.

4.2. Coastline, Hydrography and Bridges

A visual comparison of the two documents verifies that the coastline is similar, and more approximate on the Spanish map. The same is true of the fluvial hydrography: the rivers in the *Atlas* meander less, especially the Douro and the Tagus. The Portuguese map is a printed document offering a carefully worked image, with rounded lines, somewhat exaggerated when compared with a present day map. The courses of the most important rivers coincide significantly and both exaggerate the same curves and bends, although the course in the *Atlas* is more linear and closer to the real one. The courses of the secondary rivers are similar and at some points on the Spanish map a greater number of watercourses are represented, but these are only minor tributaries which were then eliminated on the printed map so as not to make it too dense overall. On the Álvaro Seco map all the rivers are drawn with a double line, while on the maps in the *Atlas de El Escorial* only the Douro and the Tagus are drawn like this, and inside Spain they are drawn only with a single line. All the rivers shown are preceded by the letter "*R.*", on both the Portuguese map and in the *Atlas,* but there are some differences: the Portuguese map uses *Douro, Minius* or *Teio,* but in the *Atlas* the names of the rivers appear several times, spelt differently in Spain and in Portugal (*Duero-Doyro, Miño-Minho* and *Tajo-Taxo).* The Spanish-Portuguese border should coincide with the hydrography, but it has been shifted back into Spanish territory in an attempt to establish a clear differentiation between the two kingdoms. This adjustment to the frontier line produces minor differences and increases the land area of Portugal, but in general the features are very similar.

The two maps show bridges over the rivers and coincide to a considerable extent, although some of those shown on the Portuguese map do not appear in the *Atlas* and vice-versa. The location of bridges was basic strategic information as their presence on the maps provided extremely useful data for travellers and military campaigns, but this was not the main aim of either of these maps: their contents were more useful for administrative tasks and as a territorial description. On the Portuguese map the bridges are shown mainly on the smaller rivers, rather than on the main rivers (only the big bridge crossing the Mondego in Coimbra is shown). The bridges are shown as two parallel lines cutting across the river, whereas in the *Atlas* the same image does not interrupt the river. On the Portuguese map some 60 bridges can be counted: 40 in the north and 20 in the south, while in Portuguese territory in the *Atlas* the total number is 135, with the following map distribution: Nº 1: 2 bridges; Nº 4: 16; Nº 8: 20; Nº 12: 72; Nº 13: 12. In southern Portugal hardly any bridges are shown, but there is a higher density in the north, on map Nº 12, with more than 50 bridges shown between the Spanish border and Porto.

Figure 3. A comparison between the general map of the *Atlas* and the Portuguese map shows the existing similarities in the coastline, frontier and hydrography

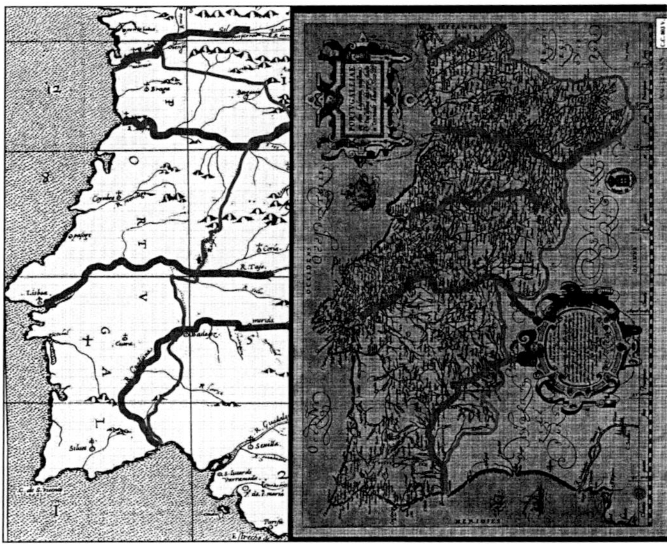

Figure 4. When the mosaic formed by the pages of the *Atlas de El Escorial* is compared with the map by Álvaro Seco, the similarities can seen between the shape of the coastline, the river hydrography and the frontier

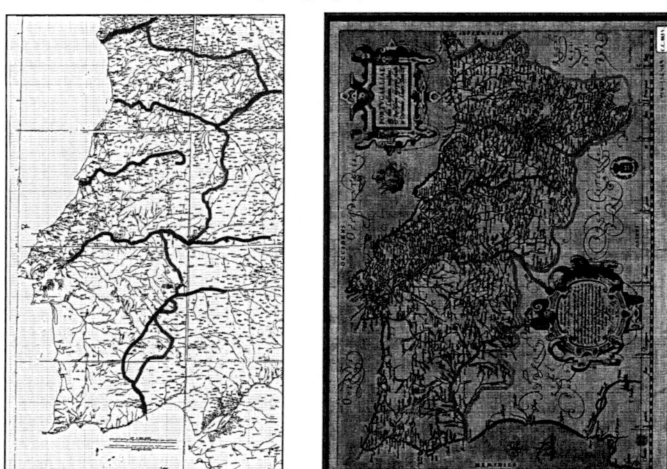

Figure 5. The bridge in the city of Coimbra is the only one on a main river and is represented cutting across the lines of the river. On the Spanish map the bridge appears as double line over the river

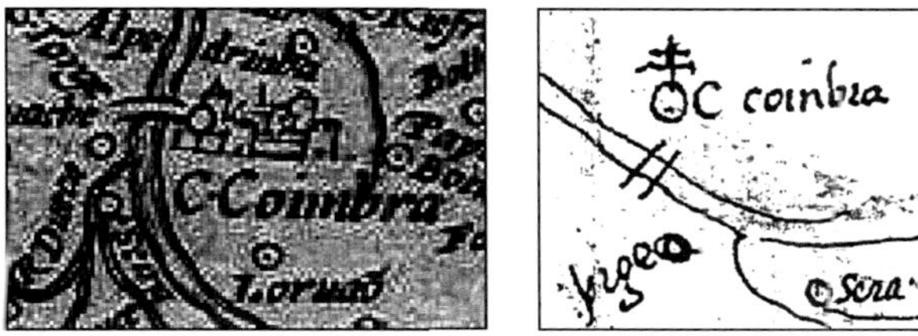

Figure 6. In the Algarve, the Portuguese map only shows one bridge for the town of "S. luzia", but in the *Atlas de El Escorial* there are two

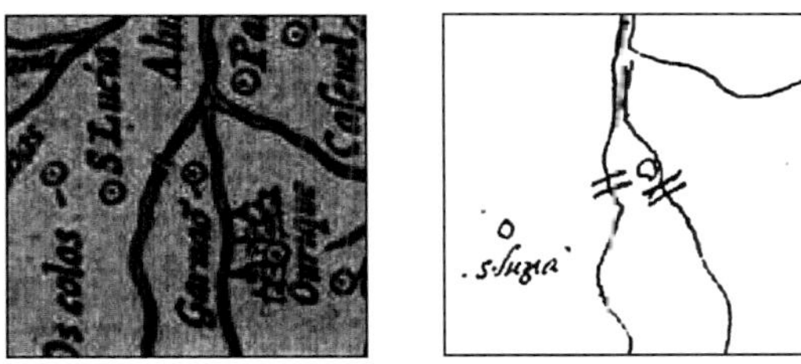

4.3. Representing the Relief

The mountains are represented in quite different ways on the two maps. In the index to the *Atlas* the most important mountain ranges are shown in schematic profile, corresponding acceptably to the text in capital letters on the Portuguese map. However, part of the drawings of the mountains in the *Atlas* have no equivalent, and some of the sierras drawn on the Portuguese map do not appear in the *Atlas* ("*Sera de Sioa*", "*Sera Do Lor*", etc.). The coincidences in the position of symbols and texts show the relationship between the two, with the Portuguese map more complete in its indications of the relief. Other texts related to natural regions, kingdoms or districts ("*Reino do Algarve", "Antreteio e Guadiana", "Riba de Coa",* etc.) have no counterpart in the index to the *Atlas*, which has fewer toponyms as it was a draft version of the map.

Figure 7. Similarities in the representation of the relief between the general map of the *Atlas* and the map of Portugal

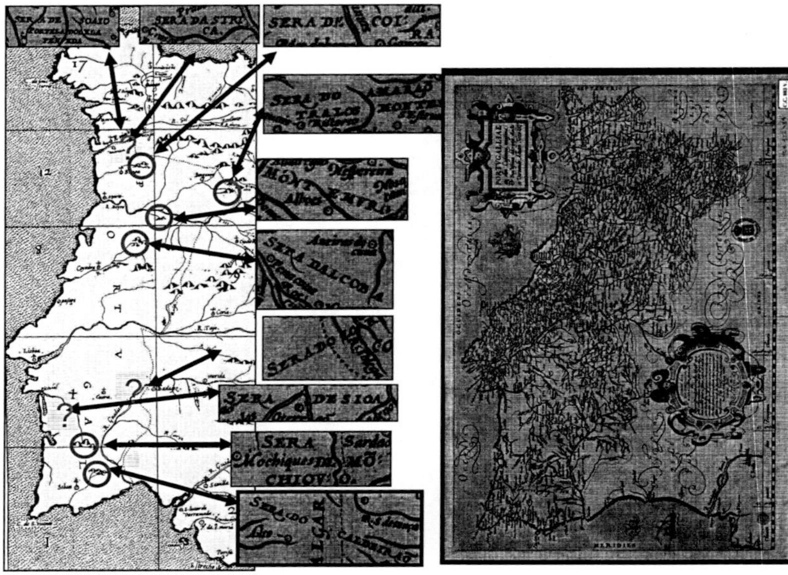

A comparative study of the orography shows little similarity between the two maps. Although in the index there was some coincidence in the location of the sierras and mountains, in the following pages this is not the case: in fact, this is where the greatest differences are found. On the Portuguese map the relief is shown with texts in capitals but on the maps in the *Atlas*, which is more modern in its cartographic representation of the features, conventional signs for mountains are used with their schematic profile, sometimes accompanied by symbols representing trees. The only exception on the printed map is a small detail to the north of Lisbon, where a group of mountains is drawn above the text "*monte iunto*". Examples of this divergence are: the conventional sign used in the *Atlas* on page 4 near "*Serpa*", beside the River Guadiana, which does not coincide with the text on the Portuguese map; or the sierra named "*Sera do lor*", which in the *Atlas* appears near Badajoz and on the Portuguese map is located further south, near Olivenza. On the general map there was a greater coincidence in the position of the orography because the cordilleras were marked in general terms, but on the pages of the *Atlas* this equivalence does not exist: schematic profiles are used and texts to mark the mountain passes, cordilleras and mountain systems.

Figure 8. On the pages of the *Atlas* the relief is represented by schematic profiles, sometimes accompanied by a text with the name of the mountain or pass. Forests may also be drawn, but are not shown on the Portuguese map

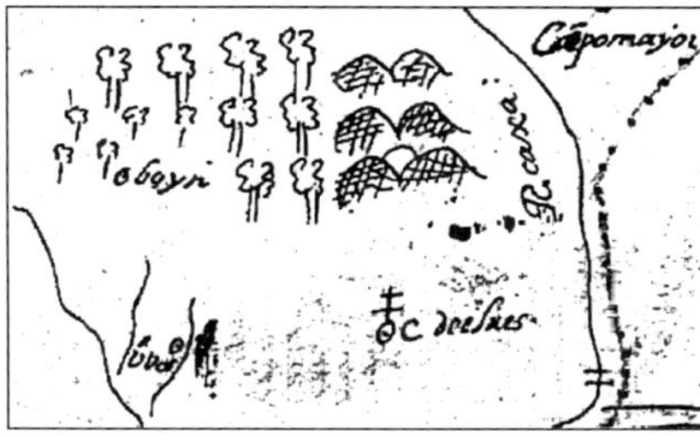

Figure 9. The only representation of mountains with schematic profiles on the map of Portugal is to the north of Lisbon. The rest of the relief is represented by text (almost always in capital letters) showing the name of the mountains or ranges.

Figure 10. In this example – near Cape St Vincent – mountains are shown on both maps, but the position of the sierras does not coincide

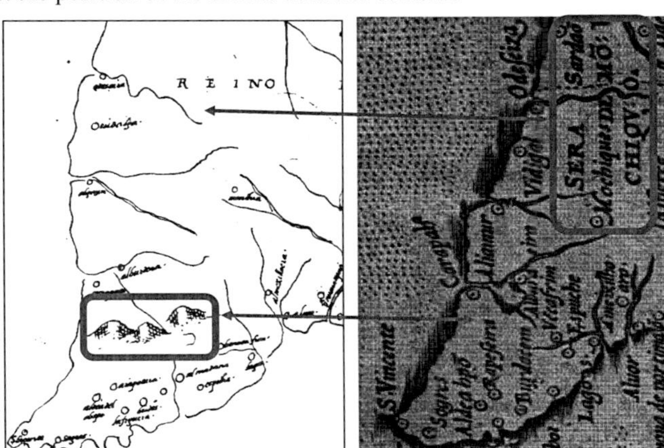

4.4. The Border and Spanish Territory

The similarity of the index in relation to the Spanish-Portuguese frontier is even more evident in the pages of the *Atlas*: similar indentations and projections, forms and peaks which the engraver rounded and shifted eastwards. The lack of information within Spain and the displacement of the actual line of the border means that Portugal is shown as isolated, separated from the neighbouring country. The information on both maps contains significant coincidences in references to the hydrographic network, the coastline, the delineation of the border, or the number and location of the cities and towns, but there are important differences in the representation of Spanish territory, which the Portuguese do not appear to have been particularly interested in. Only the coastal towns are marked – clearly evidencing the influence of the portolan charts – and there is no information at all on inland Spain (where Seville is the only city marked), some distortion of the coastline of Andalucia and an even greater deformation of the coast of Galicia.

Figure 11. The first printed map of Portugal (1561) represents Galicia with considerable distortion. The map gives the imprecise location of some cities and ports on the coast of Galicia taken from portolan charts

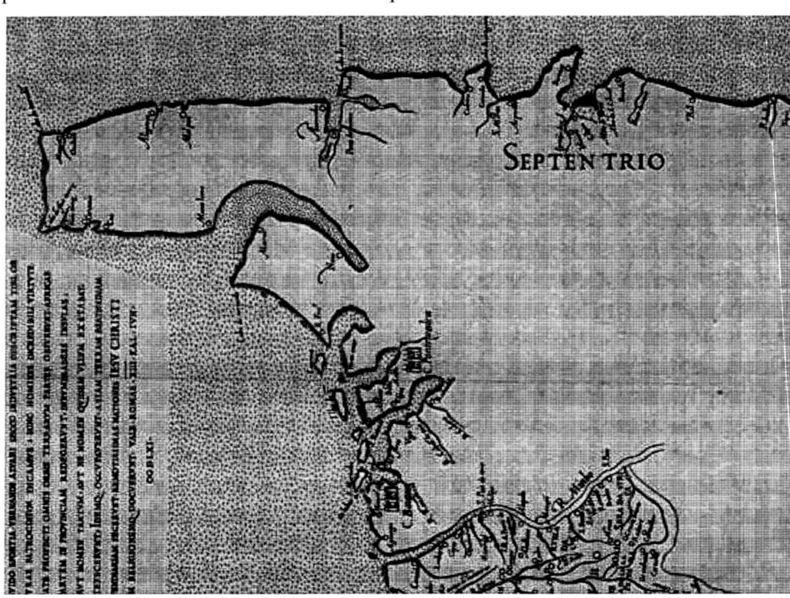

In spite of the significant similarities between the two cartographic representations, these are not identical copies of an original model. Each has specific traits which differentiate it from the other, reinforcing the hypothesis put forward from the outset: both maps are based on the same document but each cartographer included more of his own information from different sources. Alonso de Santa Cruz may have collected the data on Portugal shown in the *Atlas* during the time he spent in Lisbon in 1545, when he went there to purchase nautical charts as described in his *Libro de las Longitudes*.[8] Some historians of Portuguese cartography suspect that Santa Cruz may have plagiarized the work of the Portuguese cosmographer Pedro Nunes, but it has been shown that both of them – each on his own account – translated and extended Werner's *Commentary on Ptolemy* (1515) (Vicente Maroto, 2002: 347–381).

4.5. The Graticule and Rotation of the Map

The features which differentiate the *Atlas* are the detailed scale of the maps (1:400,000) and the presence of a real graticule to measure the coordinates. The beautiful Portuguese map, reproduced at 1:1,300,000, has no longitude grid and only includes a scale (but not a graticule) in the margin divided into degrees of 17½ leagues. It includes between 37° 30′ and 43° 30′, apparently erroneous latitude values, but rotating the graticule 9° obtains latitudes similar to those in the *Atlas*.[9] It does not seem very likely that this was drawn using topographic procedures, but like the Spanish manuscript, it must have been produced from itineraries compiled from written sources and travellers' accounts.[10]

---

[8] Santa Cruz, A, (c. 1567): *Libro de las longitúdines*, 228 fol. BNE, Mss. 9941, p. 227. En Mancho Duque, M. J. and Quirós García, M. 2005 [CD].

[9] The corrected latitudes are between approx 37° at Cape St Vincent and 43° at Bayonne, and coincide significantly with the Atlas, which also uses a degree of 17½ leagues.

[10] Cortesão considers that the Portuguese map was not drawn from triangulations supervised by Pedro Nunes, but that geometric methods were used. Cortesâo, 1960: 178.

Figure 12. To test the rotation of Álvaro Seco's map, we have drawn the parallels from the *Atlas* on it. Both maps coincide in latitude if the Portuguese version is rotated 9° in latitude, which is how much the engraver displaced the drawing to adapt it to the margins of the paper

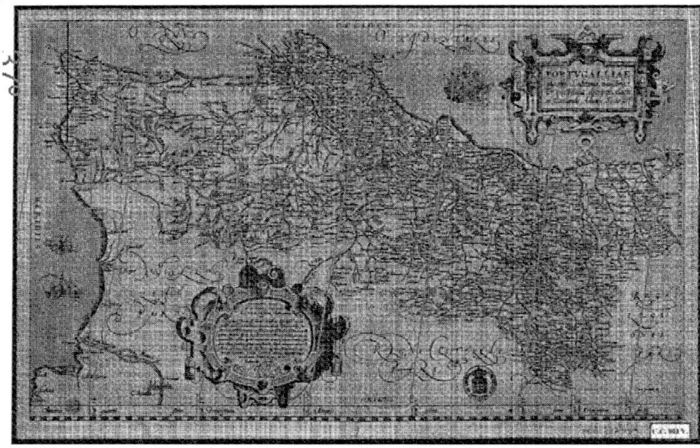

Figure 13. The texts referring to the relief are found on this map: from the Portuguese map with a green line and from the pages of the *Atlas* with red dots. Specific features are represented in the Atlas (mountain passes) and on the map of Portugal (sierras or mountain ranges). The positions of the texts and symbols do not coincide

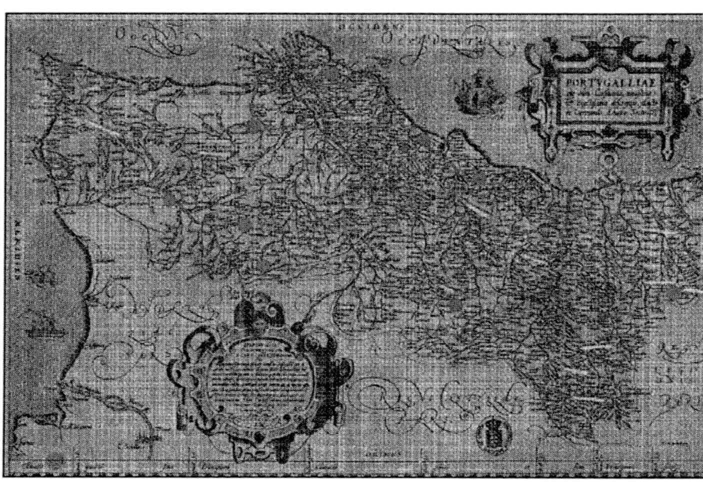

The rotation is the result of how the map was fitted to the page (instead of orienting it to the north) and the division of the degrees was superimposed on the drawing: this means that to measure latitudes the axes must be rotated about 9° with respect to the horizontal. The scale of latitudes is just an adornment and not a precise tool, as shown by the only relative coordinate measured where it gives latitude 40° for Lisbon. If we discard the rotation detected, a very similar reading to that of the *Atlas* (39° 22′) is obtained, and correcting the error caused by choosing a small degree, this approximates to the actual latitude of 38° 50′ (Crespo Sanz, 2008: 492–534). A mistake was detected (an important error due to inadequate revision before printing) on the margin of the Portuguese map where 43° 30′ is marked, although the value should appear as one degree less, but these details are commonly found on manuscript and printed maps from the Renaissance.

Figure 14. The *Atlas de El Escorial* and Álvaro Seco's map superimposed, showing the rotation of the Portuguese map

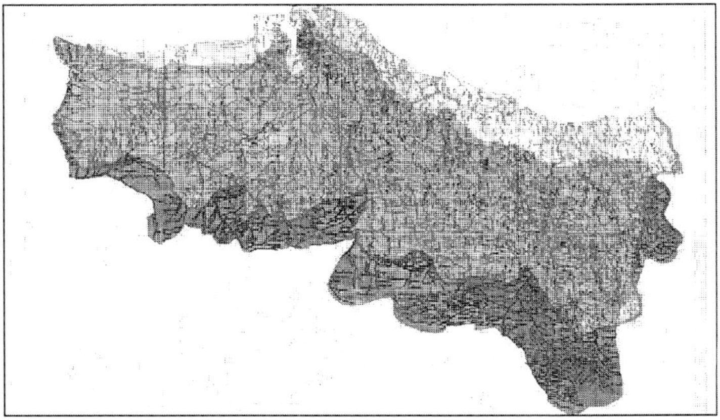

Figure 15. The maps superimposed after rotation of Alvaro Seco's map, taking the Cabo de Roca as the origin

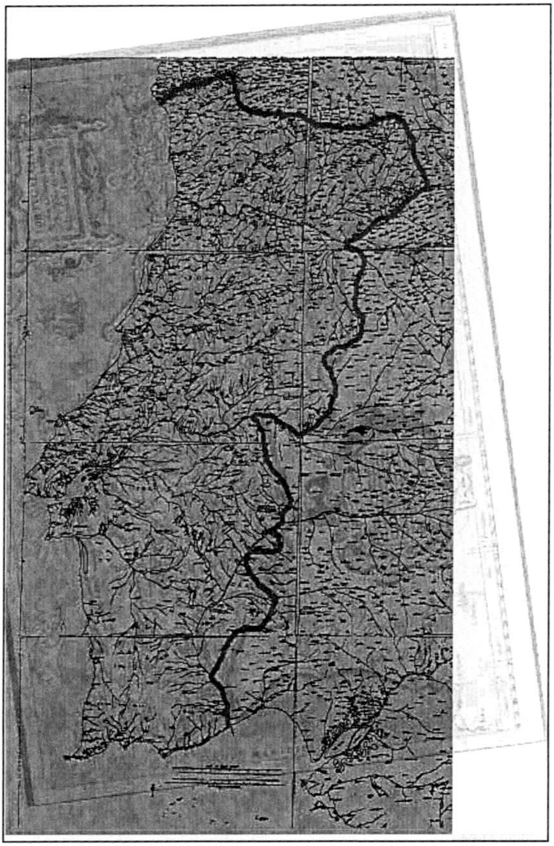

When the two maps are superimposed (using Photoshop), the results obtained confirm the hypotheses put forward: the Portuguese map is rotated by around 9° relative to the Spanish one, which makes it difficult to establish its absolute value, since different results are obtained depending on the point taken as the origin of the rotation. Once the rotation has been applied as well as a slight displacement to make both maps coincide at Cabo de Roca and the Algarve coastline, we discovered that they are not completely identical. The Portuguese map is broader in the north, with an important distortion in its prolongation towards Spain and although the frontier is similar, there are discrepancies. Both maps seem to be from a common source, but they were drawn in different ways, each with its own peculiarities: each mapmaker used his own criteria and added his own data. The author of the *Atlas de El Escorial* fitted the map of Portugal onto the graphic information for Spain he had already drawn: he adapted the scale, gave continuity to elements common to both (the frontier and the hydrography), completed some data from direct and indirect sources and added details which did not figure in the original. The rotation and the differences between the two maps cannot be attributed to the use of two different cartographic projections, but to a device of the printer to improve the presentation of the map to the detriment of the metric characteristics.

Gonzalo Reparaz and Armando Cortesao, two great experts in old maps, had already discovered this similarity, although they considered that the *Atlas de El Escorial* had been copied faithfully from Álvaro Seco's cartography (Reparaz Ruiz, 1950: 75–82; Cortesâo, 1960: 178). Cortesao considers – and we agree with him – that the map of Portugal was not drawn from "topographical triangulations" under the supervision of Pedro Nunes (as some experts claim), but instead using itineraries with estimated distances. Some contemporary historians, including Geoffrey Parker, consider that the Portuguese part of the *Atlas* is taken from a better quality map, arguing that rivers are shown more clearly on Álvaro Seco's map, but in fact they are only drawn with a double line and their course is not more accurate (Buisseret, 1992: 130–134). The main difference between the two maps is in the aesthetic aspects; the *Atlas* uses conventional symbols to show an ecclesiastical hierarchy or fortifications in the main cities; it includes a large number of bridges, adds woods, and includes relief represented by schematic profiles and other details to improve it. The Portuguese map uses a more traditional form of representation, with conventional, classic signs; it does not give information on woods and there is a surprising absence of some data which is included in the *Atlas*, such as the names of the islands in the estuary of the Tagus. Neither of the maps shows roads, a feature which was not commonly included in maps until a later date.

Figure 16. The Spanish map shows the names of the islands at the mouth of the Tagus, but these are not identified on the Portuguese map, perhaps from lack of space

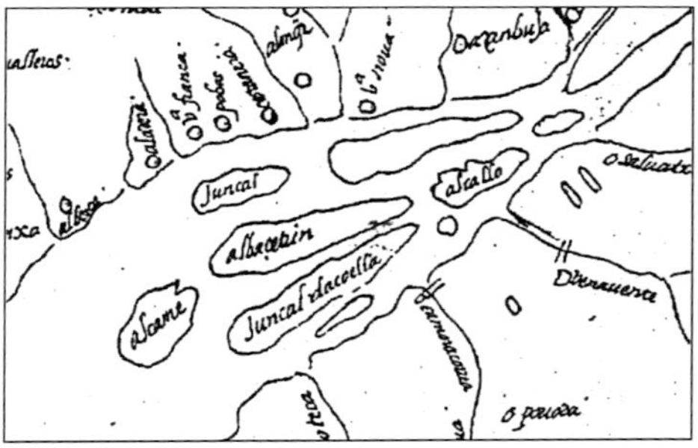

Both maps pioneer a new way of representing the terrain which could be called "terrestrial cartography" or land mapping to differentiate it from nautical charts. Neither of the two includes the compass rose; instead they have coordinate graticules (incomplete on the Portuguese map) or the scales of leagues (only on the *Atlas*) to measure distances. There are no cartographic fantasies depicted on these maps, such as the *sacro promontorio* and other fictitious places which were traditionally included on maps of the peninsula. The first edition of the printed map contains the coats of arms of Portugal, Galicia, Leon, and Andalucia on the sea and on Spanish land, to fill in blank spaces where there is no information given. The maps by Ortelius eliminate this political and administrative information and return to the more decorative features – exuberant lettering, ships, and sea monsters – characteristic of the Dutch school, which sacrificed some data to achieve a more artistic effect.

Figure 17. The *Atlas* and Álvaro Seco's map eliminate fantastic information, such as the *Sacro Promontorio*, which appears on early maps of the Peninsula. However even the prestigious mapmaker Ortelius continued to include these fantasies in his *Theatrum*. Shown from left to right: images of the *Atlas*, the map by Álvaro Seco and the map by Ortelius

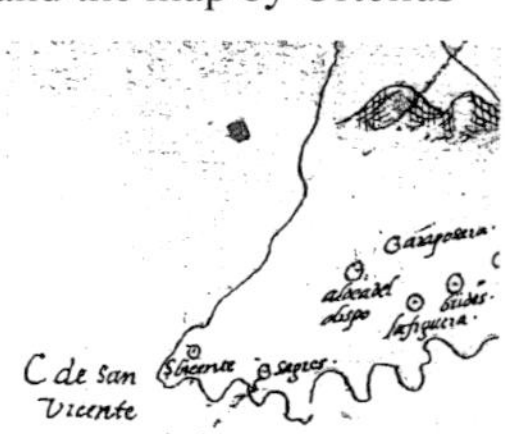

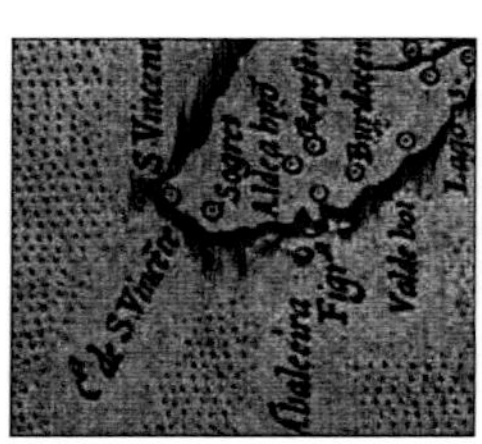

Figure 18. The similarity between the two maps is evident, but in one of them the text is horizontal and in the other vertical, there are differences in the spellings and there are some variations between the two, which may all lead to some confusion: e.g. on the fourth page of the *Atlas* chosen here, the text reads "*omote*", "*unarcos*", "*tavarede*" and "*maiora*", with one circle left blank. On Álvaro Seco's map the text appears as "*mote*", "*buarcos*", "*tavarede*", "*burobem*" and "*maiora*"

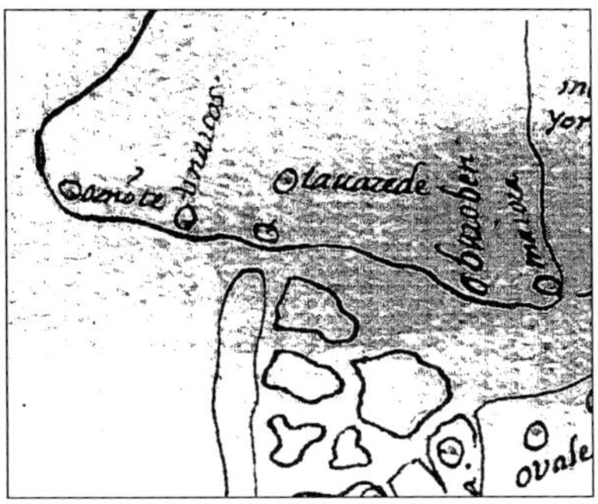

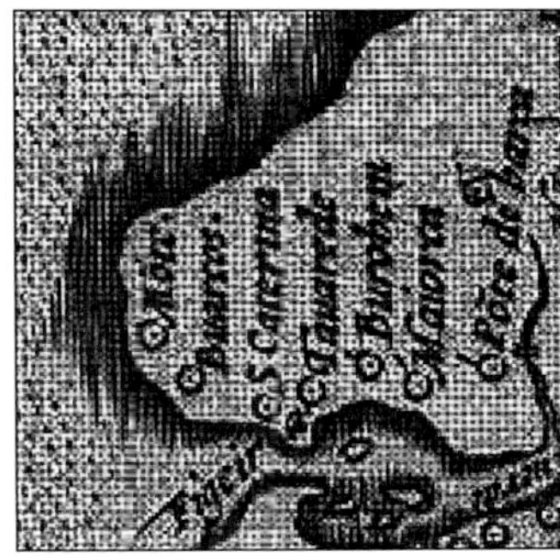

Figure 19. On the map by Álvaro Seco some cities are duplicated, the result of careless revision: e.g. "Cezimbra" which is shown on both sides of a small river. In the *Atlas* it appears only once, which shows that the Escorial manuscript is not just a mere copy of the Portuguese map. On the Cabo Espichel, Álvaro Seco names the church as "*nosa sura do cabo*", but on the fourth page in the *Atlas de El Escorial* the mapmaker seems to be unsure of the name, noting it as "*mi senora?*"

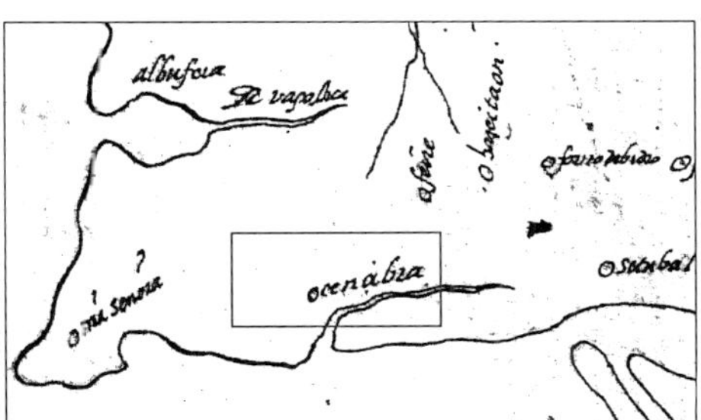

That there is a common source for the two maps comes as no surprise, as the relationship between Spain and Portugal was very close in many ways and the ruling families were related. At the same time, however, there was fierce political and territorial rivalry to control the land use (Oliveira Marques and Alves Dias, 1995: 262–263). This meant that the bilateral relationships were complex, although the constant two-way flow of people facilitated the interchange – a blend of trade, personal relationships, and espionage – of cartographic information, in spite of the efforts of each country to keep this information secret.

Figure 20. In the *Atlas de El Escorial* "Miranda" is given a special symbol to distinguish it from other towns. On the left, different symbols used on the map of Portugal; (above) on the map by Ortelius and (below) on the map in the Royal Palace in Madrid

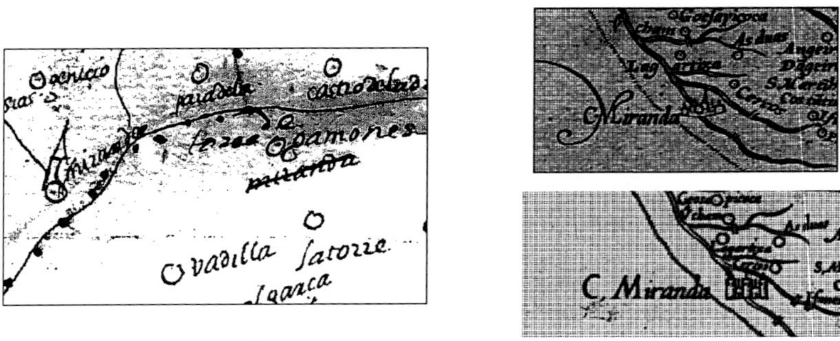

4.6. The Hamburg Codex and the Parchment in the Spanish Royal Academy of History

To consider the hypothesis that the *Atlas de El Escorial* and the map of Portugal by Álvaro Seco both come from an earlier map which has now disappeared, a manuscript conserved in the Hamburg Library (c.1526) must be included in the discussion. Known as the *Hamburg Codex*, this document contains a list of the geographical coordinates of places in Portugal. Dedicated to the fourth son of King Manuel of Portugal, this document contains more than 1500 toponyms corresponding mainly to towns (with 29 related to the relief and 8 to regions) arranged in alphabetical order and giving their geographical co-ordinates. Any doubts as to the origin and content of this document have been clarified by Suzanne Daveau: the co-ordinates were taken from an existing map, carefully measured and transferred to this sumptuous codex (Daveau, 2001a: 7–39; 2001b: 97–127; 2000: 11–38; 2008: 3–17; 2010). It was a gift for the Infante D. Alfonso (1509-1540) on the day he was made a cardinal in 1526 and must have had an accompanying map, although this has now disappeared. The number of towns and the coordinates noted in the Hamburg Codex are very similar to those which appear on Álvaro Seco's map and to those in the *Atlas de El Escorial*, confirming that a source map existed. Any relationship can be ruled out with other crown inventories produced at that time, such as the *Numeramento* of the Portuguese regional districts drawn up between 1527 and 1532. It should also be noted that the *Atlas de El Escorial* is not related either to other similar inventories carried out in Spain between 1525 and 1540 for fiscal and statistical reasons (Carretero Zamora, 2008).

Figure 21. Page from the *Hamburg Codex* (*Codex in scrinium* 136, Fol. 2 v. and 3 r.) or *Longitudo et latitudolusitaniae*. Source: DAVEAU, S. (2010): *Um antigo mapa corográfico de Portugal*

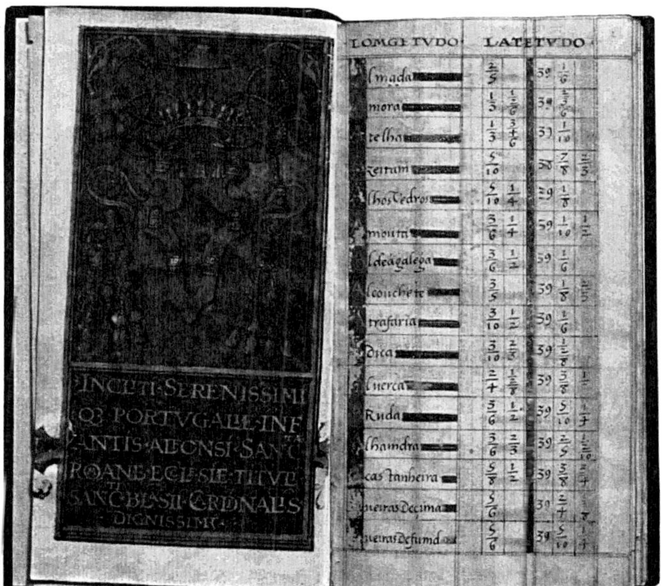

The parchment conserved in the Spanish Royal Academy of History (*Real Academia de la Historia*) – discovered by chance during the restoration of a book – contains a manuscript map linked to the map by Álvaro Seco, to the *Atlas de El Escorial,* and to the Hamburg Codex (Manso Porto, 1999: 30). It belonged to the Royal Academy collection and it occurred to someone to reuse it: it was cut up and used to cover one of the books in the library. Some time later, when the book was being rebound, it was discovered that on the reverse side of the vellum there was a map on a scale of 1:400,000 covering the south of Oporto, the Laguna de Aveiro and the Viseu region. It shows compass roses and courses, but it is not a nautical chart: it is a geographical land map. Existing references link it to a document which Jorge de la Barbuda presented to Philip II on entering his service, but we consider that it is earlier. The information it contains is almost identical to the same area in the *Atlas de El Escorial*: with the same scale, equivalent distances between towns, the same hydrographic network, bridges over rivers located in analogous positions, a very similar coastline (coinciding in the particular representation of the estuary of the Aveiro and the mouth of the Douro in Oporto), almost the same towns, in similar positions in relation to the rivers. There is no room for doubt: the data represented in the *Atlas de El Escorial* and in the fragment of the map found in the Royal Academy are very similar. The parchment is coloured, the most important cities are represented in schematic profiles and there are more bridges over the rivers, but there is no graticule of coordinates: instead there is a dense network of courses – added when the drawing was complete – covering all the features. A pencilled graticule can also be made out (with each side measuring 2.5 cm) using the directions of the cardinal points, which may have been used to place the data from another map or to transfer the data to another document. The toponymic characteristics show that it is Portuguese in origin and it may have been the source used by the mapmakers of the *Atlas de El Escorial* and the map of Portugal to draw their maps. There are some significant differences between this parchment and Álvaro Seco's map (names of rivers, the number of towns shown) and some place names have been added in different hand, with more careless lettering and more diluted ink. The map was initially dated to around 1600, but we consider that it was drawn between 1554 and 1561, and this is now reflected in the new index card in the digital map library of the Royal Academy of History. It is oriented to the east and we have calculated exactly the same scale as is used in the *Atlas de El Escorial*, 1:400,000. The map must have been larger and was cut up to reuse as the cover for another book. It is in very poor condition where it was sewn to the cover, and at the points where some pieces of the paper used as filling in the binding process are stuck to it.

Figure 22. Map of the Aveiro region conserved in the Royal Academy of History. The area shown has the same scale and features as the equivalent page in the *Atlas de El Escorial*. Source: *Real Academia de la Historia*

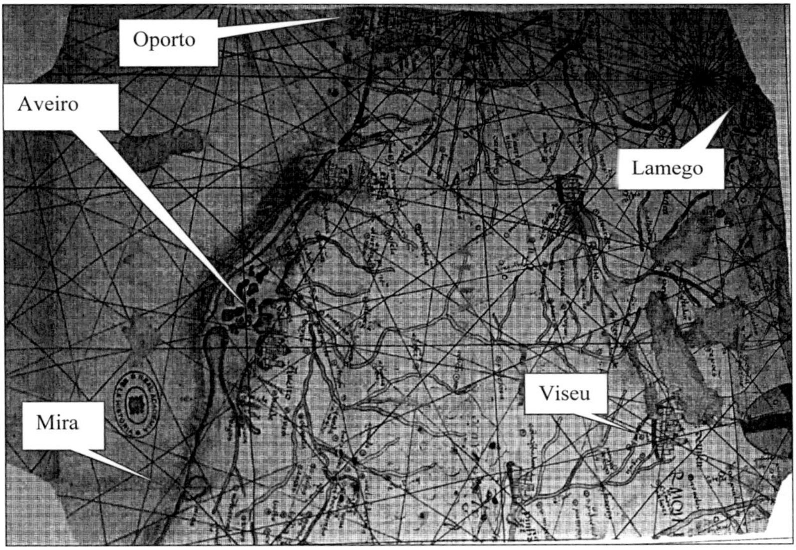

5. GIS Based Methods and Tools for the Analysis and Comparison of Old Maps

The metric analysis of a map enables an understanding of how precise and exact it is, the distortions or the changes made by the engravers over time and allows it to be compared with other documents to establish similarities and differences. From the wide-ranging possibilities of mathematics, computer systems and GIS, we have used three complementary procedures to investigate the characteristics of these cartographic documents: a traditional system with tables of coordinates; comparing the maps using homologous points with MapAnalyst; and the superimposition of images using Google Earth. The classic method is comparing the coordinates, which consists in drawing up a table recording in columns the coordinates measured on the old map in question, those of other contemporary maps and those of a present day map. The differences, means, averages, modes and other statistical parameters are calculated, a Gaussian curve is constructed and the out-of-tolerance values are obtained.[11] Erroneous numbers are discarded and for greater clarity the numerical differences are drawn on the map. This procedure was used here for the metric analysis of the *Atlas de El Escorial*, and the comparisons between the two maps were adjusted using Photoshop, where each image can be given a different level of opacity to highlight the differences.

MapAnalyst offers another possible way of checking the characteristics of a map and comparing them with other cartographic representations. With this free software homologous points are established on both maps and the distortions between the two are obtained. This programme is easy to use, but has minor drawbacks in the design and program algorithms, which make it less useful for some analyses: when comparing two maps with the same projection the results are excellent, but if an old map is compared with a modern one, the visual comparison is not totally correct as the present-day map (OpenStreetMap) is shown in UTM.[12] In spite of this disadvantage, very interesting results were obtained when the *Atlas de El Escorial* and the map by Fernando Álvaro Seco were compared, as this programme does not distort the maps, and displays the displacement vectors or distortion meshes respecting the original documents. First we compared the mosaic formed by the pages of the *Atlas de El Escorial* with the map of Portugal (1561) conserved in the Library of the Royal Palace in Madrid. We used 34 homologous points; this immediately showed up a rotation in the Portuguese map of around 8° and some distortions which warned us that the maps are not identical.

[11] All these operations can be easily carried out using Microsoft Office Excel.

[12] Although the programme internally transforms to geographical co-ordinates in order to compare them, the final view of current maps is shown in UTM projection.

Figure 23. Comparison between the mosaic of maps from the *Atlas de El Escorial* and the 1561 map of Portugal. MapAnalyst shows the distortion mesh rotated 8° and some points – shown as gray circles – where there are significant differences in the position of towns

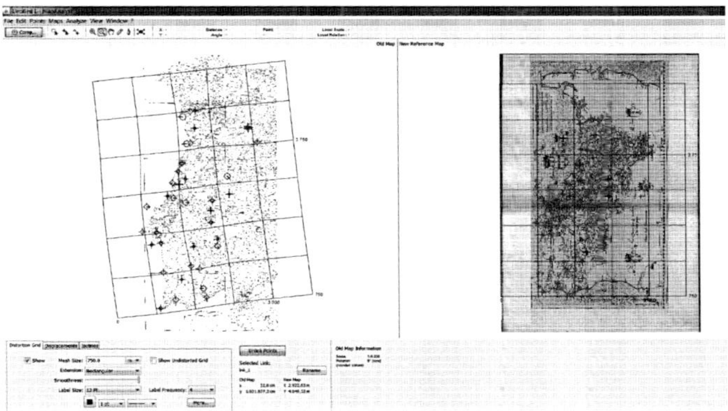

The following image shows the comparative study (using here 84 common points) between the page mosaic from the *Atlas de El Escorial* and the present-day map using MapAnalyst. The Portuguese part of the Atlas is displayed rotated and some points, especially in the south of Portugal and in the Cabo de Roca area, present the largest displacement, all with similar direction and size.

Figure 24. Results obtained with MapAnalyst comparing the Portuguese part of the *Atlas de El Escorial* with the present-day map from OpenStreetMap (http://www.openstreetmap.es/) defined as a free world map, a database with global coverage using information provided by users. This is an example of collaborative mapping which nowadays seems a very successful initiative

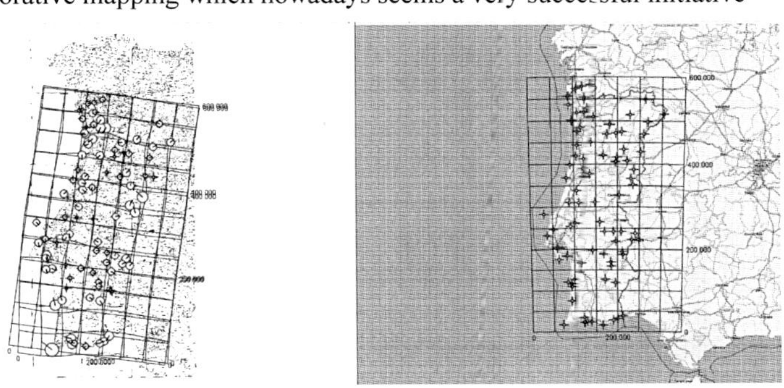

Inevitably, the similarity between the two maps of Portugal we were using had to be tested: the one from the National Library of Portugal and the one drawn by Ortelius for his *Theatrum*. Using around 30 homologous points we were able to prove that the two maps are identical and that Abraham Ortelius copied Fernando Álvaro Seco's 1561 map in its entirety. He only changed the cartouches and omitted the escutcheons of the kingdoms, but the rest of the data – including the graticule and even the errors – are the same.

Figure 25. Comparison of the two maps by Álvaro Seco. One is conserved in the Library of the Royal Palace in Madrid and the other in the National Library of Portugal. The latter, drawn by Ortelius and published in his *Theatrum Orbis Terrarum* (1579), is an identical copy of the 1561 model

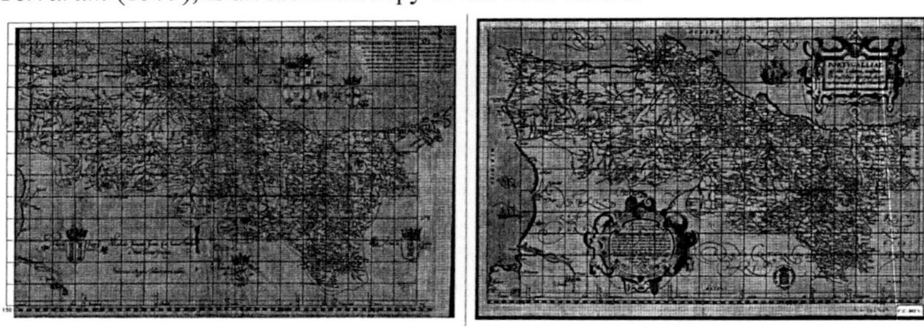

The parchment found in the Academy of History was our next objective and this was compared to the corresponding sector from the pages in the *Atlas de El Escorial*. This case was more difficult, as this area appears on two different pages and these are not a perfect match, as the reproductions of the pages have been obtained by photographic processes without corrections. To be able to compare them correctly the original conserved in the Library of the Monastery in El Escorial would have to be scanned directly, which is currently not possible. On the other hand, the parchment in the Academy of History had been cut, stitched and stretched, and used to cover another book, which has added further distortions to those expected in old vellum with the passing of time and changes in temperature and humidity. Taking these problems into consideration, the comparison was made and we discovered that the distortion mesh is not affected by rotation, as in the previous case study, and that the vectors (represented by gray circles) present minor displacements of localities grouped by sectors. Using MapAnalyst enabled us to prove that there is a close relationship between the *Atlas de El Escorial* and the parchment in the Academy, and all that was left to do was to compare the Portuguese map with the above parchment.

Figure 26. Homologous points between the parchment in the Academia de la Historia and the *Atlas de El Escorial*

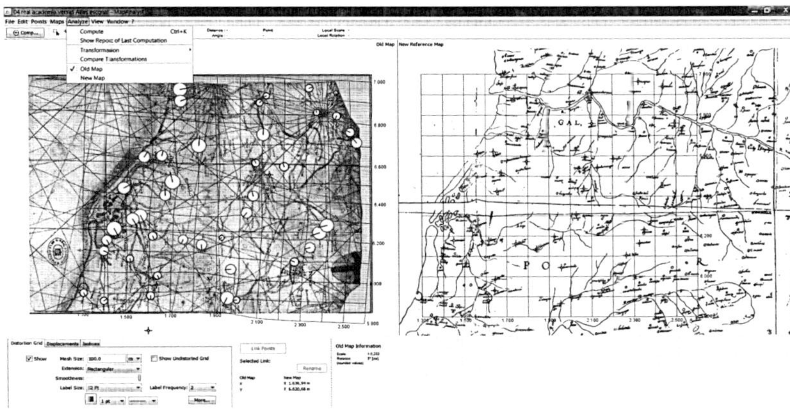

When we compared the parchment from the Academy of History with the map by Álvaro Seco, we discovered that the distortion vectors are very large, much larger than in the previous case, which led us to consider that there is a closer relationship than we had supposed between the parchment from the Academy and the *Atlas de El Escorial*; the proof is that it has a pencilled grid drawn which may have been used to transfer information to the *Atlas*.

Figure 27. Homologous points between the parchment from the Real Academia de la Historia and the map of Portugal included in the *Theatrum* of Ortelius. Important discrepancies can be seen in the position of the localities shown

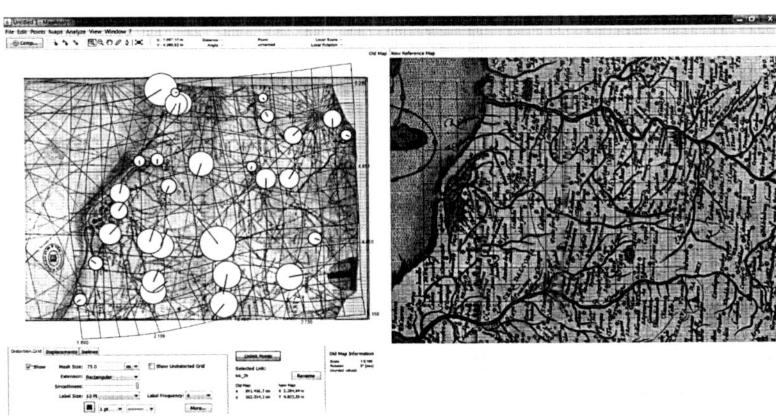

MapAnalyst offers a range of possibilities for comparing metric qualities of maps, evaluating distortions, common characteristics, and differences, but one map cannot be overlapped with another to compare two cartographies directly. This solution is provided by Google Earth, which since it appeared in 2005 has offered an option where a map in .jpg format can be superimposed on currently available images (satellite, map, or orthophotos). The image of our old map can be displaced along X and Y, rotated and the scale adjusted until it is correctly superimposed on the current map, to show the distortion directly and immediately. The transformation parameters and co-ordinates to locate, scale and geo-reference our old map are saved in a .kmz file which must be linked to the initial image. Google Earth is straightforward and user-friendly, with the advantage that it works with geographical co-ordinates (latitude and longitude) like most old maps. In our case it was useful to corroborate the conclusions obtained from the initial examination of the map and to confirm the values calculated in the metric analysis. The first superimposition was of the *Atlas de El Escorial* (Fig. 30), where parallel 36° near Gibraltar was taken as the reference line. From this point the image was rotated and amplified to find a similar scale. The result was obvious: the areas can be clearly seen where the profile of the index map from the Atlas de El Escorial does not coincide with the cartography provided by Google Earth (in gray).

Figure 28. Superimposition of the index map from the *Atlas de El Escorial* on Google Earth and the dialog box with transparency option. This program was used for a quick and easy graphic application of a Helmert transformation (translation along x, along y, rotation and scaling)

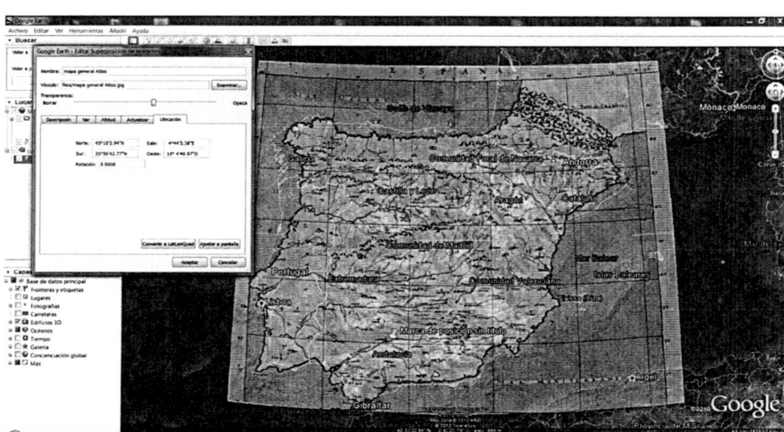

The same process was used for the map of Portugal by Fernando Álvaro Seco, although in this case the city of Lisbon was taken as the main point. Without losing sight of this reference we rotated the map of Portugal and scaled it until we considered that the dimensions of both were similar. Geo-referencing 15th–18th-century maps is very problematic and there are no fixed rules. In some cases a parallel and a meridian have to be taken as the coincidence point, in other cases two or three cities have to be used in an attempt to maintain the scale, while in others it is best to keep the coastline balanced to achieve the superimposition. These problems disappear in maps which use geodesic methods and graticules with precise co-ordinates, but this did not occur until the late 18th century. Figure 31 shows the rotation of the Portuguese map (around 9°, although the exact value cannot be given), as well as the distortion in the Algarve and in the north of the country. In relation to geo-referencing, the work carried out by the members of the CartoVIRTUAL project (2009–2011) must be mentioned, to create a distributed Virtual Historic Mapping Resource (http://www.cartovirtual.es) developed on a website which meets the requirements to form part of the Spanish Spatial Data Infrastructure (*Infraestructura de Datos Espaciales de España* (IDEE). Part of the DIGMAP proposal, where six European libraries joined together with the aim of creating a Digital Library; this now allows access to collections worldwide. CartoVIRTUAL aims to create a *Cartoteca Nacional Histórica* [National Cartography Library] within the framework of the IDEE. It includes the tools needed to set up a map server based on OGC services, with a search engine to locate old maps and a browser for geographical searches and queries. This is a flexible programme with the requisite tools, conforming to international standards, which incorporates MapAnalyst to study geometric and projective map characteristics (Fernández Wyttenbach, Bernabé Poveda, 2010).

Figure 29. Superimposition of Álvaro Seco's map of Portugal on Google Earth, where the rotation undergone by the map to adjust it to the paper can be seen. Once the map is geo-referenced, the parameters are saved in .kmz format

6. Conclusion

Research into old maps has been enriched by the use of the tools included in GIS-based programmes. In this present case study – comparing the Portuguese territory depicted in the *Atlas de El Escorial* with the map of Lusitania attributed to Fernando Álvaro Seco– these complementary tools have been used to confirm some hypotheses put forward during the traditional cartographic analysis and as a result they have revealed particular details which had previously gone unobserved. Both maps have many similarities: the coastline, the number of towns shown and their location, the border, the rivers and, to a lesser extent, the relief. However, there are minor differences which suggest that these are two different versions of an earlier map, now lost. There are two documents which may be related to this source map. The first is the Hamburg Codex (c. 1526) which contains a list of geographical co-ordinates for places in Portugal which coincides significantly with the localities found on the Álvaro Seco map and in the *Atlas de El Escorial*. It is highly probable that this list of co-ordinates was taken from a map which also provided the source data for the Spanish and Portuguese maps. The second document is a parchment conserved in the Royal Academy of History which contains a sector of Portugal (between Aveiro and Oporto) which shows extraordinary coincidences with the two maps which are the subject of our research here. The metric analyses were carried out using traditional methods (measuring map co-ordinates, drawing up tables and statistical calculations) and tested with MapAnalyst. For the comparative studies of the maps initially Photoshop was used, although in fact Google Earth provides the simplest and quickest option for map superimposition. The use of these complementary systems confirmed our initial hypothesis and thanks to MapAnalyst we discovered that the parchment in the Royal Academy of History is especially similar to the *Atlas de El Escorial*.

CHAPTER TWELVE

PORTING NATURAL LANGUAGE TIME EXPRESSIONS INTO TEMPORAL DATABASES: A CORPUS-BASED METHODOLOGY[1]

MARTA GUERRERO NIETO[2]
AND ROSER SAURÍ[3]

1. Introduction

When we think about time (e.g. when does something happen, for how long does it last, how are different events temporally related, etc.), we understand it in terms of sequentiality and inclusion: events can be ordered one before the other, they can be included in, or include, another one, or they can overlap. Moreover, we use calendars, which sort days, months, and years linearly, and clocks cyclically marking seconds, minutes, and hours.

However, beyond this simplistic conception of time that we have at an intuitive level, reasoning on time-related aspects involves complex information processing, ambiguity resolution, and inferencing processes. Other operations involved are:

- Reasoning using different levels of granularity, like minutes vs. weeks:

(1) a. *The meeting started 10 **minutes** late.*
    b. *These last few **weeks** it has been really hot.*

---

[1] Acknowledgements: This work has been partially funded by the Spanish Government project Holopedia, TIN2010-21128-C02-02 and the Marie Curie Reintegration Grant PIRG04-GA-2008-239414.

[2] Universidad Politécnica de Madrid.
[3] Barcelona Media.

- Reasoning with imprecision or vague information (fuzziness):

(2) a. The *Gulf War ended* ***a long time ago.***
b. The workers met ***during the summer***.

How long ago does "*a long time ago*" denote in (2a)? Similarly, the time expression in (2b) does not specify the year nor the precise temporal point (or points) when the meeting(s) took place. What is more, the use of "*summer*" is ambiguous – a strict interpretation (from summer solstice to autumn equinox) differs from a more relaxed one (during the hot period of the year in the northern hemisphere).

- Referencing to implicit temporal information. For instance, the expression "*in May this year*" cannot be computed without access to the specific year being referred to. Similarly, the expression "*three days ago*" requires us to know the point in time when it is uttered in order to interpret the calendar date that it denotes.

The system of time expressions in natural language provides speakers with the necessary tools for communicating temporal information effectively in our everyday life. We are as ambiguous as we need to be, choose the level of granularity that we want, or establish different temporal referents.

In addition, we can always use precise measurement systems for anchoring certain time expressions to a timeline, or for quantifying the duration of events. However, time expressions in natural language do not naturally fit into formal systems for temporal representation and reasoning. Natural language is flexible, ambiguous and often times underspecified, allowing us to conceptualize the world around us from different perspectives, whereas formal systems are rigid in nature and thus impose constraints on aspect such as levels of granularity and vagueness.

There are currently a variety of models for temporal information representation and processing. Pelekis et al. (2005) and Johnston & Weis (2010) offer a good overview of temporal databases (DBs) recent and current research. Furthermore, Geographic Information Systems (GIS) have been subject to numerous investigations on how to store and reason on time (see for example Yuan, 2009). Moreover, in the last decade or so, time information has become a central topic in the field of Natural Language Processing (NLP), resulting in the creation of temporal models at the basis of tools for extracting, representing and reasoning on temporal information as expressed in natural language text. This interest is attested, for example, in Mani et al. (2004), or by the increasing number of events in the NLP research community that revolve around time information processing. For instance, the approval of an ISO specification scheme for representing temporal information in natural language (ISO-TimeML, ISO DIS 24617-1: 2008), and the several editions of the TempEval competition, one of the tasks in the SemEval International Workshop on Semantic Evaluations.[4]

This chapter addresses the issues of porting time expressions (also known as "timexes") from natural language text into temporal DBs (or by extension also GISs), and puts forward a working methodology for carrying out the process. The work relies on an analysis of corpus data obtained from ModeS TimeBank (Guerrero & Saurí, 2012), a corpus of Modern Spanish (17th and 18th centuries) annotated with temporal and events information according to the specification language TimeML (Pustejovsky et al., 2005), and spatial information following the SpatialML scheme (Mani et al., 2008). This corpus was created within the framework of the DynCoopNet project, dedicated to the study of trade network cooperation over the 14th to 19th centuries. In the next section we explain the main features characterizing time expressions in natural language. Section 3 presents some common concepts employed to describe different approaches to time, whereas section 4 overviews the main formal systems addressed in this chapter: Geographic Information Systems (GISs), databases (DBs) and the markup language TimeML. Finally, we put forward our porting methodology in section 5.

2. Natural Language Time Expressions

Time expressions in natural language tend to have a closer connection to our subjective perception of time than to the objective, physical temporal processes. Thus, underspecified, vague or fuzzy expressions like "*for some time*" or "*in days*", are common in natural language. One can distinguish between **external time** (i.e., time references that can be associated with physical processes and have accurate measurement methods) and **internal time**, also called subjective or perceptual time.

Vagueness is another feature of temporal information in natural language. Time references can be precise (e.g., "*at 9am on September 11, 2001*") or more vague ("*one of these days*"). Imprecision also affects interval durations, which in some cases may not specify one or both of its boundaries. For example, "*until May 3rd*" expresses a duration that ends on the third day of May but starts at an underspecified point in time,

[4] http://aclweb.org/aclwiki/index.php?title=SemEval_Portal.

whereas "*for long time*" does not encode either its starting or its ending point.

Time expressions in natural language also appeal to different levels of **granularity**. Time can be measured in minutes, hours, days, weeks, months, or decades. In many Western cultures we employ the units in the Gregorian calendar and the clock, but other calendars or measurement systems are used as well, such as the astronomical calendar, which sets different year periods or seasons (*summer, fall*), the school calendar, where the division of time is in quarters, semesters, etc., the working schedule, the Catholic liturgical calendar (*Lent, Christmas, etc.*), among many others.

Natural language time expressions can express different types of **reference**: absolute (e.g., "*in May '68*") or relative ("*before summer*"). Relative references are usually expressed by adverbs ("*now*", "*after*"), prepositional phrases ("*for three days*") or noun phrases ("*last year*").

Finally, there are also natural language expressions indicating **frequency**. Like other time expressions, these can be precise ("*the first Monday of each month*") or vague ("*often*"), and can have referential character ("*every Monday of the month*"), etc. They can also provide different levels of granularity ("*every three minutes*" *vs.* "*every spring*") and be based on different calendars. What is most distinctive of these expressions is that they structure time along the temporal axis in a cyclical way, thereby allowing the representation of recurring events in time.

3. Formal Concepts

The different formal approaches to time have resulted into various models of temporal knowledge representation. Each of them establishes the units to be considered, their properties and the possible operations they can support. In what follows we present the main concepts employed for modeling time. We base the presentation on the classification by Frank (1998), which distinguishes different types of time. Aigner (2007) groups them in the following aspects: scope (or extension), structure, scale (or density), metric (or granularity), and viewpoint (or reference framework). The last two are the ones appearing the most in GISs.

Scope or extension (open, closed). Property applied to time units, which classify into *instant* (time point unit, without duration, thus of length zero), and *interval* (unit of time with length). The extension of instants and intervals may be *open* or *closed*, depending on whether their location in the timeline (in the case of instants) or limits (intervals) are defined. An instant is undefined if it is included in a duration or frequency of undefined location ("*They met* ***in May***"), and an interval is undefined if its beginning and ending times are not specified ("*The war lasted* ***for 3 years***").

Figure 1. Scope dimensions

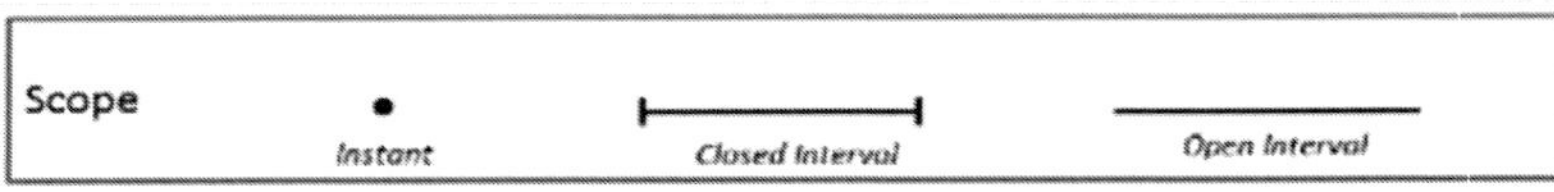

Structure (linear, cyclic, spiral). A *linear* time treatment considers that events have a sequential development in the timeline. It understands time as having a starting point and a forward direction, while in a *cyclical* treatment events are recurring. Finally, the *spiral* conception represents events in a timeline where cycles are repeated, as in "***Every Monday*** *I go to the gym*".

Figure 2. Structure dimensions

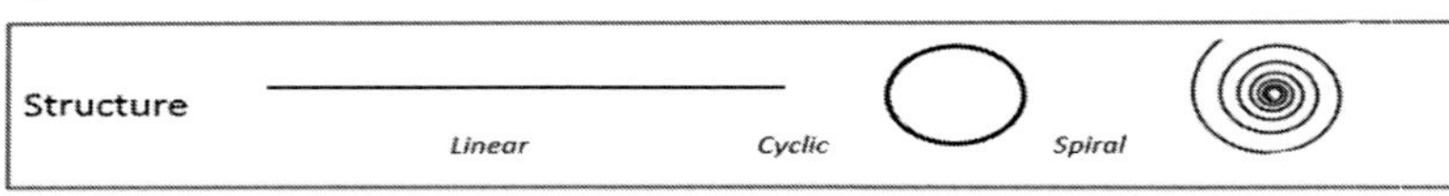

Scale or density (ordinal, discrete, continuous). Time can be conceived as *ordinal*, according to which only the ordering of events is known but not their duration or absolute position. The *continuous* model assumes that there is always a further instant between two points in time. The *discrete* model is used in maths and computer science and considers that every point in time has a preceding and a successor point.

Figure 3. Scale dimensions

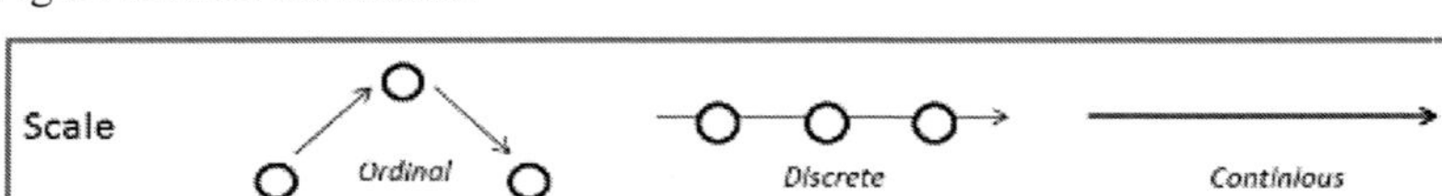

Metric or granularity (zero, single, multiple). Determining the level of division of time units (granules) based on a measurement system which, at the same, time establishes a temporal order. The calendar and clock systems have a well-known set of granules (understood as established and bounded time units): minutes, hours, days, months, etc. Granules are conceived according to a number of features: their duration is fixed, they are finite, repeated cyclically and do not allow further units between them.

In formal systems that handle temporal information, a chronon is set as the minimum unit to measure time. All time expressions, regardless of their granularity, will be normalized and expressed in chronon-based terms.

Figure 4. Granularity dimensions

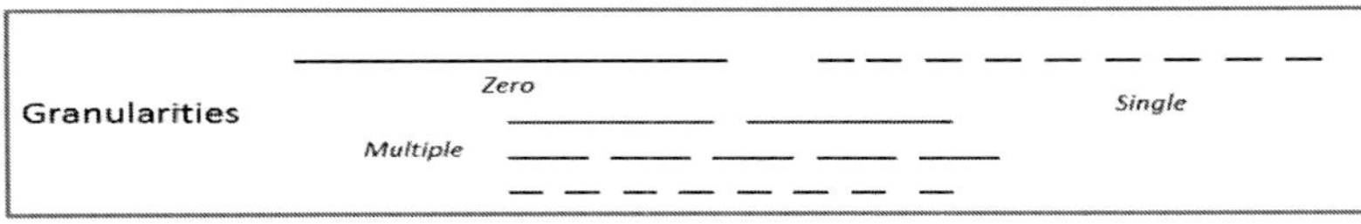

Viewpoints or reference framework (absolute, relative, branching). In an *absolute time* every instant or interval is set to a specific value, whereas in a *relative time* events are related among them without necessarily knowing their position in the universal time. The relative time is more subjective than the absolute time, because it assumes a more flexible structure which is defined in terms of the relations holding among events. By contrast, the absolute time does not involve a critical (o comparing) view. A *branched time* (*branching*) describes possible states of the world and supports alternative versions of future or past times. From this perspective, each time entity will have several predecessors and successors. The combined perspective of linear past/branched future allows representing certain and known past and an open future (Schreiber, 1994).

Figure 5. Viewpoint dimensions

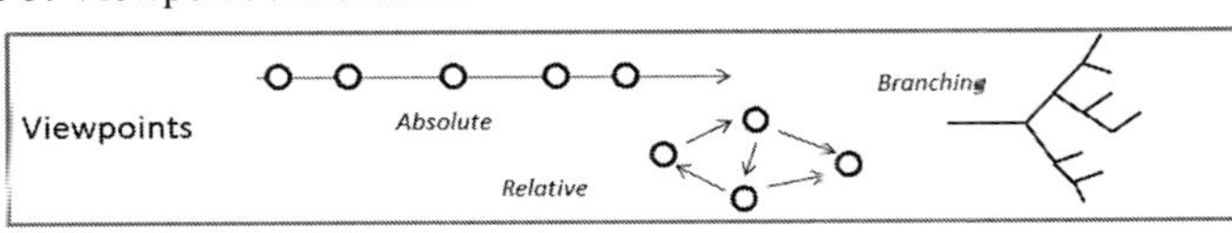

4. Time Information Modelling Systems

This section introduces the different conceptions of time underlying the main formal systems targeted in this chapter: GISs and temporal DBs, on the one hand, and TimeML (Pustejovsky et al., 2005), the markup language for representing time expressions in natural language, on the other.

4.1. Geographic Information Systems (GISs)

Starting from the 80s, there are several GIS models addressing the representation of the time component, such as the Snapshot Model (Armstrong, 1988), the Time-Stamping Model, the Event based SpatioTemporal Data Modelling, ESTDM (Peuquet, 1995), the Spatio-Temporal Entity-Relationship Model, STER (Tryfona & Jensen, 1999), the Spatio-Temporal object-oriented Model (Worboys, 1994; Wachowicz, 1999), the Moving Object Data Models (Güting, 2000), etc. In most of them, the conception of time is linear, discrete, and absolute, and therefore are not capable of handling cyclic and relative models of time.

4.2. Databases (DBs)

The component for analyzing and visualizing time in GISs is based on the formal structure of DB management systems (DBMS). Currently, most DBs includes the field date for storing time information (such as in Oracle or SQL). The SQL-92, standard for DBMSs, supports completely specified time expressions. In general, time is conceived as linear, of discrete density, and of absolute reference. However, cyclic time is partially allowed as well (Jensen, 2000). With respect to granularity, DBMSs are greatly constrained, since it has to be established previously. There are two types of unit, DATE and TIME, with two minimal granularities: day and second.

In the DATE unit, time is expressed with the format YYYY-MM-DD, where the Ys, Ms, and Ds respectively express the year, month and day digits. The TIME unit satisfies the format HH:MM:SS, where the Hs, Ms and Ss express the hour (in 24-hour cycles), minute, and second digits. In the former type, all values must be specified. In the latter, there is the option of adding the time zone or move to a level of sub-second granularity (Gulutzan & Pelzer, 1999). DBMSs set a time chronon. Coarser-grained granularities must be defined through interval equivalences based on the reference chronon.

4.3. Natural Language Processing (NLP)

Within the the field of NLP, most work on time information processing focuses on creating specification languages and information extraction tools. For example, Setzer (2001) or TIDES (Ferro et al., 2005). In this work we assume TimeML (Time Markup Language), a specification scheme for annotating temporal information expressed in text, in

particular: event mentions, time expressions, as well as the time relations holding among these entities (Pustejovsky et al., 2005). TimeML was adopted as an ISO standard in 2007 (ISO WD 24617-1:2007) and since then its use has naturally spread through the NLP community, resulting in the creation of basic resources (natural language corpora in several languages annotated with TimeML) and tools (e.g., Verhagen et al., 2005; Llorens et al., 2010; Robaldo et al., 2011).
In TimeML there are 4 basic types of time expressions (or *timexes*):

1. DATEs, which include time units of length equal to or greater than one day ("*August 2010*", "*the first week of June*", "*last summer*", "*today*"),
2. TIMEs, for expressions referring to time units smaller than one day ("*at noon*", "*4 hours*"),
3. DURATIONs, indicating the duration of something ("*the next five days*", "*many years*"), and
4. SETs, indicating event frequency ("*every day*").

Each of these timex types can be further characterized with the following attributes:

- **Value.** Encoding the normalized value referred to by the time expression, which follows the ISO 8601 standard. Hence DATEs are expressed with eight digits, YYYY-MM-DD; TIMEs with the letter T (for time) followed by hours (H)minutes (M), and possibly seconds (S): THH:MM:SS, and DURATIONs with the letter P (for period) followed by the digits for the relevant time units, sorted from greater to smaller granularity: PYYYYMMDDTHHMM. For example, the time expression "*during two days*" is represented as P2D.
- **Mod.** Optional attribute to express the semantics of any modifier, if present. For example: BEFORE ("*before summer*"), AFTER ("*after midnight*"), LESS THAN ("*almost 24 hours*"), MORE THAN ("*over three hours*"), START ("*beginning 2009*") MID ("*mid June*"), END ("the end of the month").
- **AnchorTimeId.** Optional attribute identifying the time expression that is needed in order to complete the interpretation of the current timex. For example, the time value of "*yesterday*" requires knowing the utterance time in order to compute the calendar point it refers to. This second timex is called the anchor (or anchoring) time.

- **BeginPoint.** In a timex of type DURATION, marking the beginning point of the interval. For example, "*in three days*" expresses a duration of value=P3D, which has the time of utterance the as its beginning point.
- **EndPoint.** As above but in this case indicating the ending time of a duration. For example, in "*two days ago*" (of value=P2D) the end point corresponds to the utterance time.
- **Quant.** Attribute only for timexes of type SET which allows to quantify iterations. For example, "*three days a week*": value=P1W, quant=EVERY.
- **Freq.** For expressions of type SET. Expressing the cardinality of sets of temporal entities. For example, "*three days a week*": value=P1W, freq=3D.

5. Porting Timexes from Natural Language to DBs

As argued above, the different nature of natural vs. formal languages (flexibility, ambiguity, and underspecification *vs.* rigidity and univocity), pose challenges to the porting of timex information from natural language text to DB systems. In this section we start by identifying the aspects of natural language expressions that have no immediate representation in DBs, and move then towards putting forward a methodology for porting timex information from one system to the other.

5.1. Corpus-Based Data Analysis

The data analysis is carried out over the Modern Spanish (ModeS) TimeBank corpus (Guerrero and Saurí, 2012), a corpus of Modern Spanish (17th–18th centuries) annotated with both temporal and event information according to TimeML (Pustejovsky et al., 2005), and spatial information following the SpatialML scheme (Mani et al., 2008). ModeS TimeBank contains 102 documents reporting a sea crossing cruise by a ship called La Princesa, which took place from December 1768 to April 1769. Despite these domain and period constraints of the corpus, the porting methodology put forward next can be applied to other TimeML annotated corpora.

An analysis of temporal information in ModeS TimeBank shows that the concept of time in this corpus is both linear and cyclical, with multiple granularities, open and closed extensions, and both absolute and relative references. Nevertheless, the time expressions in the corpus have been normalized according to TimeML (by means of the attribute value) without any problem, since this markup language is flexible enough to account for these features. By contrast, DBMSs pose limitations on multiple granularity, cyclic and relative times. Table 1 exemplifies the main differences observed, which are explained next.[5]

Table 1. ModeS TimeBank granularities

Granularity	Timex (Spanish)	Timex (English)	Format	Value
Minutes	*a las 5 y media*	*at 5:30 am*	THH:MM	T5:30
Hours	*a las 4 de la mañana*	*at 4 in the morning*	THH	T4
	el 4 de agosto a las 6 de la tarde	*on August 4 at 6 pm*	YYYY-MM-DD THH:MM	XXXX-08-04T18:00
Days	*el 16 de henero*	*on January 16*	YYYY-MM-DD	XXXX-01-16
	el dia miercoles 28	*on Wednesday 28*	YYYY-MM-DD	XXXX-XX-28
Years	*el año de 1692*	*the year 1692*	YYYY-MM-DD	1692-XX-XX
Part of the day	*amaneció este dia*	*that day dawned*	YYYY-MM-DDT$$	1769-02-27TMO
	a mediodia	*at midday*	T$$	TMI

Vagueness. Some timexes in ModeS TimeBank have the attribute value fully instantiated (for example, value=*T5:30* corresponding to the natural language expression "*at 5.30am*"), whereas others are only partially specified (e.g., value=XXXX-XX-28 for the timex "On *Wednesday 28*"). Since DBs do not support partial values, it will be necessary to establish a treatment for indexing them. This aspect will be taken into account when making the transfer.

[5] Being a corpus of a historical variety of Spanish, the orthographic conventions applied by the writer differ from the current ones (e.g., *henero* instead of *enero*). We keep the original spelling of the exemplified timexes.

Temporal extension. The time expressions found in the corpus can be open or closed. An interval is open when at least one of its boundaries is not specified. When neither boundaries are defined (e.g., "*for ten years*"), durations express a length which nevertheless cannot be placed on a timeline. Some examples from the ModeS TimeBank corpus are illustrated in Table 2.

Table 2. Timexes with open extensions in ModeS TimeBank

Granularity	Timex (Spanish)	Timex (English)	Format	Value
Hours	*estas 24 oras*	*these 24 hours*	PXH	P24H
	algunas oras	*A few hours*	PXH	PXH
Days	*havia tres dias*	*for three days*	PXD	P3D
Part of the day	*toda la noche*	*Overnight*	PX$$	P1NI

TimeML encodes time extensions boundaries by means of the attributes EndPoint and BeginPoint, which take timex IDs as value. In expressions like "*2 days ago*", there is a clear ending boundary corresponding to the utterance time. In other cases, the begin and end points of the temporal extension can be identified from the context. The following examples illustrate the usage of these attributes for annotating some of the timexes in Table 2.

(3) toda la noche "*all night*"
value=P1NI
beginPoint=t1
endPoint=t2

(4) todos estos días "*all these days*"
value=PXD
endPoint=t2

These attributes should be taken into account when indexing time expressions in DBs. The major challenge we foresee here is that TimeML does not impose any restriction on the granularity of these values, and therefore there may be applying different granules.

Temporal structure. A feature of the Modes TimeBank corpus is that most of the event it refers to are anchored to a time reference and project forward (hence, assuming a linear conception of time). However, it also contains examples of cyclical time. Most cases are in terms of frequency:

(5) De ora en ora "*from hour to hour*"
value=P1H
freq=XH

DBMSs do not support this kind of timex. Hence, porting this type of temporal information is not direct and should apply some sort of time linearization process. For example, setting a time interval and anchor.

Reference framework. Generally speaking, the reference framework of most timexes in the ModeS TimeBank corpus is absolute (they can be anchored to a timeline). However, the corpus presents also some expressions under a relative framework perspective since they refer to information expressed by other timexes.

(6) habia tres dias "*for three days*"
value=P3D

Example (6) refers to an imprecise duration. It is only possible to place it on a timeline if the reader appeals to the point in time when (6) was uttered. We refer to the relation between the underspecified duration and the expression referring to the utterance time as an anchoring relation.

TimeML-derived differences. Finally, some of the identified challenges from porting timex information from natural language text to DBs has to do with the TimeML scheme. The format of the attribute value in TimeML is not completely compliant with the format supported by most DBMSs. For instance, the values MO for expressing "morning" (7) or MI for expressing "midday" are not supported by the SQL-92 standard.

(7) Amaneció este día "*this day dawned*"
value=1769-01-12TMO
mod=START

Similarly, indexing the information encoded in the TimeML attribute mod is not trivial since this feature is not contemplated in SQL-92. In example (7), mod completes the information in the attribute value in order to specify that the timex refers to the beginning (START) of the morning period (1769-01-12TMO).

5.2. Data Porting Methodology

This section puts forward a methodology for porting and indexing natural language timexes (as encoded in TimeML) to temporal DBs or GISs. It aims at maintaining the expressiveness of natural language timexes, even though this aspect relies also on DB format adaptation issues that are beyond the scope of the current chapter. Figure 6 summarizes the proposed methodology.

The first step classifies timexes according to their *type* (DATE, TIME, DURATION, SET). Then, each of these will follow a different procedure depending on whether they refer to direct (*precise accuracy*) or vague (*imprecise accuracy*) times, as indicated in the value attribute. Accuracy in timexes depends on the DBMS chronon. Those expressions not conforming to the chronon (i.e., having a greater or lesser granularity) will be considered vague (or imprecise). For instance, the normalized expression value=2011-XX-XX (corresponding to the timex "*the year 2011*") will be considered imprecise in case that the DBMS chronon does not correspond to the year.

Precise timexes will be directly incorporated into the DB, whereas imprecise timexes will be further subclassified based on the origin of their imprecision:

a) Due to linguistic underspecification. For example, the timex does not explicit all the relevant temporal units (as in "*last Thursday*" or "*May 3rd*"). We refer to this imprecision type as *Timex-based*.
b) Due to the TimeML scheme (hence, *TimeML-based*). For instance, the fully normalized expression value=2011-04-26TMO (corresponding to the natural language timex "*the morning of April 26, 2011*") presents a TimeML-based imprecision because it contains the value MO (for "morning") which is not supported by DB standards.

This methodology aims at respecting natural language expressiveness (as normalized by the markup language TimeML) during the transfer to DB systems. Therefore, it is devised for being integrated into any potential porting tool.

Figure 6. Data porting methodology

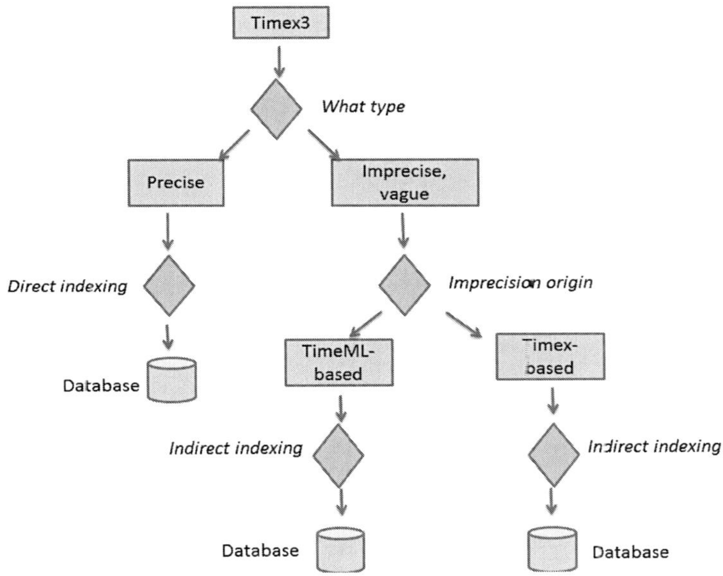

5.3. Timex Statistics

Given the formal differences between TimeML and DB standards (section 5.1), and considering the proposed methodology for porting timex information (section 5.2), we present here the statistics on different timexes in ModeS TimeBank: those requiring *direct* vs. *indirect transfer*. In this way, it can be appreciated how many timexes can be incorporated into DBs without further modification and how many must be subject to more complex processing. In the corpus, the tag TIMEX3 corresponds to 33.91% of the labeled entities (with EVENT occupying the 47.94% and PLACE the 18.14%) (Guerrero 2010, Guerrero & Saurí 2011). The percentages for each type of TIMEX3 are: DATE: 26.90%, TIME: 48.54%, DURATION: 24.44%, SET 0.11% (Table 3).

The following table provides an account of the proportion of each timex type that would be directly ported to a DB vs. the proportion that would have to undergo some kind of treatment.

Table 3. TIMEX3 classification based on type and manner of porting into a DB

Type	Classification	Number	Percentage
DATE	Direct	215	24.10%
	Indirect	25	2.80%
TIME	Direct	103	11.55%
	Indirect	330	36.99%
DURATION	Direct	101	11.33%
	Indirect	117	13.11%
SET	Direct	0	0.00%
	Indirect	1	0.11%

Imprecise DATEs are due to two causes: either the granularity is greater than the DB chronon or, although there is agreement between the chronon and the timex granularity, there is imprecision in some of the units provided by the normalized expression.

DURATIONs are normalized based on the time unit expressed by the expression (e.g., years in the case of "*three years*" or minutes in the case of "*a few minutes later*"). The porting to a DB is straightforward if the natural language timex refers to years, months, days, or seconds. However, if it refers to other types of granularity (e.g., parts of the day) it will require further treatment. Similarly, indirect duration will be necessary in case that the transfer type is nonspecific, even if the granularity agrees with the one set by the chronon.

Table 4. TIMEX3 classification based on manner of porting into a DB

Timex3	Number	Percentage
Direct	419	46.97%
Indirect	473	53.02%
Total	892	100%

The results in Table 4 justify the need of a post-processing step for porting natural language timexes (normalized according to TimeML) into a DB; otherwise half of the timexes in the corpus could not be indexed in a time field. This post-processing would allow a mapping of temporal information from the corpus to any DB without losing information.

Imprecise timexes are distinguished based on the cause of their imprecision, as reflected in the value of the value attribute. Timexes with at least one X in their normalized value attribute (X being the placeholder for underspecified information) are considered as having a Timex-based imprecision, while timexes which normalized value presents features not supported by the DB standard, will be considered as suffering a TimeML-based imprecision (Table 5).

Table 5. Indirect TIMEX3 classification

Indirect Timex3	Number	Percentage
Timex-based	34	7.18%
TimeML-based	439	92.81%
Total	473	100%

Table 5 shows that most normalized expressions that we have classified as "indirect TIMEX3s" have a format non-compliant with DB standards (92.81%).

6. Conclusions

The corpus that we analyzed, ModeS TimeBank, contains forms of both external and internal time, therefore providing rich and complex data that make the porting of this information to DBMSs challenging. The markup language TimeML allows to reconcile the varied nature of natural language timexes into a more stable and rigid format, but only to a certain extent. In this article, we analyzed and compared both at the conceptual and descriptive level, the treatment of temporal expressions in TimeML and in standard DB systems and put forward a systematization of the differences between the two formats. We concluded that in order to port natural language time expressions into DBs or GISs a post-processing capable of maintaining the original expressiveness is necessary. Future work derived from the present research involves the creation of automated tools incorporating the transfer methodology proposed here.

LITERATURE AND WORK CITED

Accarias de Sérionne, J. 1772, *Historia y descripción general de los intereses del comercio de todas las naciones de Europa en las quatro partes del mundo, traducida del francés por Domingo de Marcoleta*, Madrid: Imprenta de Miguel Escribano.

A-Hussein, A. & Hassan, F. 1986, "Las quiebras de los hombres de negocio castellanos", E. Lorenzo Sanz, ed., *Historia de Medina del Campo y su tierra*, Valladolid: Ayuntamiento de Medina del Campo, Consejeria de Educacion y Cultura de la Junta de Castilla y Leon, Diputacion Provincial de Valladolid; Caja de Ahorros Provincial de Valladolid.

Aigner, W., Bertone, A., Miksch, S., Tominski, C. & Schumann, H. 2007, "Towards a conceptual framework for visual analytics of time and time-oriented data", *Proceedings of the 39th Conference on Winter simulation: 40 years! The best is yet to come*, New Jersey: IEEE Press Piscataway: 721–729.

Akinobu, K. 2009, "The Eurasian silver century 1276–1359: commensurability and multiplicity", *Journal of Global History*, 4: 245–269.

Alegría, M. F., Daveau, S., García, J. C. & Relaño, F. 2007, "Portuguese Cartography in the Renaissance", *Cartography in the European Renaissance. History of Cartography*, Vol. III. Chicago; London: University of Chicago Press: 1034–1061.

Allami, A., *The Ā'īn-i Akbarī,* 1977–1978, 1873, 3 vols. Calcutta: Asiatic Society of Bengal; New Delhi: Oriental Book Corporation.

Allingham, M. 2002, *Choice theory: a very short introduction*, Oxford: Oxford University Press.

Alonso García, D. 2006, "Ducados entre dos dinastias. La circulación de capital entre Castilla y Flandes a comienzos del siglo XVI", C. Sanz Ayán and B. García García, eds., *Banca, crédito y capital: la Monarquía Hispánica y los antiguos Países Bajos (1505–1700)*, Madrid: Fundacion Carlos de Amberes.

—. 2007a, *El erario del reino. Fiscalidad en Castilla a principios de la Edad Moderna*, Valladolid: Junta de Castilla y León.

—. 2007b, "Guerra, hacienda y política. Las finanzas militares en los inicios de la Edad Moderna", Jiménez Estrella, A. & Andújar Castillo,

F. eds., *Los nervios de la Guerra. Estudios sociales sobre el ejército de la Monarquía Hispánica (siglos XVI-XVIII): nuevas perspectivas*, Granada: Comares: 37–57.

—. 2011, "Crear espacios, cobrar impuestos. Los partidos fiscales en Castilla a principios de la Edad Moderna", L. Salas Almela, ed., *Los ámbitos de la fiscalidad: fronteras, territorio y percepción de tributos en los imperios ibéricos (siglos XV-XVIII)*, Madrid: Instituto de Estudios Fiscales: 41–56.

—. 2012, "El fenómeno del arrendamiento de rentas: nuevas vías de análisis", *Viator Multilingual*, nº 43: 339–358.

Alvarez, A. 2003, "Geomagnetism and the Cartography of Juan de La Cosa", *Terrae incognitae*, 35: 1–15.

Álvarez Nogal, C. 2011, "Mercados o redes de mercaderes: el funcionamiento de las ferias de Portobelo", Böttcher, N., Hausberger, B. & Ibarra, A. eds, *Redes y negocios globales en el mundo ibérico, siglos XVI-XVIII*, México: El Colegio de México: 53–87.

Amith, J. 2005, *The Mobius strip: a spatial history of colonial society in Guerrero, México*, Stanford: Stanford University Press.

Anderson, Ph. 1999, "Complexity Theory and Organization Science", *Organization Science*, 10 (3): 216–232.

Andersson, Å. E., Anderson, W. P. & Johansson, B. 2003, *The economics of disappearing distance*, Aldershot: Ashgate.

Andrade, A. de, 2006, *Os senhores do desterro de Portugal. Judeus portugueses em Veneza e Ferrara em meados do século XVI*, in "Veredas-Revista da Associação Internacional de Lusitanistas", 6: 65–108.

—. 2011, *De Antuérpia a Ferrara: o caminho de Amato Lusitano e da sua família*, in *Medicina na Beira Interior da Pré-História ao século XXI*, "Cadernos de Cultura", nº. 25 Castelo Branco, Novembro: 5–17.

Anhai D., Domingos, P. & Halevy, A. 2003, "Learning to match the schemas of data sources: A multistrategy approach," *Machine Learning* 50: 279–301.

Antunes, C. 2009, *Lisboa e Amesterdão, 1640–1705. Um caso de globalização na História Moderna*. Lisboa: Livros Horizonte.

Aram, B. 2008, *Leyenda negra y leyendas doradas en la conquista de América. Pedrarias y Balboa*, Madrid: Marcial Pons.

Arasaratnam, F. 1958, *Dutch power in Ceylon 1658–1687*, Amsterdam: Djambatan for the Netherlands Institute for International Cultural Relations.

Armitage, D. 2004, "Tres conceptos de historia atlántica", *Revista de Occidente*, 281: 7–28.

Armstrong, M. P. 1988, "Temporality in spatial databases", *Proceedings GIS/ LIS'88*, 2: 880–889.

Arthur, W. Brian. 1999, "Complexity and the economy", *Science*, 284/5411: 107–109.

Artola Renedo, A. 2011, "El patrocinio intraclerical en el Antiguo Régimen: curias y familias episcopales de los arzobispos de Toledo (1755–1823)", *REDES- Revista Hispana de redes sociales,* 21.

Axelrod, R. 1984, *The evolution of cooperation*. New York: Basic Books.

Azevedo e Silva, J. M. 1998, "A cartografía hispano portuguesa no tempo de Filipe II". *Actas del Simposium: Felipe II y su Época,* 2 vols. Madrid: Estudios superiores de El Escorial: 317–342.

Bachman, Ch. 1973, "The Programmer as Navigator. ACM Turing Award lecture", *Communications of the ACM* 16/11.

Badia, A. & Lemire, D. 2011, "A call to arms: revisiting database design," *Association for Computing Machinery (ACM) Special Interest Group on Management of Data (SIGMOD) Record* 40 (3): 61–69.

Bailyn, Bernard. 2005, *Atlantic history: concepts and contours*. Harvard University Press.

Barabási, A-L. & Réka, R. 1999, "Emergence of scalling in random networks", *Science* 286: 509–512.

Barabási, A-L. 2003, *Linked: how everything is connected to everything else and what it means*, New York: Plume.

Barendse, R. J. 2002, *The Arabian Seas: the Indian Ocean world of the seventeenth century*, New York: M. E. Sharpe.

Barkley R. Jr. 2000, *From catastrophe to chaos: A general theory of economic discontinuities*, Vol. I, 2nd ed. Boston: Kluwer Academic.

Barrientos Martínez, M. Á. *Network Analyst. El análisis de redes desde ArcGis 9.2.*

Barros, A. J. 2012, *Minas e açúcares. Notas sobre a evolução da economia brasileira (1554–1573)*, in *Livro de Homenagem ao Professor Romero Magalhães*, Luís Miguel Duarte, Leonor Freire Costa and Álvaro Garrido eds, Coimbra: Almedina: 311–336.

—. 2013, *Os negócios e a aritmética. Bento Fernandes e as redes cristãs-novas do Porto no século XVI*, in *Humanismo, Diáspora e Ciência*, Porto: BPMP, 2013, pp. 297–322.

—. 2006, *Oligarquia política e elite económica no Porto dos séculos XV e XVI*, in *Estudos em homenagem ao Professor Doutor José Amadeu Coelho Dias*, vol. I, Porto: Faculdade de Letras: 49–70.

Barua Bahadur, L. 1966, *Early history of Kamrupa from the earliest times to the end of the sixteenth century*, 2nd ed. Gauhati: Lawyer's Book Stall.

Basset, D.K. 1989, "British 'Country' trade and local trade networks in the Thai and Malay states, c. 1680–1770", *Modern Asian Studies*, 23 (4): 625–643.

Beckman, C. M., Haunschild, P. R. & Phillips, D. J. 2004, "Friends or Strangers? Firm-specific uncertainty, market uncertainty, and network partner selection", *Organization Science* 15 (3): 259–275.

Beller, A. T., Giblin, K. V., Le, S., Litz, K. T. & Schimel, D. 1991, "A temporal GIS prototype for global change research", *GIS/LIS'91*: 752–765.

Benavides Quecán, J. 2004, "El uso de métodos históricos de levantamiento topográfico y cartográfico en la elaboración de mapas confiables de ciudades. Caso de estudio: Zwolle (Holanda), *Scripta Nova, Revista Electrónica de Geografía y Ciencias Sociales,* VIII (170): 65.

Benneworth, R. L., Bishop, C. D., Turnbull, C. J. M., Holman, W. D. & Monette, F. M. 1981, "The implementation of GERM, an entity-relationship data base management system", *Very Large Data Bases, 7th International Conference, September 9–11, 1981, Proceedings*. IEEE, Cannes: Computer Society: 478–484.

Bernard, C. 1994, *Descubrimiento, conquista y colonización de América a quinientos años*, México: Fondo de Cultura Económica.

Bhattacharya, Bh. 2005, "Armenian European relationship in India, 1500–1800: no Armenian foundation for European empire?", *Journal of the Economic and Social History of the Orient*, 48 (2): 277–322.

Bhattacharya, Fr. 2007, *La Victoire de Manasa: Traduction francaise du Manasavijaya, poeme bengali de Vipradasa (XVe)*, Pondicherry: IFP, EFEO, Collection Indologie 105.

Blaut, J. M. 1961, "Space and Process", *The Professional Geographer*, 13: 1–7.

Blockmans, W. & Antheun Janse (eds), 1999, *Showing status: representation of social positions in the Late Middle Ages*, Turnhout: Brepols.

Bodenhamer, D. J., Corrigan, J. & Harris, T. M. 2010, *The spatial humanities: GIS and the future of humanities scholarship*, Bloomington: Indiana University Press.

Bodenhamer, D. J. 2013, "Beyond GIS: geospatial technologies and the future of history", Lünen, A. & Travis, C. eds, *History and GIS*, Netherlands: Springer: 1–13.

Boisselier, J. 1971, Review: [Untitled], Reviewed Work(s): Kamalakanta, G. 1967, *Copper-Plates of Sylhet, vol. I (7th-11th Century A.D.)*,

Sylhet, in: *Journal of the Economic and Social History of the Orient*, 14 (1): 90–94.

Bonabeau, E. 2002, "Agent-based modeling: methods and techniques for simulating human systems", *Proceedings of the National Acadmey of Sciences of the United States of America,* 99 (14), Suppl. 3: 7280–7287.

Borgatti, S. P., Everett, M. G. & Freeman, L.C. 2002, *Ucinet for Windows: Software for Social Network Analysis.* Harvard: Analytic Technologies.

Bose, S. 2006, *A hundred horizons: the Indian Ocean in the age of global empire*, London; Cambridge: Harvard University Press.

Bosque González, I. del, Fernández-Freire, C., Martín-Forero, L. & Pérez Asensio, E. 2012, *Los sistemas de información geográfica y la investigación en ciencias humanas y sociales*, Madrid: CECEL-CSIC.

Brading, D. A. 1993, *Orbe Indiano: de la monarquía católica a la República criolla, 1492–1867*, México: Fondo de Cultura Económica.

Braudel, F. 1983, *O Mediterrâneo e o Mundo Mediterrânico na Época de Filipe II*, 3 vols., Lisboa: Publicações Dom Quixote.

—. 1983, *The wheels of commerce: civilization and capitalism 15th–18th century*, Berkeley: University of California Press.

—. 1984, *Civilización material, economía y capitalismo*, Madrid: Alianza Editorial, 3 vols.

—. 1985, *Civilização material, economia e capitalismo, sécs. XV-XVIII*, vol.: *Os Jogos das Trocas*, Lisboa: Edições Cosmos.

—. 2002, "Geohistoria: la sociedad, el espacio y el tiempo", recogido en *Las ambiciones de la historia*, Barcelona: Crítica, 2002: 53–87.

Britnell, R. H. 1996, *The commercialisation of English society, 1000–1500,* Manchester.

Brodie, M. L. & Liu, J. T. 2010, "The power and limits of relational technology in the age of information ecosystems," Seminar, Vrije Universiteit Brussel.

Brown, A.P. G. 1975, "Modelling a real-world system and designing a schema to represent it", Douqué, B. C. M. & Nijssen, G. M. (eds), *Data base description*, Amsterdam: North-Holland Publishing Company: 339–347.

Brunelle, G. 1991, *The New-World merchants of Rouen, 1559–1630.* Kirksville.

Bryan, T. 2013, "Literature survey of graph databases," www.systap.com/pubs/graph_databases.pdf.

Buisseret, D. 1992, *Monarchs, ministers, and maps: the emergence of cartography as a tool of government in early modern Europe*, Chicago: University of Chicago Press.

Bustos Rodríguez, M. 2005, *Cádiz en el sistema atlántico. La ciudad, sus comerciantes y la actividad mercantil (1650–1830)*, Madrid: Sílex Ediciones; Universidad de Cádiz.

Campbell, B. M. S. & Britnell, R. H. (eds), 1995, *A commercialising economy: England 1086 to c. 1300*, Manchester.

Caquard, S. 2013, "Cartography I Mapping narrative cartography", *Progress in Human Geography*, 37: 135–144.

Carayol, N. & Roux, P. 2005, "Self-organizing innovation networks: when do small worlds emerge?", *Journal of Economic and Social Systems*, 18 (2): 307–332.

Cardim, P., Herzog, T., Ruiz Ibáñez, J. J. & Sabatini, G. (eds), 2012, *Polycentric monarchies: how did Early Modern Spain & Portugal achieve and maintain a global hegemony?* Portland: Sussex Academic Press.

Carlos Morales, C. J. de. 2000, *Carlos V y el crédito de Castilla. El tesorero general Francisco de Vargas y la Hacienda Real entre 1516 y 1524*, Madrid: SECCFC.

—. 2006, "Felipe II y sus banqueros", Sanz Ayán, C. & García García, B. (eds), *Banca, crédito y capital: la monarquía Hispánica y los antiguos Países Bajos (1505–1700)*, Madrid: Fundación Carlos de Amberes.

Carmagnani, M. 2011, *El otro occidente. América Latina desde la invasión europea hasta la globalización*, México: Fondo de Cultura Económica; El Colegio de México.

Carretero Zamora, J. M. 1999, "Los arrendadores de la Hacienda de Castilla a comienzos del siglo XVI (1517–1525)", *Studia Historica. Historia Moderna*, 21: 117–152.

—. 2008, *La averiguación de la Corona de Castilla (1525–1540): los buenos vecinos pecheros y el dinero del Reino en época del emperador Carlos V*. 3 vols. Valladolid: Junta de Castilla y León, Consejería de Cultura y Turismo.

Carrington, P., Scott, J. J. & Wasserman, S. (eds). 2005, *Models and methods in social network analysis*, Cambridge: Cambridge University Press.

Carter, P. & Malouf, D. 1989, "Spatial History", *Textual Practice*, 2 (3): 173–183.

Cartografia impresa dos Seculos XVI e XVII. Imagens de Portugal e Ilhas Atlánticas. VI centenario do nascimento do infante D. Henrique, 1994.

Porto: Comissao nacional para as comemoraçoes dos descubrimientos portugueses.

Carvalho, J. R. de. 1999, "O social emergente: sistemas adaptáveis complexos e interpretação histórica", *Revista de História das Ideias* 20: 623–688.

Casado Alonso, H., ed. 1995, *Castilla y Europa: comercio y mercaderes en los siglos XIV, XV y XVI*, Burgos: Diputación Provincial.

Casado Alonso, H. 1997, "Crecimiento económico y redes de comercialización en la Castilla septentrional (Siglos XV y XVI)", in *Imágenes de la diversidad. El mundo urbano en la Corona de Castilla (S. XVI - XVIII)*, ed. J. I. Fortea Perez, Santander: 283–322.

—. 1998, "Le Rôle des marchands castillans dans la commercialisation international du pastel toulousain (XVe et XVIe siécles)", in *2nd International Symposium "Woad, Indigo and others Natural Dyes: Past, Present and Future*, ed. D. Cardon, H. E. Müllerott, B. Bemjelloun, F. Brumont & M. Delmas, Arnstadt: 65–70.

—. 2001, "Medina del Campo fairs and the integration of Castile into 15th and 16th century European economy", Cavaciocchi, S. (ed.), *Trentaduesima Settimana di studi*, Prato: Le Monnier.

—. 2008, "Los flujos de información en las redes comerciales castellanas de los siglox XV y XVI", *Investigaciones de Historia Económica*, 10: 35–68.

—. 2012, "Crecimiento económico, redes de comercio y fiscalidad en Castilla a fines de la Edad Media", Bonachía Hernando, J. A. & Carvajal de la Vega, D. (eds), *Los negocios del hombre. Comercio y rentas en Castilla. Siglos XV y XVI*, Valladolid: Ediciones Castilla: 17–35.

—. 2013, "El papel de las colonias mercantiles castellanas en el imperio hispánico (siglos XV y XVI)", in *Las vecindades de las Monarquías Ibéricas*, coord. de José Javier Ruiz Ibáñez. Madrid: FCE: 355–374.

Casas, Bartolomé de las, 1983, *Brevísima relación de las destrucción de las Indias*, Madrid: Anjana Ediciones.

Casey, E. S. 1993, *Getting back into place: toward a renewed understanding of the place-world*, Bloomington: Indiana University Press.

—. 1997, *The fate of place: a philosophical history*, Berkeley: University of California Press.

Casson, L. (ed. & tr.) 1989, *Periplus Maris Erythraei*, Princeton: Princeton University Press.

Céspedes del Castillo, G. 1991, *La exploración del Atlántico*, Madrid: Mapfre.

Chamberlin, D. D. & Boyce, R. F. 1974, "SEQUEL: A Structured English Query Language," Randall, R., ed., *Proceedings of the 1974 Association for Computing Machinery SIGMOD Workshop on Data Description, Access and Control*, New York: ACM vol. 1: 249–264.

Chaunu, P. 1955–1960, *Séville et l'Atlantique (1504–1650)*, 12 vols. Paris, SEVPEN.

—. 1973, *L'Espagne de Charles Quint*, 2 vols., París: SEDES.

Chen, P. 1976, "The entity-relationship model – toward a unified view of data," *Association for Computing Machinery (ACM) Transactions on Database Systems* 1 (1): 9–36.

—. 1985, "Database design base on entity and relationship," S. B. Yao (ed.), *Principles of database design*, New Jersey: Prentice-Hall: 174–210.

—. 2002, "Entity Relationship Modeling: Historical Events, Future Trends, and Lessons Learned," M. Broy and E. Denert, eds., *Software Pioneers: Contributions to Software Engineering*, Berlin: Springer-Verlag, Lecturing Notes in Computer Sciences: 100–114.

Chen, J. & Jiang, J. 2000, "An event-based approach to spatio-temporal data modeling in land subdivision systems", *GeoInformatica*, 4: 387–402.

Chittolini, G. 1995, "The 'Private', the 'Public', the 'State': The origins of the state in Italy, 1300–1600", *Journal of Modern History*, 67: 34–61.

Choudhury, P. C. 1988, *Assam–Bengal relations: from the earliest to the twelfth century A.D.*, Guwahati; Delhi: Spectrum Publications, 1988.

Choudhury, R. (ed.) 2007, *Sri Sri Padmapuran, Silchar*, Assam: Choudhury Library.

Choudhury, S. 1995, *Folklore and history: a study of the Hindu folkuts of the Barak valley*, New Delhi: K. K. Publishers.

Claramunt, C. & Thériault, M. 1995, "Managing time in GIS and event-oriented approach", *Recent advances in temporal databases*, Springer: 23–42.

Codd, E. F. 1970, "A relational model for large shared data banks," *Communications of the Association for Computing Machinery (ACM)* 13 (6): 377–387.

—. 1972a, "Normalized data base structure – a brief tutorial," Dean, A. L. (ed.), *Proceedings of the 1971 Association for Computing Machinery SIGFIDET Workshop on Data Description, Access, and Control*, San Diego: Association of Computing Machinery.

—. 1972b, "Further normalization of the data base relational model", Randall J. Rustin (ed.), *Data base systems: Courant computer science symposia series 6*. New Jersey: Englewood Cliffs; Prentice-Hall: 33–64.

—. 1974, "Recent investigations into relational data base systems," IBM Research Report RJ1385, *Proceedings, IFIP 1974 Congress, Stockholm, Sweden,* Amsterdam: 1017–1021.

—. 1979, "Extending the database relational model to capture more meaning," *Association for Computing Machinery (ACM) Transactions on Database Systems (TODS)* 4 (4): 397–434.

—. 1981, "1981 Turing Award Lecture – Relational database: a practical foundation for productivity," *ACM Annual Conference in Los Angeles,* http://awards.acm.org/images/awards/140/articles/2485527.pdf (accessed 23 March 2013).

—. 1990, "Serious flaws in SQL," *The Relational Model for Database Management: Version 2,* Boston: Addison-Wesley: 371–389.

Colás, A. 2007, *Imperio*, Madrid: Alianza Editorial.

Corpo Diplomatico Portuguez, tomo 8: 311–312.

Cortesâo, A. 1960, *Cartografía portuguesa antiga*, Lisboa: Comissâo executiva das comemoraçoes do quinto centenario da morte do Infante D. Enrique.

Cortesâo, A. & Teixeira Mota, 1960, *Portugaliae Monumenta Cartographica*, 6 vols. Lisboa: Commemoraçoes do V Centenário da morte do Infante D. Enrique.

Costa Pinto Gomes, S. M. 2013, *A Companhia de Simón Ruiz. Análise especial de uma rede de negócios no século XVI*, Porto: Universidade Porto (unpublished PhD thesis).

Crespo Sanz, A. 2008, *El Atlas de El Escorial*, Valladolid: Universidad de Valladolid (unpublished PhD thesis).

Crespo Solana, A. 1996, *La Casa de la Contratación y la Intendencia General de Marina, 1717-1730*, Cádiz: Universidad de Cádiz.

—. 2000, *El comercio marítimo entre Amsterdam y Cádiz (1713–1778),* Madrid: Banco de España.

—. 2001, *Entre Cádiz y los Países Bajos: Una comunidad mercantil en la ciudad de la Ilustración,* Cádiz: Cátedra Adolfo de Castro, Fundación Municipal de Cultura.

—. 2006, *América desde otra frontera: la Guayana holandesa (Surinam), 1680–1795,* Madrid: CSIC.

—. 2009, *Mercaderes atlánticos: Redes del comercio flamenco y holandés entre Europa y el Caribe*, Córdoba: Universidad de Córdoba, Caja Sur.

—. 2010a, "Legal strategies and smuggling mechanisms in the trade with the Hispanic Caribbean by foreign merchants in Cadiz – the Dutch and

Flemish case (1680–1750)," *Jahrbuch für Geschichte Lateinamerikas* 47: 181–212.

—. (ed.), 2010b, *Comunidades transnacionales. Colonias de mercaderes extranjeros en el mundo atlántico (1500–1830)*, Madrid: Doce Calles.

Crespo Solana, A. 2011, "Geostrategy of a system? Merchant societies and exchange networks as connection centres in the Spanish Atlantic trade in the First Global Age," Rila Mukherjee, ed., *Networks in the First Global Age, 1400–1800*, Delhi: Primus Books: 11–35.

—. 2012a, "Self-organizing: the case of merchant cooperation in the Hispanic Atlantic economy (1680–1778)," *Journal of Knowledge Management, Economics and Information Technology*, Special Issue: Crespo Solana, A. & Alonso García, D. (eds), *Self-organizing networks and GIS tools. Cases of use for the study of trading cooperation (1400–1800)*, June: 191–224.

—. 2012b, "El sistema portuario del Atlántico español: aproximaciones metodológicas para el estudio de redes mercantiles y puertos", *V Colóquio Internacional: Redes Portuárias Atlânticas (séculos XVI a XX) Balanço e prospectiva historiográfica*. Porto.

Crespo Solana, A. & Alonso García, D. 2012, *Self-organizing networks and GIS tools. Cases of use for the study of trading cooperation (1400–1800)*, Scientific Papers: 145–189.

Crespo Solana, A. & Herrero Sánchez, M. (eds), 2002, *España y las 17 Provincias de la Países Bajos: una revisión historiográfica (XVI-XVIII),* Córdoba: Universidad de Córdoba; Madrid: Ministerio de Asuntos Exteriores, Fundación Carlos de Amberes.

Cunqueiro, Á. 1998, *Fábulas y leyendas de la mar*, Madrid, Tusquets.

Curtin, P. D. 1984, *Cross-cultural trade in world history*, Cambridge: Cambridge University Press.

Curtis, J. W. 2012, "Interacting sketch maps with GIS to explore fear of crime in the urban environment: a review of the past and prospects for the future", *Cartography and Geographic Information Science*, 39: 175–186.

Curto, D. R. & Molho, A. (eds) 2002, *Commercial networks in the Early Modern world*, Firenze: EUI Working Paper.

Danvers, F. C. & Foster, W. 1899, *Letters received by the East India Company*, Vol. IV, London: Sampson Low, Marston & Co.

Daveau, S. 2000, "A rede hidrográfica no mapa de Portugal de Fernando Álvaro Seco (1560)" *Finisterra: Revista portuguesa de Geografia*, 35 (69): 11–38.

—. 2001, "A descrição territorial no numeramento 1527–32". *Penélope: Revista de história e ciencias sociais,* 25: 7–39.

—. 2008, "O fragmento de mapa corográfico de Portugal da Real Academia de la Historia de Madrid". *Cadernos de Geografia*, 26–27: 3–17.

—. 2010, *Um antigo mapa corográfico de Portugal (c. 1525) Reconstituição a partir do Códice de Hamburgo*, Lisboa.

David J. S. 2003, *Computers, visualization and history,* Armonk: M. E. Sharpe.

Dedieu, J.-P., Marzagalli, S., Pourchasse, P. & Scheltjens, W. 2012, "Navigocorpus at work: a brief overview of the potential of a database", *International Journal of Maritime History*, XXIV (1): 331–359.

Denègre, J. & Salgé, F. 2004, *Les systèmes d'information géographiques*. 2nd ed. Paris: PUF, 2004 (collection Que sais-je?).

Desfeuilles, P. 1959, "La Navigation à travers le Sund en 1784: d'après les registres de la douane danoise," *Annales: Histoire, Sciences Sociales*, XIV (3): 492–520.

—. 1961, "La Navigation à travers le Sund et la guerre russo-suédoise de Finlande (1788–1790)," *Vierteljahrschrift für Sozial- und Wirtschaftsgeschichte*, XLVIII (1): 56–74.

Diago Hernando, M. 1991a, "Arrendadores arandinos al servicio de los Reyes Católicos", *Historia. Instituciones. Documentos*, 18: 71–95.

—. 1991b, "Los Velázquez de Cuéllar, Tenentes de Arévalo, en el horizonte político a fines de la Edad Media", *Cuadernos Abulenses*, 16: 11–40.

—. 1994, "La alta nobleza en la vida política de las ciudades castellanas en las décadas precomuneras: el ejemplo de Cuenca", *Cuadernos de Historia Moderna*, 15: 121–141.

Dias, G. C. 1983, *O Cabido da Sé do Porto e a Comuna dos Judeus. Por uma dobra e um açougue*, in "Humanística e Teologia", Tomo IV (Set./Dez): 356–357.

Dias, Licenciado F. 1937, *Memórias quinhentistas dum Procurador Del Rei no Porto*, edição de Artur de Magalhães Basto. Porto: Câmara Municipal/Gabinete de História da Cidade.

Dolman, A. & Ettinger, J. van. 1990, *Ports as nodal points in a global transport system: proceedings of Pacem in Maribus*, Pergamon Press.

Dopfer, K.; Foster, J. & Potts, J. 2004, "Micro-meso-macro", *Journal of Evolutionary Economics* 14: 263–279.

Düring, M., Bixler, M., Kronenwett, M. & Stark, M. 2011, "VennMaker para historiadores: fuentes, redes sociales y programas informáticos" *REDES_ Revista Hispana de Redes Sociales,* 21: http://revista-redes.rediris.es/html-vol21/vol21_8.htm.

Earle, P. 2004, *Piratas en Guerra*, Barcelona: Editorial Melusina.

Echevarría Bacigalupe, M. A. 2013, *En los orígenes del espacio global. Una historia de la Mundialización*, Madrid: Ediciones Catarata.

Eklund, R. B. Jr., Herbert, R. F. & Tollison, R. D. 2004, "The economics of the Counter-Reformation: incumbent-firm reaction to market entry". *Economic Inquiry,* 42 (4).

Elden, S. 2002, *Mapping the present: Heidegger, Foucault and the project of a spatial history*, Continuum International Publishing Group.

Elliott, J. H. 2002, *La Europa dividida (1559–1598)*, Barcelona: Crítica.

—. 2006, *Empires of the Atlantic World: Britain and Spain in America, 1492–1830*, New Haven; London: Yale University Press.

Eltis, D. 2012, "Some implications from the transatlantic slave trade for maritime databases", *International Journal of Maritime History*, XXIV (1): 257–264.

Eltis, D. & Richardson, D. 2008, *Extending the frontiers: essays on the new transatlantic slave trade database,* Yale University Press.

Elwood, S. & Leszczynski, A. 2012, "New spatial media, new knowledge politics", *Transactions of the Institute of British Geographers*. Doi: 10.1111/j.1475-5661.2012.00543.x.

Emmer, P. C. 2008, "The myth of early globalisation: the Atlantic economy, 1500–1800", Nuevo Mundo, Mundos Nuevos (consultado el 27 junio 2013). URL: http://nuevomundo.revues.org/42173; DOI: 10.4000/nuevomundo.42173.

Emmer, P. C., Pitri-Grenouilleau, O. & Roitman, J. V. (eds) 2006, *A deus ex machina revisited: Atlantic colonial trade and European economic development*, Leiden: Brill.

Engerman, S. 2009, "War, colonization, and migration over five centuries", in *Migration, trade, and slavery in an expanding world. Essays in honor of Pieter Emmer*, ed. Wim Klooster, Leiden: Brill.

Epstein, S. R. 2009, *Libertad y comercio. El desarrollo de los estados de los mercados en Europa, 1300–1750*, Valencia: Publicaciones de la Universitàt de Valencia.

Erikson, W. & Bearman, P. 2004, *Routes into Newtwors. The structure of English trade in the East Indies, 1601–1833*, New York: Columbia University Press.

Ethington, P. J. 2007, "Placing the past: 'Groundwork' for a spatial theory of history", *Rethinking History,* 11 (4): 465–493.

Falola, T. & Roberts, K. D. (eds) 2008, *The Atlantic world, 1450–2000*, Indiana: Indiana University Press.

Fernández-Armesto, F. 2004, *Las Américas*, Barcelona: Debate.

—. 2007, *Amerigo: the man who gave his name to America,* New York: Random House.

Ferro, L., Gerber, L., Mani, I., Sundheim, B. & Wilson, G. 2005, *TIDES 2005 standard for the annotation of temporal expressions*. MITRE Technical Papers.

Finnegan, D. 2008, "The spatial turn: geographical approaches in the history of science", *Journal of the History of Biology*, 41: 369–388.

Fitzball, Edward, 1826, *The Flying Dutchman.*

Frank, A. G. 1998, *ReORIENT: Global economy in the Asian age*. Berkeley and Los Angeles: University of California Press.

—. 2008. *Re-Orientar: La economía global en la era del predominio asiático*, Valencia: Universidad de Valencia.

Frank, A. I. 1998, "Differents types of 'Times'", Egenhofer, M. J. & Golledge, R. G. (eds), *GIS: Spatial and temporal reasoning in GIS*, New York: Oxford University Press: 40–61.

Fraser, M. 2006, "Event", *Theory, Culture and Society*, 23: 129–132.

Freeman, L. C. 2004, *The development of social network analysis: a study in the sociology of science*. Vancouver: Empirical Press.

Fruchterman, T. M. J. & Reingold, E. M. 1991, "Graph drawing by force-directed placement", *Software: Practice and Experience*, 21 (11): 1129–1164.

Fujita, M., Krugman, P. & Venables, A. J. (1999). *The spatial economy: cities, regions, and international trade.* Cambridge, MA, and London: MIT Press.

Fujita, M., Krugman, P. & Venables, A. J. 1999, *The spatial economy: cities, regions and international trade*, Cambridge; London: MIT Press.

Gaddis, J. L. 2002, *The landscape of history: how historians map the past*, New York: Oxford University Press.

Games, A. 2006, "Atlantic history: definitions, challenges and opportunities", *American Historical Review*, 111 (3): 741–757.

García del Pino, C. 2001, *El corso en Cuba, siglo XVII: causas y consecuencias*, La Habana: Editorial de Ciencias Sociales.

García-Baquero González, A. 1992, *La Carrera de Indias, Suma de la Contratación y océano de negocios*, Sevilla: Sociedad Estatal para la Exposición Universal.

García-Molina, H., Ullman, J. D. & Widom, J. 2008, *Database systems: the complete book.* 2nd ed. Upper Saddle River: Pearson/Prentice Hall.

Gascon, R. 1971, *Grand commerce et vie urbaine au XVIe siècle: Lyon et ses marchands*, Paris: SEVPEN.

Gaulin, J. L. et Rau, S. (eds), 2009, *Lyon vue d'ailleurs (1245–1800): échanges, compétitions et perceptions*, Lyon: Presses Universitaires de Lyon.

Getis, A. 2008, "What holds us together" *Essays on Geography and GIS. GIS best practices*, 3–8. ESRI.

Ghosh, J. C. 1984, "Hayagriva-Worship in Assam", *Readings in the History and Culture of Assam*, Guwahati: Kamrupa Anusandhana Samiti: 83–86.

Gibbon, E. 1993, *The Decline and Fall of the Roman Empire*, London: Everyman's Library.

Gil, J. 2000–2003, *Los conversos y la Inquisición sevillana: ensayo de prosopografía*, 3 vols. Sevilla: Fundación El Monte.

Giménez Fernández, M. 1984, *El plan Cisneros-Las Casas para la reformación de las Indias. I Bartolomé de las Casas. Delegado de Cisneros para la reformación de las Indias (1516–1517)*, Madrid: CSIC.

Ginalsky, S. 2011, "Las empresas familiares en la red interempresas: el caso de la industria suiza de máquinas y metales en el siglo XX" *REDES_ Revista Hispana de Redes sociales,* vol. 2.

Girard, A. 2006, *El comercio francés en Sevilla y Cádiz en tiempos de los Habsburgo: contribución al estudio del comercio extranjero en la España de los siglos XVI al XVIII*, Sevilla: Editorial Renacimiento.

Gøbel, E. 2010, "The Sound Toll Registers Online Project", *International Journal of Maritime History*, XXII (2): 305–308.

Goldenberg, A., Zheng, A. X., Fienberg, S. F. & Airoldi, E. M. 2010, "A survey of statistical network models," *Foundations and Trends in Machine Learning* 2 (2): 129–233.

González Jiménez, M. 1995, "Las relaciones entre Portugal y Castilla en el siglo XV". *El Tratado de Tordesillas y su época. II/ Congreso Internacional de Historia*, Valladolid: Sociedad V Centenario del Tratado de Tordesillas.

González Torga, J. M. 2005, "Simón Ruiz, um mercader banquero: volcado a Europa desde Medina del Campo". *Revista de la Hermandad del Valle de los Caídos*, 97 (40).

Goonewardena, K. W. 1958, *The Foundation of Dutch Power in Ceylon 1638–1658*, Amsterdam: Djambatan.

Grabher, G. 2006, "Trading routes, bypasses, and risky intersections: mapping the travels of networks between economic sociology and economic geography", *Progress in Human Geography* 30: 163–189.

Granovetter, M. 1973, "The strength of weak ties" *American Journal of Sociology,* 78 (6): 1360–1380.

—. 1985, "Economic action and social structure: The problem of embeddedness" *American Journal of Sociology*, 91: 481–510.

Granovetter, M. 2005, "Business groups and social organization", Smelser, R. N. J. S. (ed.), *The handbook of economic sociology*, Princeton: Princeton University Press.

Greene, J. P. & Morgan, P. D. 2009, *Atlantic history: a critical appraisal,* New York: Oxford University Press.

Gregory, I. 2005, *A place in history: A guide to using GIS in historical research*, 2nd edition (http://www.ccsr.ac.uk/methods/publications/ig-gis.pdf).

Gregory, I. N., Bennet, C., Gilham, V. L. & Southall, H. R. 2002, "The Great Britain historical GIS project: from maps to changing human geography", *Cartographic Journal*, 39: 37–49.

Guerrero Elecalde, R. 2012, *Las élites vascas y navarras en el gobierno de la monarquía borbónica. Redes sociales, carreras y hegemonía en el siglo XVIII (1700–1746)*. Bilbao: Servicio Editorial, Universidad del País Vasco.

Guerrero Nieto, M. & Saurí, R. 2012, "ModeS TimeBank 1.0. Linguistic Data Consortium, Philadelphia", available at: http://www.ldc.upenn.edu/Catalog/catalogEntry.jsp?catalogId=LDC2012T01.

Guerrero Nieto, M. 2010, *Creación de recursos lingüísticos para la incorporación del tiempo en los Sistemas de Información Geográfica*. Masters thesis.

Guerrero, M., Saurí, R. & Bernabé Poveda, M. A, 2011, "ModeS TimeBank: A Modern Spanish TimeBank Corpus", *Revista de la Sociedad Española del Procesamiento del Lenguaje Natural,* 47: 259–267.

Guldi, J. "The spatial turn in history", *Spatial humanities: a project of the Institute for Enabling Geospatial Scholarship* (http://spatial.scholarslab.org/).

Gupta, A. Das. 1979, *Indian merchants and the decline of Surat, c. 1700–1750*, Wiesbaden: Franz Steiner Verlag.

Gurevic, Aron, 1989, "O Mercador", in *O homem medieval*, dir. Jacques Le Goff. Lisboa: Editorial Presença.

Güting, R. H., Böhlen, M. H., Erwig, M., Jensen, C. S., Lorentzos, N. A., Schneider, M. & Varzirgiannis, M. 2000, "A foundation for representing and querying moving objects", *ACM Trans.Database Syst.*, 25 (1): 1–42.

Habermas, J. 1995, *Conocimiento e Interés*, València: Universitat de València.

Hägestrand, T. 1970, "What about people in regional science?" *Regional Science*, 24: 7–24.

Haggerty, S. 2006, *The British-Atlantic trading community, 1760–1810: men, women, and the distribution of goods*, Atlantic World 6 Series, Leiden: Brill Academic Publishers.

Haken, H., 1983, *Advanced synergetics: instability hierarchies of self-organizing systems and devices*, Berlin: Springer-Verlag.

Hamilton, W. D. 1964a, "The genetical evolution of social behavior I", *Journal of Theoretical Biology* 7 (1): 1–16.

—. 1964b, "The genetic evolution of social behavior II", *Journal of Theoretical Biology* 7 (1): 17–52.

Hancock, D. J. 1997, *Citizens of the world: London merchants and the integration of the British Atlantic community, 1735–1785*, Cambridge University Press.

—. 2009, *Oceans of wine: Madeira and the emergence of American trade and taste*. Yale University Press.

Harris, R. 2005, "The formation of the East India Company as a cooperation-enhancing institution", *Social Science Research Network*: 9–47.

Hayes, D. W. 2001, *Atlas histórico del Océano Pacífico Norte: mapas de descubrimiento y exploración científica, 1500–2000*, Douglas & McIntyre.

Haythornthwaite, C. 1996, "Social network analysis: an approach and technique for the study of information exchange", *Library and Information Science Research*, 18 (4): 323–342.

Heimann, J. 1980, "Small change and ballast: Cowry trade and usage as an example of Indian Ocean economic history", *Southasia: Journal of Southasian Studies*, 3 (1): 48–69.

Helland, P. 2011, "If you have too much data, then 'good enough' is good enough," *Association for Computing Machinery (ACM) Queue* (23 May).

Helland, P. & Campbell, D. 2009, "Building on quicksand", *Conference on Innovative Data Systems Research (CIDR) '09*, Asilomar, California (4–7 January).

Henrich, J. 2004. "Cultural group selection, coevolutionary processes and large-scale cooperation", *Journal of Economic Behavior and Organization* 53: 3–35.

Herrero Sánchez, M. 2000, *El acercamiento hispano-neerlandés (1648–1678)*, Madrid: CSIC.

Herrero Sánchez, M. & Pérez Tostado, I. 2010, "Conectores del mundo atlántico: los irlandeses en la red comercial internacional de los Grillo y Lomelin", Pérez Tostado, I.. & García Hernán, E. (eds), *Irlanda y el Atlántico Ibérico: movilidad, participación e intercambio cultural*, Valencia: Albatros Ediciones: 307–323.

Hewings, G. J. D., Sonis, M. & Boyce, D. (eds) 2002. *Trade, networks, and hierarchies: integrated approaches to modeling regional and interregional economies*. Springer Verlag.

Hill, L. 2006, *Georeferencing: the geographic associations of information*. Cambridge: MIT Press.

Hill, R. & Dunbar, R. 2002, "Social network size in humans", *Human Nature* 14 (1): 53–72.

Holland, J. H. 1998, *Emergence: from Chaos to Order*, Oxford: Oxford University Press.

Hunter, G. J. & Williamson, I. P. 1990, "The development of a historical digital cadastral database", *International Journal of Geographical Information Systems*, 4: 169–179.

Husserl, E. 1990, *Las crisis de las ciencias europeas y la fenomenología trascendental*, Barcelona: Crítica.

Imfeld, S. 2000, *Time, points and space: towards a better analysis of wildlife data in GIS*. PhD thesis. http://www.geo.unizh.ch/~imfeld /diss/.

Imízcoz Beunza, J. M. 2009a, "Familia y redes sociales en la edad moderna", Lorenzo Pinar, F. J. (ed.), *La Familia en la Historia*, Salamanca: Ediciones Universidad de Salamanca: 135–186.

—. 2009b, "Las redes sociales de las élites. Conceptos, fuentes y aplicaciones" Soria Mesa, E. and Caro Bravo, J.M., ed., *Las élites en la época moderna: la monarquía española, Tomo I: Nuevas perspectivas*, Córdoba: Universidad de Córdoba: 77–111.

Imízcoz Beunza, J. M. & Arroyo Ruíz, L. 2011, "Redes sociales y correspondencia epistolar: del análisis cualitativo de las relaciones personales a la reconstrucción de redes egocentradas" *REDES_ Revista Hispana de Redes Sociales*, vol. 21.

Imízcoz, J. M. & Oliveri Korta, O. 2010, "Economía domestica y redes sociales: una propuesta metodológica", Imízcoz Beunza, J. M. & Korta, O. Oliveri (eds), *Economía doméstica y redes sociales en el Antiguo Régimen*, Madrid: Silex: 15–51.

Irving, W. 1855, T*he Flying Dutchman on Tappan Sea.*

Jackson, G. 2001, "The significance of unimportant ports", *International Journal of Maritime History*, XIII: 1–17.

Jackson, M. O. 2003, "A strategic model of social and economic networks", *Journal of Economic Theory* 71: 44–74.

Jantschs, E. (ed.) 1980, *The self-organizing universe: scientific and human implications of the emerging paradigm of evolution*, New York: Pergamon.

Jenny, B.; Weber, A. & Hurni, L. 2007, "Visualising the planimetric accuracy of historical maps with MapAnalyst", *Cartographica*, 42 (1): 89–94.

Jensen, C. S. 2000, *Introduction to temporal databases*, PhD thesis, http://www.cs.auc.dk/csj/Thesis/pdf/chapter1.pdf.

Jesus, P. De & Triba, L. 2009, "Cooperação", Cattani, J.-L., Gaiger, L. I. & Hespanha, P. (eds), *Dicionário internacional da Outra economia*, Coimbra: Almedina.

Johansen, H. C. 1983, *Shipping and trade between the Baltic and Western Europe, 1784–95,* Odense: Odense University.

Johnson, N. 2007, *Simply complexity: a clear guide to complexity theory*, Oxford: Oneworld.

Johnston, T. & Weis, R. 2010, *Managing time in relational databases: how to design, update and query temporal data*, Amsterdam: Elsevier.

Kantabutra, V. 2007, "A new type of database system: intentionally-linked entities – a detailed suggestion for a direct way to implement the entity-relationship data model," *Proceedings of the 2007 International Conference on E-Learning, E-Business, Enterprise Information Systems, and E-Government (EEE),* Las Vegas, Nevada: Computer Science Research, Education and Applications [CSREA] Press: 258–263.

Kantabutra, V., Owens, J. B., Ames, D. P., Burns, C. N. & Stephenson, B. 2010, "Using the newly-created ILE DBMS to better represent temporal and historical GIS data," *Transactions in GIS* 14 (1): 39–58.

Kathirithamby-Wells, J. 1986, "The Islamic City: Melaka to Jogjakarta, c. 1500–1800", *Modern Asian Studies*, 20 (2): 333–351.

Kauffman, S. 1993, *Origins of order: self-organization and selection in evolution*, New York: Oxford University Press.

Kearney, M. 2004, *The Indian Ocean in world history*, London: Routledge.

Khair, T., Edwards, J. D., Leer, M. & Ziadeh, H. (eds) 2006, *Other routes: African and Asian travel writings from before 1900*, Bloomington: Indiana University Press.

Knoke, D. & Yang, S. 2008, *Social network analysis*. London: Sage.

Knowles, A. K. 2000, *Historical GIS: the spatial turn in social science history*, Duke University Press.

—. 2005, "Emerging trends in historical GIS", *Historical Geography*, 33: 7–13.

—. 2008, *Placing history: how maps, spatial data and GIS are changing historical scholarship,* New York: ESRI.

Knox, H., Savage, M. & Harvey, P. 2006, "Social networks and the study of relations: networks as a method, metaphor and form", *Economy and Society* 35 (1): 113–140.

Kraus, M. 1949, *The Atlantic civilization: eighteenth-century origins,* Ithaca: Cornell University Press.

Krebs, V. "A brief introduction to social network analysis" www.orgnet.com/sna.html.

—. 2002, "Uncloaking terrorist networks" *First Monday*, 7 (4).

Krugman, P. 1996, *The self-organizing economy.* Mitsui Lectures in Economics. Cambridge, MA, and Oxford: Blackwell Publishers.

Kunz, A. 2000, "El impacto de la info-cartografía en la historia económica del transporte europeo (siglos XIX-XX)", Aranda Pérez, F. J., Fernández Izquierdo, F. & Sanz Camañes, P. (coords), *La Historia en una nueva frontera*, Cuenca: Universidad Castilla-La Mancha.

Kwan, M.-P. 2004, "GIS methods in time-geographic research: geocomputation and geovisualization on human activity patterns", *Geografiska Annaler Series B: Human Geography*, 86: 267–280.

Kwan, M.-P. & Ding, G. 2008, "Geo-narrative: extending geographic information systems for narrative analysis in qualitative and mixed-method research", *The Professional Geographer*, 60: 443–465.

Lamikiz, X. 2010, *Trade and trust in the eighteenth century Atlantic world. Spanish merchants and their overseas networks*, London: Royal Historical Society; Boydell Press.

Lanero Fernández, J. & Ortega Montes, E. 2007, "De aprendiz a mercader: El factor en el comercio internacional inglés del siglo XVI", in *Pecvnia*, 5: 146–148.

Lang, D. M. 1981, *The Armenians: a people in exile*, London: Allen & Unwin.

Lang, M. F. 1998, *Las flotas de la Nueva España (1630–1710): despacho, azogue, comercio*, Muñoz Moya.

Langran, G. & Chrisman, N. R. 1988, "A framework for temporal geographic information", *Cartographica*, 25: 1–14.

Langran, G. 1989, "A review of temporal database research and its use in GIS applications", *International Journal of Geogrpahical Information Systems*, 3: 215–232.

Lapeyre, H. 1953, *Simón Ruiz et les asientos de Philippe II*, Paris: SEVPEN.

—. 1955, *Une famille de marchands: les Ruiz. Contribution à l'étude du commerce entre la France et l'Espagne au temps de Philippe II*, Paris: Librairie Armand Colin.

Latham, A. 2002, "Retheorizing the scale of globalization: topologies, actor-networks, and cosmopolitanism", Wright, M. W. (ed.), *Geographies of power, placing the scale*, Oxford: Blackwell.

Latour, B. 2005, *Reassembling the social: an introduction to actor-network-theory*, Oxford: Oxford University Press.

Law, J. & Hassard, J. (eds), 1999, *Actor-network theory and after*, Oxford: Blackwell.

Lawson, P. 1993, *The East India Company: a history*, London: Longman.

Leavitt, N. 2010, "Will NoSQL databases live up to their promise?" *Computer (IEEE Computer Society),* 43 (2): 12–14.

Lefebvre, H. 1991, *The production of space*, Cambridge: Blackwell.

Leider, J. P. 2002, "On Arakanese territorial expansion: origins, context, means and practice", Gommans, J. L. & Leider, J. (eds), *The maritime frontier of Burma: exploring political, cultural and commercial interaction in the Indian Ocean world, 1200–1800*, Leiden: Koninklijke Nederlandse Akademie van Wetenschappen: 127–149.

Lemercier, C., Guzzi-Hebb, S. & Bertrand, M. 2011, "Introducción al análisis de redes e historia: herramientas, aproximaciones, problemas" *REDES_Revista Hispana de Redes Sociales*, 21 (8).

Li Yu-ju, 2010, *From supercargo to mutual commissioned trade: traditional junk trade between Lugang Guild merchants and China proper during the Qing dynasty*, in Schottenhammer, A. (ed.), *Trading networks in Early Modern East Asia*, Wiesbaden: Otto Harrassowitz Gunbtl & Co.: 195–220.

Li Zh. 2001, "Migration and privatization of space and power in late Socialist China", *American Ethnologist* 28 (1): 179–205.

Lippe, P. von der, 1993, *Deskriptive Statistik*, Stuttgart; Jena: Gustav Fischer Verlag.

Livro em que se contem toda a Fazenda e Real Património dos Reinos de Portugal, India e Ilhas Adjacentes, ordenado por Luiz de Figueiredo Falcão, secretário de El-Rei Filippe II, Lisboa: Imprensa Nacional, 1859.

Llorens, H., Saquete, E. & Navarro, B. 2010, "TIPSem (English and Spanish): Evaluating CRFs and Semantic Roles in TempEval-2", *Proceedings of the 5th International Workshop on Semantic Evaluation*, Association for Computational Linguistics: 284–291.

López Beltrán, M. T. 2012, "Redes familiars y movilidad social en el negocio de la renta: el tandem Fernando de Córdoba-Rodrigo Álvarez de Madrid y los judeoconversos de Málaga", *Revista del Centro de Estudios Históricos de Granada y su Reino*, 24: 33–72.

López de Mariscal, B. 2004, *Relatos y relaciones de viaje al nuevo mundo en el siglo XVI*, Madrid: Polifemo.

Lorenzo Sanz, E. 1986, *Comercio de España con América en la época de Felipe II*, 2 vols., Valladolid: Diputación Provincial.

Ludden, D. 2003a, "Investing in Nature around Sylhet: An Excursion into Geographical History", *Economic and Political Weekly*, 29: 5080–5088.

—. 2003b "The first boundary of Bangladesh on Sylhet's northern frontiers", *Journal of the Asiatic Society of Bangladesh*, 48 (1): 1–54.

Luke, D. A. & Harris, J. K. 2007, "Network analysis in public health: history, methods and applications" *Annual Review of Public Health:* 69–93, doi: 10.1146/annurev.publhealth.28.021406.144132.

Lynch, J. 2007, *España bajo los Austrias*, 2 vols., Barcelona: Crítica.

McCrindle, J. W. (tr. & ed.) 1879, *The Commerce and Navigation of the Erythraean Sea*.

McMaster, R. B., Lindberg, M., & Riper, D. van. 2003, "The National historical geographic information system (NHGIS)", *Proceedings 21st International Cartographic Conferencia*: 821–828.

McPherson, K. 1993, *The Indian Ocean: a history of people and the sea*, New Delhi: Oxford University Press.

Maestre Martínez, R. 2010, "snGraph. Software óptimo para manipulación de redes libres de escala" http://digital.csic.es/handle/10261/27556.

Magnani, M. & Montesi, D. 2010, "A survey on uncertainty management in data integration," *Journal of Data and Information Quality* 2 (5): 1–32.

Mani, I., Hitzeman, J., Richer, J., Harris, D; Quimby, R. & Wellner, B. 2008, "SpatialML: Annotation scheme, corpora, and tools", *Sixth International Language Resources and Evaluation, LREC'08. Marrakesh,* Morocco.

Mani, I., Pustejovsky, J. & Sundheim, B. 2004, "Introduction to the special issue on temporal information processing", *ACM Transactions on Asian Language Information Processing (TALIP),* Special Issue on Temporal Information Processing, 3 (1): 1–10.

Manso Porto, C. 1999, *Cartografía histórica portuguesa. Catálogo de manuscritos (siglos XVII-XVIII)*, Madrid: Real Academia de la Historia.

Marcos Martín, A. 2000, *España en los siglos XVI, XVII y XVIII: economía y sociedad*, Barcelona: Crítica.

Marin Fotache, 2006, "Why normalization failed to become the ultimate guide for database designers", *Social Science Research Network, working papers series,* http://papers.ssrn.com/so13/papers.cfm?abstract_id=905060. (accessed 28 March 2013).

Marryat, F. 1837, *The Phantom Ship.*

Marsilio, C. 2009, "'Four times a year for so many years'. The Italian Exchange Fairs during the XVIth-XVIIth centuries", Paper read at *XVth World Economic History Congress*, 3–7 August, Utrecht.

Martens, F. 1925, *Hispanische Reise-Beschreibung de anno, 1671*, Junk, W. ed., Berlin.

Martínez Millán, J. & Carlos Morales, J. de, 1990, "Conversos y élites de poder en Castilla durante la primera mitad del siglo XVI: Rodrigo de Dueñas, consejero de Hacienda de Carlos V", in *Jornadas Sefardíes y Seminario de las Tres Culturas "Las tres culturas en la Corona de Castilla y los sefardíes*. Valladolid: Consejería de Cultura y Bienestar Social: 149–161.

Martínez Shaw, C., & Oliva Melgar, J. M. (eds) 2005, *El sistema atlántico español (siglos XVII-XIX)*, Madrid: Marcial Pons.

Mauro, F. 1990, "Merchant communities, 1350-1750", in *The Rise of Merchant Empires. Long-distance trade in the early modern world, 1350-1750*, ed. James D. Tracy, Cambridge: Cambridge University Press, 1990: 255–86.

—. 1997, *Portugal, o Brasil e o Atlântico: 1570–1670*, Lisboa: Editorial Estampa.

Mazlish, B. 2006, *The new global history*, New York; London: Routledge.

Meinig, D. W. 1986, *The shaping of America: a geographical perspective on 500 years of history,* New Haven: Yale University Press.

Mena García, C. 1998, *Sevilla y las flotas de Indias. La Gran Armada de Castilla del Oro (1513–1514)*, Sevilla: Universidad de Sevilla; Fundación El Monte.

Mennis, J., Mason, M. J. & Cao, Y. 2012, "Qualitative GIS and the visualization of narrative activity space data", *International Journal of Geographical Information Science,* 27(2): 267–291.

Middell, M. 2010, "Global history and the spatial turn: from the impact of area studies to the studies of critical junctures of globalization", *Journal of Global History*, 5: 149–170.

Milgram, S. 1967, "The small-world problem", *Psychology Today* 1: 61–67.

Mires, F. 2001, *La revolución permanente: las revoluciones sociales en América Latina*, Madrid: Siglo XXI editores.

Modelski, G., Devezas, T. & Thompson, W. R. (eds) 2008, *Globalization as evolutionary process: modeling global change*, Abingdon; New York: Routledge.

Modelski, G. 1990, "Is world politics evolutionary learning?", *International Organization* 44 (1): 1–24.

—. 2000, "World System Evolution", Denemark, F., Gills, B. & Modelski, G. (eds), *World system history: the social sciences of long-term change*, New York: Routledge.

Modelski, G. 2004, "Self-organization in the world system", *Encyclopedia of life support systems (EOLSS)*, Oxford: UNESCO, Eolss Publishers.

Molho, A. & Ramada Curto, D. (eds) 2003, 'Réseaux marchands', in *Annales: Histoire, Sciences Sociales*, 3: 567–672.

Mollat, M. 1995, *A Europa e o mar*. Lisboa: Editorial Presença.

Moreno, H. B. 1978, "Reflexos na cidade do Porto da entrada dos conversos em Portugal nos fins do século XV", in *Revista de História*, vol. I. Porto: INIC/Centro de História da Universidade do Porto: 7–38.

Morineau, M. 2006, "Le système atlantique au péril de l'histoire", *Jahrbuch für Geschichte Lateinamerikas*, 43: 301–316.

Moutokias, 1989, *Contrabando y control colonial. Buenos Aires y el Espacio peruano en el siglo X VII*. Buenos Aires, 1989.

Mukherjee, R. (ed.) 2011, *Networks in the First Global Age: 1400–1800*, Delhi: Primus Books.

Mukherjee, R. (ed.) 2013, *Oceans connect: reflections on water worlds across time and space,* Delhi: Primus Book.

Mukherjee, R. [forthcoming], *Between land and sea: the integration of "small" polities into a Northern Bay of Bengal economic system between the 9th and 19th centuries*, Singapore: Nalanda Sriwijaya Centre/ISEAS.

Mullins, N. C. & Mullins, C. J., 1973, *Theories and theory groups in contemporary American sociology*, New York: Harper & Row.

Murdoch, J. 1998, "The spaces of actor-network theory", *Geoforum*, 29: 357–374.

Nalini, S. A. P. 2005, *The court chronicle of the Kings of Manipur: the Cheitharon Kumpapa*, New York: Routledge.

Neale, W. J. 1840, *The Flying Dutchman, a legend of the high sea*, 2 vols., Philadelphia.

Newman, M. E. J. 2010, *Networks: an introduction*, Oxford: Oxford University Press.

Noë, R. 2011, "Introduction", Dixon, B., Hoogland, E. & Rustat-Flinton, C. (eds), *The evolution of cooperation and trading (TECT): final report*, Strasbourg: European Science Foundation: 3–4.

North, D. C. 1985, "Transaction costs in history", *Journal of European Economic History*, 14 (3): 557–576.

Northup, D. 2006, "Globalization and the Great Convergence: rethinking world history in the long term", *Journal of World History*, 16 (3): 249–267.

Novoa, J. N. [forthcoming], *A sixteenth century voyage of legitimacy: the paths of Jácome and António da Fonseca from Lamego to Rome and beyond*.

Núñez Espinoza, M. 2008, "Manual en español para la representación y análisis de rede de desarrollo rural con Visone. Un acercamiento a una herramienta de apoyo para la evaluación de proyectos", *revista-redes.rediris.es* http://revista-redes.rediris.es/webredes/.

Ogilvie, Sheilagh, *Institutions and European trade: merchant guilds, 1000–1800*, Cambridge, 2011.

Olivares García, J. M. 2006, "Cartografía Catastral en Google Earth". *CT: Catastro*, 59.

Oliveira Marques, A. H. & Alves Dias, J. J. 1995, "A populaçao portuguesa nos seculos XV e XVI", *El Tratado de Tordesillas y su época. I/Congreso Internacional de Historia*, Valladolid: Sociedad V Centenario del Tratado de Tordesillas.

Oliveira, M. & Gama, J. 2012, "An overview of social network analysis" *WIRE's Data Mining Knowledge Discovery*, 2: 99–115 doi: 10.1002/widm.1048.

Olle, T. W., 1978, *The Codasyl approach to data base management*, Chichester; New York: Wiley.

Ortego Rico, P. 2012, *"Estrategias financieras y especulación en torno al arrendamiento "por mayor" de rentas regias ordinarias en Castilla: aproximación a partir del caso de Castilla la Nueva (1462–1504)"*, Bonachía Hernando, J. A. & Carvajal de la Vega, D. (eds), *Los negocios del hombre. Comercio y rentas en Castilla, siglos XV y XVI*, Valladolid: Castilla Ediciones: 235–260.

Owens, J. B. 2007, "Towards a geographically-integrated, connected world history: employing Geographic Information Systems (GIS)" *History Compass*, 5 (6): 2014–2040.

—. 2008a, "A multi-national, multi-disciplinary study of trade networks and the domains of Iberian monarchies during the First Global Age, 1400–1800", *Society for Spanish and Portuguese Historical Studies*, Bulletin, XXXII (2): 23–30.

—. 2008b, "What historians want from GIS". *Essays on Geography and GIS: GIS best practices*, ESRI: 35–46.

—. 2008c, "Dynamic complexity of cooperation-based self-organising networks in the First Global Age (DynCoopNet)", Noe, R., Klein, R., Bowman, J. & Rustat-Flinton, C. (eds), *The evolution of cooperation and trading (TECT)* (Strasbourg, FR: European Science Foundation, 2008): 23–35. www.esf.org/activities/eurocores/programmes/tect.html.

—. 2009, "Space, connections, and place in the First Global Age", *Sixteenth Century Journal*, XL (1): 190–192.

—. 2012, "Dynamic complexity of cooperation-based self-organizing commercial networks in the First Global Age (DynCoopNet): What's in a name?", *Journal of Knowledge Management, Economics and Information Technology*, Special Issue: Crespo Solana, A. & Alonso García, D. (eds), *Self-organizing networks and GIS tools: cases of use for the study of trading cooperation (1400–1800)*, June: 25–51.

Owens, J. B., Yuan, M., Wachowicz, M., Kantabutra, V., Coppola Jr, E., Ames, D. P. & Gangemi, A. 2009, "Visualizing historical narratives: geographically-integrated history and dynamics GIS" *Position paper for the National Endowment for the Humanities workshop "Visualizing the Past: Tools and Techniques for Understanding Historical Processes"*, 20–21 February 2009, University of Richmond, Virginia.

Padgett, J. F. 2011, "Introducción a matrimonio y estructura de la élite en la Florencia del Renacimiento" *REDES_ Revista hispana de Redes Sociales,* 21, http://revista-redes.rediris.es/html-vol21/vol21_2.htm.

Padmanath, B. 1920, "To the East of Samatata (On the six countries mentioned but not visited by Yuan Chwang)", *Journal of the Royal Asiatic Society of Great Britain and Ireland*, 1: 1–19.

—. (ed.) 2003, *Heramba Rajyer Dandabidhi ebong Heramba Rajyer Hrinadanbidhi (Penal Codes and Financial Rules of Hedamba Kingdom)*, Silchar, Assam: Barak Upatyaka Banga Sahitya o Sanskriti Sammelan.

Pagden, A. P. 1995, *Lords of all the worlds: ideologies of empire in Spain, Britain and France, c.1500–c.1850*, New York: Yale University Press.

Paladini Cuadrado, A. 1996, "Sobre el Atlas de El Escorial y su posible autor", *Exposición de Cartografía IV Centenario ciudad Valladolid 1596–1996*, Valladolid: Ayuntamiento de Valladolid: 35–50.

—. 2000, "La formación de la carta moderna de España en el siglo XVI". *9º Congreso Jornadas Nacionales de Historia Militar. El emperador Carlos y su tiempo*, Madrid: Deimos.

Parello, V. 1999, "Sociología conversa en los siglos XV y XVI: la dinámica de las familias manchegas", *Sefarad,* 59 (2): 391–418.

Parry, J. H. 1990, *The Spanish seaborne empire*, Berkeley: University of California Press.

—. 1991, *El descubrimiento del mar*, México: Consejo Nacional para la cultura y las artes.

Pedreira, J. 1998, "*As consequências económicas do império: Portugal (1415–1822)*", in *Análise Social*, XXXII (2–3): 146–147.

Pelekis, N., Theodoulidis, B., Kopanakis, I. & Theodoridis, Y. 2005, "Literature review of spatiotemporal database models", *Knowledge Engineering Review*, 19: 235–274.

Pérez Asensio, E., Bosque González, I. del, Maestre Martínez, R., Crespo Solana, A. & Sánchez-Crespo Camacho, J. M., 2010, "Integración de bases de datos históricas en una IDE. Comercio mundial y redes de cooperación en la primera Edad Global (1400–1800)" *Contribución a las JIIDE 2009*.

Pérez Asensio, E., Bosque González, I. del, Maestre Martínez, R., Crespo Solana, A. & Sánchez-Crespo Camacho, J. M., 2012, "Modelling and implementation of a spatio-temporal historic GIS", *Journal of Knowledge Management, Economics and Information Technology,* Special Issue: Crespo Solana, A. & Alonso García, D. *Self-organizing networks and GIS tools: cases of use for the study of trading cooperation (1400–1800)*, Scientific Papers: 145–189.

Pérez, J. 1997, *La revolución de las Comunidades de Castilla (1520–1521)*, 7th ed., Madrid: Siglo XXI.

Petti Balbi, G. ed. 2001a, *Comunitá forestiere e nationes nell'Europa dei secoli XIII-XVI*, Nápoles: GISEM.

—. 2001b, *Strutture del potere ed elites economiche nelle città europee dei secoli XII-XVI*, Nápoles: GISEM.

Petti Balbi, 2005, *Negoziare fuori patria: nazioni e genovesi in età medievale.* Bolonia.

Peuquet, D. J. 1994, "It's about time: a conceptual framework for the representation of temporal dynamics in Geographic Information Systems", *Annals of the Association of American Geographers*, 84: 441–461.

Peuquet, D. J. & Duan, N. 1995, "An event-based spatiotemporal data model (ESTDM) for temporal analysis of geographical data", *International Journal of Geographical Information Systems*, 9 (1): 7–24.

Peuquet, D. J. 2002, *Representations of space and time,* New York; London: Guilford Press.

Picazo Muntaner, A. 2003, "La concepción de hombre-otro en la percepción de la ampliación colonial del espacio indiano y el olvido de indígena americano", *Cauces*, 4: 29–34.

—. 2011a, "La evaluación de la percepción espacial y el peso cartográfico de los nódulos del Indico y Pacífico, siglos XVI-XVIII", *Diálogos, Revista Electrónica de Historia*, 12 (2).

—. 2011b, "Las redes comerciales en el galeón de Manila", *Delaware Review of Latin American Studies*, 12 (2): 1–15.

Pieper, R. & Schmidt, P. (eds) 2005, *Latin American and the Atlantic World /El mundo atlántico y América Latina (1500–1800). Essays in honor of Horst Pietschmann*, Köln: Böhlau Verlag.

Pieris, P. E. 1924, *Ceylon and the Hollanders*, Tellippalai: American Ceylon Mission Ppress, 1924.

Pietschmann, H. 2002, *Atlantic history: history of the Atlantic system, 1580–1830*, Kolhn: Vandenhoeck & Ruprecht.

—. 2010, "The Spanish Atlantic in an age of transition, 1648–1700", *Jahrbuch für Geschichte Lateinamerikas*, 47: 345–360.

Pinker, S. 2011, *The better angels of our nature: the decline of violence in history and its causes*, London: Allen Lane.

Polónia Da Silva, A. 2010b. "Ruling strategies and informal power of self-organising networks in the First Global Age: the Portuguese case", Sessão "The power of the commoners: informal agent-based networks as source of power in the First Global Age" in *ASSHC 2010* (Chicago, 18–21 Novembro) (http://dyncoopnet-pt.org/working-paper).

—. 2010c, "European seaports in the Early Modern age: concepts, methodology and models of analysis", *Cahiers de la Méditerranée*, 80: 17–39.

—. 2011, "Seaports as centres of economic growth: the Portuguese case, 1500–1800", Unger, R. (ed.), *Shipping and Economic Growth. 1350–1800*, Leiden: Brill Academic Publishers: 379–409.

—. 2012, "Indivíduos e redes auto-organizadas na construção do império ultramarino português", Garrido, A., Freire Costa, L. & Duarte, L. M. (eds), *Economia, Instituições e Império. Estudos em Homenagem a Joaquim Romero de Magalhães*, Coimbra: Almedina: 349–372.

—. [Forthcoming], "Portuguese Seaports as Gateways, 1500–1800", Mukherjee, R. ed., *Vanguards of globalization: port-cities from the classical to the modern times*, Delhi: Primus Books.

Polónia Da Silva, A. & Owens, J. B. 2010a, "Cooperation-based self-organizing networks in Portuguese overseas expansion in the First Global Age, 1400–1800" (www.dyncoopnet-pt.org/working-papers).

Pomeranz, K. 2000, *The great divergence: China, Europe, and the making of the modern world economy.* Princeton & Oxford: Princeton University Press.

Porto: a construção de um espaço marítimo no início da Época Moderna (in press). Lisboa: Academia de Marinha.
Poujade, P. 2008, *Une société marchande: le commerce et ses acteurs dans les Pyrénées modernes*. Toulouse: Presses Universitaires du Mirail.
Priotti, J. P. *Réseaux marchands en conflit et hiérarchisation des centres économiques: l'Atlantique franco-espagnol, 1560–1570*, available in http://www.usc.es/estaticos/congresos/histec05/b24_priotti.pdf.
—. 2004, *Bilbau et ses marchands au XVIe siècle: genèse d'une croissance*, Lille: Presse Universitaire du Septentrion.
Pugli, I. 2005, *I Piccolomini d'Aragona duchi di Amalfi (1461–1610). Storia di un patrimonio nobiliare*. Naples: Editorial Scientifica: 14–18.
Pustejovsky, J., Knippen R., Littman, J. & Saurí, R. 2005, "Temporal and event information in natural language text", *Language Resources and Evaluation*, 39 (2–3): 123–164.
Puu, T. 2003, *Attractors, bifurcations and chaos: non-linear phenomena in economics*, Berlin & Heidelberg: Springer-Verlag.
Radhamadhav D. & Panchali, M. (eds) 2004, *Amalendu Bhattacharjee*, Calcutta: Mahajati Prakashan.
Ramírez, B. R. 2001, "Krugman y el regreso a los modelos espaciales. ¿la nueva geografía?", *Terra Livre*, 16: 25–38.
Ravasz, É. & Barabási, A.-L. 2003, "Hierarchical organization in complex networks", *Physical Review* E 67 (2).
Recio Morales, O. 2012, "Redes de nación y espacios de poder en la Monarquía Hispánica: un estado de la cuestión", Recio Morales, O. (ed.), *Redes de nación y espacios de poder en España y la América española, 1600–1825*, Madrid: Ministerio de Defensa: 37–52.
Reinhard, W., 2002, "Nepotism", Levillain, P. (ed.), *The Papacy: An Encyclopedia*, Routledge: 1031–1033.
Reparaz Ruiz, G. 1950, "The topographical maps of Portugal and Spain in the 16th century", *Imago Mundi*, 7: 75–82.
Reynoso, C. 2008, "Redes libres de escala en ciencias Sociales: Significado y Perspectivas", *I Reunión Latinoamericana de análisis de Redes Sociales. Agosto de 2007, La plata, Argentina*, Universidad Nacional de la Plata.
Ribeiro, A. S. 2011, *Mechanisms and criteria of cooperation in trading networks of the First Global age: the case study of Simón Ruiz network, 1557–1597*, PhD dissertation, Universidade Porto.
—. 2012, "Nos bastidores do poder: a aproximação a lugares de poder como estratégia de negócio em redes comerciais auto-organizadas da época moderna. O caso da rede de Simón Ruiz (1557–1597)", *XXXII*

Encontro da Associação Portuguesa de História Económica e Social, ISCTE, Lisboa.

Ringrose, D. 2001, *Expansion and global interaction, 1200–1700*, San Diego: University of California; New York: Addison Wesley Longman.

Robaldo, L., Caselli, T., Russo, I. & Grella, M. 2011, "From Italian text to TimeML document via dependency parsing", *Journal on Data Semantics* (February): 177–187.

Robinson, R. 1976, "Modelling the port as an operational system: a perspective for research", *Economic Geography*, 52 (1): 71–86.

Rodríguez González, R. 1995, *Mercaderes castellanos del siglo de oro*, Valladolid: Secretariado de Publicaciones Universidad de Valladolid.

Rodríguez Pascual, A., Emilio López, E. & Abad, P. 2005, "La infraestructura de datos espaciales de España (IDEE): una realidad emergente*", Mapping*, 100: 30–32.

Rodríguez, R. 2008, "Análisis de redes sociales: el caso de Rafael Altamira" *Comunicaciones del I Encuentro de Jóvenes Investigadores en Historia Contemporánea de la AHC*, Universidad Autónoma de Madrid, http://ifc.dpz.es/recursos/publicaciones/27/15/76.rosariorodriguez.pdf.

Roitman, J. 2008, *Us and them: inter-cultural trade and the Sephardim*, Leiden: Universiteit Leiden.

Rosé, I. 2011, "Reconstitución, representación gráfica y análisis de redes de poder de la alta Edad Media. Aproximación de las prácticas sociales de la aristocracia a partir del ejemplo de Odon de Cluny († 942)", *REDES_Revista Hispana de Redes Sociales,* 21, http://revista-redes.rediris.es/html-vol21/vol21_5.htm.

Rosetti, G. (ed.), *Dentro della cittá. Stranieri e relata urbane nell'Europa dei secoli XII-XVI*, Naples: GISEM, 1989.

Ruiz Ibáñez, J. J. 2011, *Las vecindades de la Monarquía Hispánica*, México: Fondo de Cultura Económica.

Ruiz Martin, F. 1970, *El Banco de España: una historia económica*, Madrid: Banco de España.

—. 1986, "Las ferias de Castilla", Lorenzo Sanz, E. (ed.), *Historia de Medina del Campo y su tierra*, Valladolid: Ayuntamiento de Medina del Campo/ Consejeria de Educacion y Cultura de la Junta de Castilla y Leon/ Diputación Provincial de Valladolid/ Caja de Ahorros Provincial de Valladolid.

Sánchez-Albornoz, N., Lockhart, J. & Bowser, F. P. (eds), 1990. *América Latina en la época colonial*, vol. 2. *Economía y Sociedad,* Barcelona: Editorial Crítica.

Santos Pérez, J. M. & Cabral de Sousa, G. F. (eds) 2006, *El desafío holandés al dominio ibérico en Brasil en el siglo XVII*, Salamanca: Universidad de Salamanca.

Sauer, C. O. 1956, "The education of a geographer", *Annals of the Association of American Geographers*, 46: 287–299.

Schaub, J.-F. 2004, "Historia colonial de Europa. De civilización a barbarie", *ISTOR*, 16: 45–71.

Scheltjens, W. 2009, "The volume of Dutch Baltic shipping at the end of the eighteenth century: a new estimate based on the Danish Sound toll registers", *Scripta Mercaturae*, 43 (1): 74–102.

Scheltjens, W. & Veluwenkamp, J. W. 2012, "Sound Toll registers online: introduction and first research examples", *International Journal of Maritime History*, XXIV (1): 301–330.

Schmidl, U. 1903, *Viaje al Río de la Plata, 1534–1554*. Notas bibliográficas y biográficas por Bartolomé Mitre. Prólogo, traducción y anotaciones por Samuel A. Lafone Quevedo, Buenos Aires. Cabaut y Cia., editores; Buenos Aires: Librería del Colegio.

Schmidt, P. 2002, "Les minorités religieuses européennes face à l'espace atlantique à l'époque moderne", in Pietschmann, H. (ed.), *Atlantic history: history of the Atlantic system, 1580–1830,* Göttingen: 83–96.

Schoff, W. H. 1912, *The Periplus of the Erythraean Sea: travel and trade in the Indian Ocean by a merchant of the first century*, New York: Longmans, Green.

Schottenhammer, Angela (ed.) 2010, *Trading networks in Early Modern East Asia*, Wiesbaden: Otto Harrassowitz Gumbtl.

Schreiber, F. A. 1994, "Is time a real time? An overview of time ontology in informatics", *Real Time Computing*: 283–307.

Schulte-Beerbühl, M. 2007, *Deutsche Kaufleute in London: Welthandel und Einbürgerung*, München: R. Oldenbourg Verlag.

—. 2008, "Spatial construction and social norms in eighteenth century trade networks", *DyncoopNet Workshop Trust, Reputation, Defectors and Sustaining Social norms: Studying Spatially complex cooperative relationships in ways that connect TECT projects*, Porto.

Schulte-Beerbühl, M. & Vögele, J. (eds) 2004, *Spinning the commercial web: international trade, merchants, and commercial cities, c.1640–1939*. Frankfurt am Main: Peter Lang.

Scott, J. 1996, *Social network analysis: a handbook*. London: Sage Publications.

Serker, S. & Sidorova, A. 2006, "Actor-networks and business process change failure: an interpretive case study", *Journal of Management Information Systems* 23 (1): 51–86.

Serrat, O. 2009, "Social network analysis" in *Cornell University ILR School. International Publications. Key Workplace Documents*, http://digitalcommons.ilr.cornell.edu/cgi/viewcontent.cgi?article=1167&context=intl.

Serrera Contreras, R. 1992, *Tráfico terrestre y red vial en las Indias españolas*, Lunwerg.

Setzer, A. 2001, *Temporal information in newswire articles: an annotation scheme and corpus study*, PhD thesis, University of Sheffield.

Sevilla Soler, R. 1991, "Santo Domingo, frontera franco-española. Consecuencias de la presencia francesa en la isla Española", Solano, F. de & Bernabeu, S. (eds), *Estudios (Nuevos y Viejos) sobre la Frontera*, Madrid: CSIC: 163–185.

Shaw, S.-L. & Yu, H. 2009, "A GIS-based time-geographic approach of studying individual activities and interactions in a hybrid physical-virtual space", *Journal of Transport Geography*, 17: 141–149.

Shnayder, E. 2010, "A data model for spatial history: the Shaping the West geodatabase", Working paper published on 15 November 2010.

Shvaiko, P. & Euzenat, J., 2005, "A survey of schema-based matching approaches," *Journal on Data Semantics*: 146–171.

Siebert, L. J. 2000, "Using GIS to document, visualize and interpret Tokyo's spatial history", *Social Science History*, 24 (3): 537–574.

Sigrist, R., & Widmer, E. D. 2001, "Relaciones de formación y transmisión del conocimiento en la botánica del siglo XVIII: un análisis de redes sociales", *REDES_ Revista Hispana de Redes Sociales*, 21 http://revista-redes.rediris.es/html-vol21/vol21_7e.htm.

Silberschatz, A., Korth, H. F. & Sudarshan, S. 2010, *Database system concepts,* 6th ed. New York: McGraw-Hill.

Silva, J.-G. da. 1956, *Stratégie des affaires à Lisbonne entre 1595 et 1607: Lettres marchandes des Rodrigues d'Évora et veiga*, Paris: Librairie Armand Colin.

Silveira, L. 1946, *Manuscritos portugueses da biblioteca Estatal de Hamburgo*, Lisboa: Instituto para alta Cultura.

Sircar, D. C. 1990, *Studies in the geography of ancient and medieval India*, New Delhi: Motilal Banarsidass.

Sluiter, E. 1942, "Dutch maritime power and the colonial status quo, 1585–1641", *Pacific Historical Review*, 11 (1): 29–4.

Smoot, M., Keichiro, O., Ruscheninski, J., Wang, P.-L. & Ideker, T. 2010, "Cytoscape 2.8: new features for data integration and network visualization" *Bioinformatics*, 1, 27 (3): 431–432.

Soldani, M. L. 2011, *Uomini d'affari e mercanti toscani nella Barcelona del Quattrocento*. S/l: Consejo Superior de Investigaciones Cientificas, 2011.

Solé, R. 2009, *Redes complejas. Del genoma a internet*, Barcelona: Tusquets.

Steven, R., Flint, M. & George, E. 2010, "Spatializing Social Networks: Using Social Network Analysis to Investigate Geographies of Gang Rivalry, Territoriality, and Violence in Los Angeles", *Annals of the Association of American Geographers*, 100 (2): 307–326.

Subrahmanyam, S. 1996, *Merchant networks in the Early Modern world*, Aldershot: Ashgate.

Sugiyama, S. & Grove, L. (eds) 2001, *Commercial networks in modern Asia*, Richmond: Curson Press.

Tesoros de la cartografía española: exposición con motivo del XIX Congreso Internacional de Historia de la Cartografía, 2001. Madrid: Biblioteca Nacional de España.

Tesouros da Cartografia Portuguesa, 1997. Lisboa: Comissao Nacional para as comemmoraçoes dos descobrimentos portugueses.

Thrift, N. J. 2008, *Non-representational theory: space, politics, affect*, Taylor & Francis.

Tognetti, S. 1999, *Il banco Cambini: affari i mercati di una compagnia mercantile-bancaria nella Firenze del XV secolo*. Florença: Leo S. Olschki.

Tooley, R. V. 1982, *Maps and mapmakers,* London: B. T. Batsford.

Torget, A. & Wilson, J. 2009, "Visualising the past: tools and techniques for understanding historical processes", A White Paper for the *National Endowment for the Humanities*, Virginia: University of Richmond.

Tracy, J. D. 1990, *The rise of merchant empires: long-distance trade in the Early Modern world, 1350–1750*, Cambridge: Cambridge University Press.

Treppo, M. del (ed.) 1994, *Sistema di rapporti ed élites economiche in Europa (secoli XII-XVI)*, Naples: GISEM.

Treseler, M. E. (ed.) 2009, *Learning SQL,* 2nd ed. Sebastopol: O'Reilly.

Trivellato, F. 2009, *The familiarity of strangers: the Sephardic diaspora, Livorno, and cross-cultural trade in the Early Modern period*, New Haven: Yale University Press.

Tryfona, N. & Jensen, C. 1999, "Conceptual data modeling for spatiotemporal applications", *GeoInformatica*, 3 (3): 245–268.

Tutte, W. T. 2001, *Graph theory*, Cambridge: Cambridge University Press.

Um imenso Portugal: História e Historiografia. São Paulo: Editora 34, 2002.

Valenzuela Ruz, V. 2003, *Manual de análisis y diseño de algoritmos version 1.0*, Chile: Instituto Nacional de Capacitación.

Vaquero Piñeiro, M. 1999, *La renta y las casas. El patrimonio inmobiliario de Santiago de los Españoles de Roma entre los siglos XV y XVII*, Roma: CSIC.

Varela Marcos, J. 2012, "La cartografía de los puertos de partida hacia el Nuevo Mundo y su origen", González Cruz, D. (coord.), *Descubridores de América. Colón, los marinos y los puertos*, Madrid: Sílex: 17–30.

Vásquez de Prada, V. s.d, *Lettres marchandes d'Anvers*. 4 vols. Paris: SEVPEN.

Vásquez Maure, F. 1982, "Análisis y evaluación del Atlas de El Escorial". *Boletín de la Real Sociedad Geográfica*, 98: 203–215.

Verhagen, M., Mani, I., Saurí, R., Knippen, R., Littman, J. & Pustejovsky, J. 2005, "Automating temporal annotation with TARSQI. Demo session". *Proceedings of the ACL.*

Vicente Maroto, M. I. 1995, "Alonso de Santa Cruz y el oficio de Cosmógrafo Mayor del Consejo de Indias", *Mare Liberum 10, II Congreso luso español sobre descobrimentos e expansão colonial*, Lisboa: Comissão Nacional para as comemorações dos descobrimentos portugueses: 509–529.

—. 2002, "El arte de navegar". *Historia de la Ciencia y de la Técnica en la Corona de Castilla III. Siglos XVI y XVII*, Valladolid: Junta de Castilla y León, Consejería de Educación y Cultura: 347–381.

Vicente Maroto, M. I. & Esteban Piñeiro, M. 2006, *Aspectos de la Ciencia Aplicada en la España del Siglo de Oro*, Valladolid: Junta de Castilla y León; Consejera de Cultura y Turismo.

Vila Vilar, E. 2012, *Hispanismo e Hispanización: El Atlántico como nuevo Mare Nostrum*, Madrid: Real Academia de la Historia.

Vilar, P. 2004, *Pensar históricamente. Reflexiones y recuerdos*, Barcelona, Crítica.

Vries, Jan de, 2008, *The industrious revolution: consumer behavior and the household economy, 1650 to the present*, New York: Cambridge University Press.

Wachowicz, M. & Owens, J. B. 2012, "Dynamics of trade networks: the main research issues on space-time representations," *Journal of Knowledge Management, Economics and Information Technology*" Special Issue: Crespo Solana, A. & Alonso García, D. (eds), *Self-*

organizing networks and GIS tools: cases of use for the study of trading cooperation (1400–1800), Scientific Papers: 53–79.
Wachowicz, M. 1999, *Object-oriented design for temporal GIS*, London: Taylor & Francis.
Wachowicz, M. & Owens, J.B. 2013, "The role of knowledge spaces in geographically-integrated history," Lünen, A. von & Travis, C. (eds), *History and GIS: epistemologies, considerations and reflections*, Dordrecht: Springer Verlag: 127–144.
Wallerstein, I. 1984, *El moderno sistema mundial II: El mercantilismo y la consolidación de la economía-mundo europea, 1600–1750*, México: Siglo XXI.
—. 2000, "El eurocentrismo y sus avatares: los dilemas de las ciencias sociales", *New Left Review* (http://www.newleftreview.es/).
Warf, B. & Arias, S. 2008, *The spatial turn: interdisciplinary perspectives*, Routledge.
Wasserman, S. & Faust, K. 1999, *Social network analysis: methods and applications*, Cambridge: Cambridge University Press.
Watts, D. J. 2003, *Small worlds: the dynamics of networks between order and randomness*, Princeton: Princeton University Press.
—. 2004, *Six degrees: the science of a connected age*, New York: Norton.
—. 2006, *The structure and dynamics of networks*, Princeton: Princeton University Press.
Wee, H. van der. 1963, *The growth of Antwerp market and the European economy*. 3 vols. The Hague: Martinus Nijhoff.
Welling, G. M. 1998, *The prize of neutrality: trade relations between Amsterdam and North America 1771–1817. a study in computational history*, [s.l.]: [s.n.].
West, S. A., El Mouden, C. & Gardner, A. 2011, "Sixteen common misconceptions about the evolution of cooperation in humans", *Evolution and Human Behavior* 32 (4): 231–262.
White, R. 2010, "What is spatial history?" *Stanford University Spatial History Lab*. Available at: www.stanford.edu/group/spatialhistory/cgi-bin/site/pub.php?id=29&project_id=997.
Whitehead, A. N. 1960, *Process and reality: a essay in cosmology*, New York: Harper & Row.
Winius, G. D. 1983, "The shadow empire of Goa in the Bay of Bengal", *Itinerário* 7 (2): 83–101.
Worboys, M. F. 1994, "A unified model of spatial and temporal information", *Computer Journal*, 37: 26–34.
—. 2005, "Event-oriented approaches to geographic phenomena", *International Journal of Geographical Information Science*, 19: 1–28.

Worboys, M. F. & Hornsby, K. 2004, "From Objects to Events: GEM, the Geospatial Event Model", Max, J. E., Freksa, C. & Miller, H. J. (eds), *Geographic Information Science*, Adelphi: Springer: 327–344.

Yang, B. 2004, "Horses, silver, and cowries: Yunnan in global perspective", *Journal of World History*, 15 (3).

—. 2011a, "The rise and fall of cowry shells: the Asian story', *Journal of World History*, 22 (1): 1–26.

—. 2011b, "The Bay of Bengal connections to Yunnan", Mukherjee, R. (ed.), *Pelagic passageways: the northern Bay of Bengal before colonialism,* Delhi: Primus Books: 317–42.

Yu, H. & Shaw, S.-L. 2006, "Exploring potencial human activities in physical and virtual spaces: a spatio-temporal GIS approach", *International Journal of Geographic Information Science*, 22 (4): 409–430.

Yuan, M. 1996, "Temporal GIS and spatio-temporal modeling", *Proceedings of Third International Conference Workshop on Integrating GIS and Environment Modeling*, Santa Fe, NM http://loi.sscc.ru/gis/data_model/may.html (accesed 30 April 2013).

—. 1999, "Use of three-domain representation to enhance GIS support for complex spatiotemporal queries", *Transactions in GIS*, 3: 137–159.

—. 2008, "Dynamics GIS: recognizing the dynamic nature of reality," *ArcNews* 30 (1): 4–5, http://www.esri.com/news/arcnews/spring08articles/dynamics-gis.html (accessed 15 March 2013), reprinted in 2008, *GIS best practices: essays on geography and GIS,* Redlands: ESRI: 17–24.

—. 2009, "Challenges and critical issues for temporal GIS research and technologies", *Handbook of Research on Geoinformatics*: 144–153.

Yuan, M. & Stewart, K. 2008, *Computation and visualization for understanding dynamics in geographic domains: a research agenda,* Boca Raton: CRC.

Archives

ADP: Arquivo Distrital do Porto.
AGI: Archivo General de Indias, Sevilla.
AGS: Archivo General de Simancas.
AHMP: Arquivo municipal do Porto.
AHPV: Archivo Histórico Provincial de Valladolid.
ARCV: Archivo de la Real Chancillería de Valladolid.
ASR: Archivo Simón Ruiz.
BNP: Biblioteca Nacional de Portugal.

BNE: Biblioteca Nacional de España.
GAA: Gemeente Archief Amsterdam.
MNM: Museo Naval de Madrid.

ABBREVIATIONS

API	Application Programming Interface
BCE	Before Christian Era
CCHS	Centro de Ciencias Humanas y Sociales, Madrid
CE	Christian Era
CIDHEUS	Interdisciplinary Centre for History, Cultures and Societies, University of Evora
CITCEM	Centro de Investigaçâo transdisciplinar, Cultura, Espaço e Memória
CODASYL	Conference on Data Systems Languages
CSIC	Consejo Superior de Investigaciones Científicas
DBMS	database management systems
EIC	East Indian Company
FBI	Federal Bureau of Investigation
FCT	Fundaçâo para a Ciência e a Tecnologia, Portugal
GEM	Geospatial Event Model
GIS	Geographic Information System
GUI	graphic user interface
HHI	Herfindahl-Hirschman Index
IDEE	Infraestructura de Datos Espaciales de España
ILE	intentionally-linked entity
MICINN	Ministerio de Ciencia e Innovacion, España
Mss.	Manuscrit
NLP	Natural Language Processing
NSF	National Science Foundation, USA
RDBMS	relational database management schemes
RDF	Resource Description Framework
SNA	social network analysis
SQL	structured query language
STRO	The Sound Toll Registers online
TECT	The Evolution of Cooperation and Trading
VOC	Vereenigde Oost-Indische Compagnie

LIST OF IMAGES

Chapter Six

Chapter Seven

Chapter Eight

Chapter Nine

Chapter Ten

Chapter Eleven

Chapter Twelve

INDEX